Control Engineering

A Modern Approach

Control Engineering

A Modern Approach

Pierre R. Bélanger
McGill University

Saunders College Publishing

Harcourt Brace and Company

Fort Worth   Philadelphia   San Diego   Orlando   Austin   San Antonio   Toronto
Montreal   London   Sydney   Tokyo

Text Typeface: Times Roman
Compositor: Electronic Technical Publishing Services
Acquisitions Editor: Sara Tenney
Managing Editor: Carol Field
Project Editor: Laura Shur
Copy Editor: Mary Patton
Manager of Art and Design: Carol Bleistine
Art Director: Robin Milicevic
Art & Design Coordinator: Sue Kinney
Cover Designer: Ann McDonald
Text Artwork: Grafacon
Director of EDP: Tim Frelick
Senior Production Manager: Joanne Cassetti
Marketing Manager: Marjorie Waldron

Cover credit: Bettman Archive

Printed in the United States of America

CONTROL ENGINEERING: A MODERN APPROACH

ISBN 0-03-013489-7

Library of Congress Catalog Card Number:

4567890123 042 987654321

Preface

This text is intended to give students a *real* introduction to control engineering—covering material at the introductory level that is truly new and up-to-date.

This textbook is written for a one-semester first course in Control Engineering. It is intended for an audience of students with a prior course in Signals and Systems so that, in particular, Laplace transforms and Bode plots are assumed to be known.

The emphasis of the text is on Control Engineering. The "mother discipline" of Linear Systems Theory receives its share of attention, but this is not a textbook on Systems Theory. The coverage of theory is limited to that which is to be applied, and proofs are sometimes given only for special, simple cases (e.g., matrices with distinct eigenvalues). It is the intent that enough of the practical flavor should come through, so as to whet an interest to the more formal theory.

The approach is a true blend of "classical" and "modern" methods. It is recognized that "modern" control theory is not limited to the state approach, but also includes frequency-based ideas, such as those of the H^∞ theory. Indeed, the contemporary trend is to use state-based methods as a convenient means of obtaining designs with good frequency properties; linear-quadratic state design, for example, is used with Loop Transfer Recovery methods as a computationally convenient tool for frequency design.

The computer is an essential ingredient in this book. Several good packages are available for microcomputers at prices affordable to universities. MATLAB$^{\circledR}$ is used throughout, but other packages could have been used equally well. The book does require the use of the computer as a regular working tool. The pedagogical approach assumes that to be so: for example, several of the Root Locus "rules" are omitted, because the detailed construction of the Locus is left to the computer.

There is no attempt in this text to teach numerical methods. Modern application software is sufficiently robust for most purposes, so that the study of numerical methods has become a more specialized topic. A few of the problems do explore numerical aspects, in order that the student may have some awareness of that general area.

Modeling and simulation are given more prominence than in most Control texts. It can be argued that, with more and more of the actual design burden being shifted to the computers, Control engineers will tend to concentrate on modeling. In any case, simulation occupies an important place in the practical design of control systems, and, with the advent of inexpensive and powerful computing, should now be receiving more attention in Control courses.

Models for a number of systems are introduced and defined in the problems in Chapter 2. These same models recur in problems in other chapters in order to serve as continuous examples for the application of the various techniques developed. A list of these systems and where they occur throughout the text can be found following the Table of Contents. This matrix is also provided in the solutions manual.

Some of the problems are "think" problems, designed to implant theoretical concepts. Others are relatively complex, but easily within the capabilities of today's microcomputers. This opens up the possibility of studying several control scenarios (e.g., several parameter values, bandwidths) for possible tradeoffs. Those problems requiring the use of MATLAB are identified as such with an icon: **ⓜ** .

Chapter 1 introduces the control problem. Chapter 2 is given to simulation and modeling. Several modeling examples are worked out, from several technical areas; those examples are carried throughout the book, as applications of control design concepts. The problems in Chapter 2 also recur as method problems in later chapters, so that students may retain their Chapter 2 simulations for later use.

Chapter 3 is the main Systems Theory chapter. Basic concepts of Linear Systems theory are covered, including: linearity, exponential matrices, poles and zeros in relation to state equations, controllability, observability, realizations, and stability. Chapter 4 studies performance measures and limitations. It covers open-loop control, an easy initial structure, and both single- and double-degrees-of-freedom configurations. It is somewhat novel in an undergraduate text, in that sensitivity and complementary sensitivity are used as the design parameters. The interpolation conditions imposed by the dynamics are studied and used to derive performance limitations. A brief introduction to the H^∞ ("one-block") problem is given, in order to provide a yardstick of achievable low-frequency performance.

Feedback stability in terms of the loop gain is the subject of Chapter 5. Standard tools such as the Routh criterion, the Root locus, the Nyquist criterion and the Nichols chart are covered. The effect of uncertainty is addressed through the Kharitonov polynomials and, in the frequency domain, the unstructured multiplicative uncertainty. Chapter 6 applies the concepts of Chapter 5 to design. Classical performance specifications are discussed, including dc steady-state behavior, transient response, peak frequency response and stability margins. There follows single-loop design with pure gain, lag, PI, lead, PD, lead-lag, PID, feedforward, and 2-degrees-of-freedom control. The last section treats briefly systems with delay and the Smith predictor.

Chapter 7 covers state-space design. It begins with a discussion of the properties of systems with state feedback, and moves on to pole placement feedback. The matrix Lyapunov equation is used to lead into Linear-Quadratic theory. The LQ state feedback is presented for the regulator problem for stable disturbances, and also with feedforward for constant disturbances. Next comes observer theory, first using pole placement, followed by a deterministic derivation for what is basically the Kalman filter. There follows an exposition of observer-based control laws.

Chapter 8 is concerned with multivariable control and H^∞ design. The emphasis is on H^2 and H^∞ design, so concepts such as singular values and norms are given prominence. The student is referred to the literature for several proofs, because the accent is placed on setting up problems and on obtaining computer solutions.

The final chapter introduces sampled-data implementation. It is not meant to be a thorough treatment of discrete-time control, and the accent is placed on the implementation in discrete-time of continuous-time control laws. Aliasing is discussed, and the z-transform is introduced. That is followed by the study of the stability of discrete-time systems, and of some common methods used to translate an analog design into its discrete counterpart.

At McGill, we have taught the first 6 chapters to senior undergraduates in Electrical Engineering, and the first 7 chapters to first-year graduate students from Electrical, Mechanical, and Biomedical Engineering. The semester in Canada has 13 weeks of lectures. Students are assumed to have had a good course in Signals and Systems, with some notions of Complex Variables theory. A course in Linear Algebra is essential. From experience, it appears that prior experience with MATLAB is not necessary.

I would like to thank Ms. Mildred Leavitt for typing the manuscript. Her patience with my handwriting has been exemplary.

Several drafts of the manuscript were scrutinized by numerous reviewers. In particular, I want to acknowledge James G. Owen for his thorough accuracy review; Joseph J. Feeley, University of Idaho; John A. Fleming, Texas A&M; Charles P. Newman, Carnegie Mellon; M.G. Rekoff, University of Alabama at Birmingham; Chris Rahn, Clemson University; Susan Riedel, Marquette University; Zvi Roth, Florida Atlantic University; Steven Yurkovich, Ohio State University; and Byron Winn, Colorado State University.

For those who would like to review any of the material necessary for the understanding of this text, suggested readings follow. In addition, a listing of references to using MATLAB is also included.

Useful background references:
On Systems:

Kwakernaak, H., R. Sivan and R. C. W. Strijbos, *Modern Signals and Systems*, Prentice-Hall (1991).

Oppenheim, A. V. and A. S. Willsky, with I. T. Young, *Signals and Systems*, Prentice-Hall (1983).

Ziemer, R. E., W. H. Tranter and D. R. Fannin, *Signals and Systems: Continuous and Discrete*, Macmillan (1993).

Gabel, R. A. and R. A. Roberts, *Signals and Linear Systems*, Wiley (1987).

Glisson, R. H., *Introduction to Systems Analysis*, McGraw-Hill (1985).

Neff, H. P. Jr., *Continuous and Discrete Linear Systems*, Harper and Row (1984).

On MATLAB:

The MathWorks, Inc., The Student Edition of *MATLAB*, Prentice-Hall (1992).

Strum, R. D., and D. E. Kirk, *Contemporary Linear Systems Using MATLAB*, PWS Publishing Company (1994).

Kuo, B. C. and D. C. Hanselman, *MATLAB Tools for Control System Analysis and Design*, Prentice-Hall (1994).

Etter, D. M., *Engineering Problem Solving with MATLAB*, Prentice-Hall (1993).

Saadat, H., *Computational Aids in Control Systems Using MATLAB*, McGraw-Hill (1993).

Contents

Problem Matrix:
This grid outlines by chapter, where you will find the recurring case studies of eleven systems modeled in Chapter 2.

	Servo Simplified Model	Servo Flexible Shaft	Drum Speed Control	Blending Tank	Heat Exchanger	Chemical Reactor	Flow Control	Crane	High Wire Artist	Two-pendula Problem	Maglev
Chapter 2 Simulation & Modelling	2.4	2.5	2.6	2.7 2.13	2.8 2.14	2.9 2.15	2.10 2.16	2.11 2.17	2.12 2.18	2.19	2.0
Chapter 3 Linear Systems Theory	3.13 3.28 3.40	3.14 3.29 3.41 3.53	3.15 3.30 3.46	3.17 3.32 3.47	3.18 3.33 3.54	3.19 3.34 3.42 3.48		3.20 3.35	3.16 3.31	3.21 3.36	3.22 3.37 3.38
Chapter 4 Specifications, Structures, Limitations									4.31	4.32	
Chapter 5 Feedback System Stability in Terms of Loop Gain	5.15 5.42	5.16 5.22			5.17	5.18		5.20 5.47	5.19 5.46	5.21 5.48	
Chapter 6 Classical Design	6.25 6.38	6.7 6.12 6.26	6.8	6.33	6.13 6.17 6.28 6.32 6.45	6.14 6.27 6.34 6.39	6.18 6.19 6.20	6.29			6.9 6.10 6.35
Chapter 7 State Feedback	7.10 7.32 7.60	7.11 7.22 7.46 7.51 7.61 7.62	7.23 7.33 7.34 7.47 7.63	7.24 7.35	7.12 7.36 7.39 7.40 7.64	7.13 7.41 7.42 7.65		7.15 7.26 7.52 7.66	7.14 7.25 7.53 7.67	7.27 7.48 7.54 7.68	7.16 7.28 7.55 7.56 7.69 7.70
Chapter 8 Multivariable Control			8.3 8.15	8.4	8.16	8.5				8.17	8.18
Chapter 9 Sample Data Implementation	9.34 9.40 9.47	9.35 9.41 9.48	9.42	9.43	9.37 9.49	9.36		9.38 9.44		9.45 9.50	9.39 9.46

Chapter

1

Introduction

1.1 CONTROL SYSTEMS AND WHY WE NEED THEM

In many situations, it is necessary that the value of a variable be kept near some target value. The target value may be constant, in which case the objective is *regulation*, or it may vary, and the problem becomes one of *tracking*. Examples of variables to be regulated are the speed of rotation of a computer disk, the moisture content of paper in the papermaking process, and the chemical composition of the outlet of a reactor. Examples of tracking, or *servo*, problems are the control of a robotic manipulator along a preset trajectory, the control of an aircraft or missile to make it follow a specified flight path, and the tracking of a target by an antenna or a telescope.

God must have been a control engineer, for the human body is replete with clever regulation and servo systems. Our body temperature is held very near 37°C by a system that uses a multitude of sensors and several mechanisms (perspiration, circulation, and shivering) to affect temperature. The way in which we reach out to an object, grasp it, and place it represents an admirable solution to a very tough control problem, certainly a much better one than can be managed with robotics at this time.

There are many reasons why we need control systems, but the four most important ones are performance, economics, safety, and reliability. Many systems cannot achieve specified levels of performance without controls. A high-performance aircraft must have a control system to achieve the required maneuverability. A robot depends greatly on its control system to achieve good accuracy. A distillation column must be controlled if the chemical composition of its outlet stream is to be within specifications.

Economic considerations are also important, especially in *process control*, i.e., in the control of production systems. There are many so-called *continuous processes*, designed to operate in the steady state. Power plants, distillation columns, and paper machines are all examples of systems that are designed to keep constant operating conditions except during start-up and shutdown. The operating points of the units of such processes are often chosen to maximize economic returns, subject to inequality constraints imposed on certain variables by quality or safety considerations. It is often the case that the optimal solution calls for operation at a constraint boundary. For example, the basis weight (weight per unit area) of paper produced by a papermaking machine is required to be above a certain lower limit. It is clear that, the lighter the paper, the smaller the quantity of pulp required, and hence the lower the cost. Therefore, the optimal operating point calls for a basis weight exactly at the limit. However, because of basis weight fluctuations, the operating point must be moved to some value greater than the limit, so as to guarantee that specifications will be met most of the time despite fluctuations. A good control system minimizes those fluctuations and makes it possible to operate closer to the constraint boundaries. This type of analysis is often used to justify the acquisition of a control system, bearing in mind that, in high-volume industries, a small-percentage improvement may represent very substantial savings. Another economic benefit, one that is more difficult to predict, is that a process can often be run at a higher capacity with improved control.

Safety is a third justification for control. An aircraft landing under poor visibility conditions must be controlled to follow a safe glide path. A nuclear reactor must operate in such a way that key variables are kept within safe limits. Many systems have "danger zones"; it is the task of the control system to avoid them.

Finally, control is used to achieve better reliability. Physical systems are usually less prone to failure if they are operated smoothly and with low levels of fluctuation. A sheet of paper threaded through drying rolls is less likely to break if the speed of the rolls is nearly constant. Automobile manufacturers advise drivers to avoid quick starts and stops in order to lengthen the lives of their cars. Nature appears to abhor radical, quick changes, as it is said to abhor a vacuum.

1.2 PHYSICAL ELEMENTS OF A CONTROL SYSTEM

Figure 1.1 depicts the elements of a control system. The word "plant" is used generally, to denote the object under control; in this context, an aircraft is a plant.

A plant has *output variables*, some of which are the ones to be controlled. These variables are measured by *sensors*. A sensor is basically a transducer, i.e., a device that transforms one type of physical quantity to another, usually electrical. Examples of sensors are tachometers, accelerometers, thermocouples, strain gauges, and pH meters.

A plant must have *input variables*, which can be manipulated to affect the outputs. The elements that permit these manipulations are called *actuators*. Control valves, hydraulic actuators, and variable voltage sources are examples of actuators. Actuators are often themselves complete systems, with their own controls.

The role of the *control elements* is to carry out the control strategy, i.e., to derive command signals for the actuators in response to the sensor outputs. The control elements may be analog devices, but digital control has become prevalent. In the latter case, the control element includes the A/D and D/A converters and a portion of the computer software.

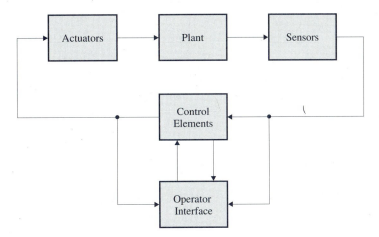

Figure 1.1   Physical components of a control system

The *operator interface* is the window to the outside world that allows human monitoring and intervention. It receives information concerning the inputs and outputs, plus certain status variables from the control elements. It activates displays (dials, strip recorder charts, computer graphics) and triggers alarm indicators (red lights, bells). It also serves as a means of altering the control strategy—for example, by changes in set points or control strategy parameters.

At this point, the reader should be warned that not all systems are physical systems, and that the blocks of Figure 1.1 do not always correspond to physical devices. For example, the plant may be the economy of a nation, with inputs such as monetary and fiscal policies and outputs such as GNP growth rate, unemployment rate, and inflation rate. It is still possible to identify elements that behave functionally like sensors and actuators, but these elements are not physical devices.

1.3   ABSTRACT ELEMENTS OF A CONTROL SYSTEM

The great power of control theory is that it is applicable to so many types of systems. That is why it is taught in so many disciplines, including electrical, chemical, and mechanical engineering; operations research; and economics. The power of control theory derives from its ability to fit a multitude of different problems into a single abstract, mathematical framework.

Figure 1.2 shows the abstract elements of a control system. The plant is represented by $\mathcal{P}$, which specifies the mathematical operations that generate the output variables $\mathbf{y}$ from the input variables $\mathbf{u}$ and $\mathbf{w}$. Vectors are used to indicate that, in general, there are several inputs and outputs. It is convenient to divide the inputs into two categories: the *manipulated inputs* $\mathbf{u}$, and the *disturbance inputs* $\mathbf{w}$.

The sensors are represented by another mathematical operator, $\mathcal{S}$, operating on the output variables $\mathbf{y}$ and another set of disturbances, $\mathbf{v}$, to produce the measured outputs, $\mathbf{y}_m$.

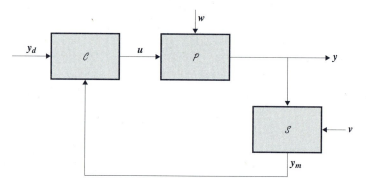

Figure 1.2   Abstract elements of a control system

Finally, the controller is represented by the operator $\mathcal{C}$, which acts on the measurements and the target (or setpoint) variables $\mathbf{y}_d$ to generate the actuator commands $\mathbf{u}$. The actuators are lumped with the plant in this representation.

1.4   THE DESIGN PROCESS

In general terms, the objective of a control system is to maintain the vector of outputs $\mathbf{y}(t)$ close to the vector of desired values $\mathbf{y}_d(t)$, in spite of the disturbances. The design of a control system is divided into two distinct portions that deal respectively with the abstract and physical levels.

The end product of the design at the abstract level is the specification of the mathematical operator $\mathcal{C}$, which acts on the measurements $\mathbf{y}_m$ and the desired inputs $\mathbf{y}_d$ to produce the command signals $\mathbf{u}$. In order to carry this out, the following information is required:

1. An approximation of $\mathcal{P}$, the plant operator; this approximation is called a *mathematical model*.

2. Characterizations of the functions $\mathbf{y}_d(t)$ and $\mathbf{w}(t)$, the desired values and disturbances. These functions may be described in the time domain as specific functions (e.g., steps) or as sets of functions. They are often characterized in the frequency domain according to properties of their spectra.

3. A set of performance specifications.

4. Mathematical models for the actuators and sensors. The latter include the characterization of the noise variables $\mathbf{v}(t)$, if such are used to represent measurement errors.

5. Restrictions on the structure of $\mathcal{C}$. For example, analog proportional, integral, derivative (PID) controllers have the mathematical structure

$$u(t) = \frac{1}{T_I} \int (y_d - y_m)dt + k(y_d - y_m) + T_D \frac{d}{dt}(y_d - y_m)$$

and the design is reduced to the choice of the constants T_I, k, and T_D. Digital systems may also be limited as to structure; in many turnkey systems, control algorithms must be built from certain given software "building blocks."

At the physical level, one is concerned with the following:

1. The choice of actuators and sensors.
2. The selection of the control elements. In most cases, the control system includes a computer, which raises issues of both hardware and software.
3. The design of the man–machine interface. This includes the definition of alarm conditions and backup procedures as well as the selection of display devices.

The design at the abstract level results in the definition of a control strategy. It is primarily a mathematical exercise, and that exercise is the subject of this book. In actual fact, control engineers spend much more time designing the physical system, where the issues are quite different and include economics, project management, availability of equipment, reliability of both equipment and vendors, interaction with clients, ergonomics, and labor relations. The two levels interact to a considerable extent, and the overall process is iterative. For example, performance specifications are seldom set a priori, but are often the result of compromises between performance, actuator, sensor, and computer costs and the needs of the end user.

The reader will readily appreciate that a control engineering project is a team undertaking. It is almost impossible for one person to be an expert in all the subdisciplines that are brought into play. This book will serve as an introduction to the design process at the abstract level. This process is general, and is applicable to control problems in all areas of technology as well as to many socioeconomic systems. The control engineer must bring to the team a high level of expertise in this kind of design. To be an effective member of the team, he or she must also have some acquaintance with instrumentation, computer engineering, and economics. The serious student is strongly encouraged to plan his or her studies so as to get at least some exposure to those subjects.

2 Simulation and Modeling

2.1 INTRODUCTION

The starting point for a control system design is a model of the plant. There are two basic approaches to modeling. When modeling from *fundamental principles*, we identify the important physical processes taking place in the system, and we write equations describing those processes and their interrelations. The *black-box* approach lies at the other end of the scale and makes no assumptions about internal mechanisms. When we use a black-box approach, we observe actual inputs and outputs and try to fit a model to the observations. For example, we may start from a transfer function of a given order and proceed to adjust its parameters so that the model response to an input approximates the plant's response. Between these two extremes lies a variety of "gray-box" methods in which the structure of the model is dictated by the physics but the parameters are identified by experiment.

The identification of parameters from experimental data is a subject unto itself and lies beyond the scope of this book. The models we shall use will be derived mostly from fundamental principles.

If modeling is the first step in control system design, simulation is the last. To simulate a control system is to solve the equations that describe it, given specific inputs and disturbances. The equations to be solved include those of the controllers, actuators, and sensors as well as those of the plant model. Sometimes the design is carried out directly by simulation, if it requires only modest adjustments of a previous design. In that case, the designer proceeds by trial and error, manipulating a few key controller parameters. More often, the simulation is the final check of the design before the physical implementation. This check is often necessary because of the common practice of using simplified models to obtain a design, which is then verified with a more realistic simulation model.

2.2 STATE EQUATIONS

We begin with the simple example of Figure 2.1. A mass M is moving on a horizontal plane under the influence of an external force, F, against a friction force, $D(v)$, where v is the velocity. From Newton's second law,

$$M\frac{dv}{dt} = F - D(v)$$

$$\frac{dv}{dt} = \frac{F}{M} - \frac{D(v)}{M}. \tag{2.1}$$

Also,

$$\frac{dx}{dt} = v. \tag{2.2}$$

Equations 2.1 and 2.2 describe the motion of the mass. Note that these two equations have very different origins: Equation 2.1 comes from the physics of the problem, whereas Equation 2.2 simply expresses the definition of velocity.

Equation 2.1 is a conservation law and expresses the rate of change of momentum in terms of force. Such rate equations are quite common in physics because of the many conservation principles, e.g., mass and energy. Rate equations are generally first-order differential equations, as are many other equations that, like Equation 2.2, express definitions. It follows that a large class of models can be expressed as sets of first-order differential equations, as follows:

$$\dot{x}_1 = f_1(x_1, \ x_2, \dots, x_n, \ u_1, \ u_2, \dots, u_r, \ t)$$

$$\dot{x}_2 = f_2(x_1, \ x_2, \dots, x_n, \ u_1, \ u_2, \dots, u_r, \ t)$$

$$\vdots \qquad\qquad \vdots$$

$$\dot{x}_n = f_n(x_1, \ x_2, \dots, x_n, \ u_1, \ u_2, \dots, u_r, \ t) \tag{2.3}$$

$$y_1 = h_1(x_1, \ x_2, \dots, x_n, \ u_1, \ u_2, \dots, u_r, \ t)$$

$$\vdots \qquad\qquad \vdots$$

$$y_m = h_m(x_1, \ x_2, \dots, x_n, \ u_1, \ u_2, \dots, u_r, \ t). \tag{2.4}$$

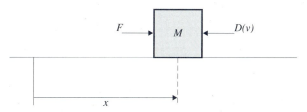

Figure 2.1   Mass moving on a plane

Here, u_1, u_2, ..., u_r are the r inputs, and y_1, y_2, ..., y_m are the m outputs. The variables x_1, x_2, ..., x_n are called *state variables*. Equations 2.3, the *state equations*, describe the dynamical behavior of the system. Equations 2.4, the *output equations*, are algebraic expressions for the outputs in terms of the state variables, the inputs, and time.

Equations 2.1 and 2.2 are in state form. If we define x and v as state variables and F as the input, then

$$x_1 = x$$

$$x_2 = v$$

$$u = F$$

and we write

$$\dot{x}_1 = x_2$$

$$\dot{x}_2 = -\frac{D(x_2)}{M} + \frac{u}{M} \tag{2.5}$$

so that, comparing with Equation 2.3,

$$f_1(x_1,\ x_2,\ u,\ t) = x_2$$

$$f_2(x_1,\ x_2,\ u,\ t) = -\frac{D(x_2)}{M} + \frac{u}{M}.$$

The output equations depend on the choices of output variables. For example, if x is considered the output of interest, then

$$y_1 = x_1 \tag{2.6}$$

is the output equation, and

$$y_1 = h_1(x_1,\ x_2,\ u,\ t) = x_1. \tag{2.7}$$

The right-hand sides in Equations 2.3 and 2.4 are explicit functions of t, which means that the equations themselves, rather than just the variables, are changing with time. The right-hand sides in Equations 2.5 and 2.6 do not depend explicitly on time; Equation 2.5 would if, for example, M represented the mass of a container that was simultaneously moving and being filled with water. Systems in which neither the f's nor the h's depend explicitly on t are called *time-invariant*; other systems are *time-varying*.

If the f's and h's are linear functions of the state and input variables, the system is called a *linear system*. Equations 2.5 and 2.6 represent a linear system if the function $D(v)$ is linear in v.

It is very cumbersome to carry so many variables and equations, and we use vector notation to reduce the clutter. Define the following vectors:

$$\mathbf{x} = \begin{bmatrix} x_1 \\ x_2 \\ \vdots \\ x_n \end{bmatrix}, \qquad \mathbf{y} = \begin{bmatrix} y_1 \\ y_2 \\ \vdots \\ y_m \end{bmatrix}, \qquad \mathbf{u} = \begin{bmatrix} u_1 \\ u_2 \\ \vdots \\ u_r \end{bmatrix}.$$

Define also the vector functions

$$\mathbf{f}(\ ) = \begin{bmatrix} f_1(\ ) \\ f_2(\ ) \\ \vdots \\ f_n(\ ) \end{bmatrix}, \qquad \mathbf{h}(\ ) = \begin{bmatrix} h_1(\ ) \\ h_2(\ ) \\ \vdots \\ h_m(\ ) \end{bmatrix}.$$

Then Equations 2.3 and 2.4 are written compactly as

$$\dot{\mathbf{x}} = \mathbf{f}(\mathbf{x}, \mathbf{u}, t) \tag{2.8}$$

$$\mathbf{y} = \mathbf{h}(\mathbf{x}, \mathbf{u}, t). \tag{2.9}$$

The system is linear if the functions $\mathbf{f}(\ )$ and $\mathbf{h}(\ )$ are expressed as linear functions of $\mathbf{x}$ and $\mathbf{u}$ (but not t). Linear expressions are expressed compactly using matrices and vectors. This leads to the following equations:

$$\dot{\mathbf{x}} = A(t)\mathbf{x} + B(t)\mathbf{u} \tag{2.10}$$

$$\mathbf{y} = C(t)\mathbf{x} + D(t)\mathbf{u}. \tag{2.11}$$

If the matrices are constant, the system is time-invariant.

The state variables $\mathbf{x}$ have an important property. A key result from the theory of differential equations is that, given $\mathbf{u}(t)$ for $t > t_0$ and given an n-vector $\mathbf{x}_0$, then, for $t \geq t_0$, there may exist a unique solution $\mathbf{x}(t)$ of Equations 2.7 with the property that $\mathbf{x}(t_0) = \mathbf{x}_0$. These existence and uniqueness results require only mild conditions on the functions $f_1(\ ), f_2(\ ), \ldots, f_n(\ )$. A consequence of this result is that, given the state at some time t_0 and the future inputs $\mathbf{u}(t), t > t_0$, the future evolution of the state variables is uniquely determined. Since the outputs $\mathbf{y}(t)$ are algebraically (i.e., instantaneously) related to $\mathbf{x}(t)$ and $\mathbf{u}(t)$, future output evolution is also determined by $\mathbf{x}(t_0)$ and future inputs. For example, the motion of the mass of Figure 2.1 for $t > 0$ is uniquely determined by the force $F(t), t > 0$, and by the position, and velocity at $t = 0$, i.e., by the initial conditions.

The state summarizes the influence of the past on the future. The mass of Figure 2.1 may have moved in many different ways prior to $t = 0$, but as far as its future motion is concerned, the only thing that matters about the past is its end result, position and velocity, at $t = 0$.

A system that can be described by a finite number of state variables is called a *lumped system*. Not all systems are of this type. For example, to predict the motion of a vibrating string, we need to know the initial position and velocity of every point along the string's length. But those are functions of the distance variable, and a function has an infinite number of points; it cannot be specified by a finite number of variables. There are many such systems in practice—for example, a flexible link of a robotic manipulator, a packed distillation column, a transmission line. They are usually represented by partial differential equations and are called *distributed systems*. Fortunately, it is almost always possible to approximate a distributed system by use of a lumped system; that is why, except for rare excursions, this book is concerned with lumped systems.

State equations have the further advantage of being in a form suitable for numerical solution. Mathematically,

$$\mathbf{x}(t) = \mathbf{x}(t_0) + \int_{t_0}^{t} \mathbf{f}[\mathbf{x}(\tau), \mathbf{u}(\tau), \tau] d\tau. \tag{2.12}$$

The integration in Equation 2.12 cannot be carried out directly, because $\mathbf{x}(\tau), t_0 \leq \tau < t$, is not known in advance; this integral is not a quadrature.

The numerical solution of differential equations has been, and continues to be, the subject of much investigation [1,2]. Today the problems are well understood, and many software packages have been developed to solve differential equations.

2.3 MODELING FROM CONSERVATION PRINCIPLES

Conservation principles lead to balance equations, in which the rate of change of a quantity (e.g., momentum, matter, or heat) is equated to other quantities, such as force and flow. This leads quite naturally to equations in state form.

This section gives several examples of modeling. They will serve throughout the book to demonstrate design principles.

Example 2.1   (dc Servo)

Description: A dc motor with constant field is driven by application of a voltage to its armature terminals. Through a set of gears, the motor drives a load with moment of inertia J, subject to an external torque (see Fig. 2.2). The control objective is to keep the load angle at some desired value.

Inputs and Outputs: The armature voltage v is the control input, and the load torque T is a disturbance. The outputs are the shaft angle θ and angular velocity ω.

Basic Principles: The gear ratio N is the ratio of angles and velocities of the two shafts; the torques have the same ratio. The load shaft is the high-torque, low-velocity shaft. The torque T_m produced by a dc motor is given by $T_m = k_1 \phi i$, where k_1 is a constant and ϕ is the field intensity. The armature circuit has resistance R and inductance L, and the motion generates a back emf $k_2 \phi \omega_m$. It can be shown that $k_1 = k_2$ in Système International (SI) units. The rotor of the dc motor has inertia J_m.

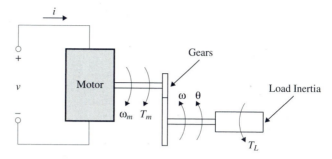

Figure 2.2   A dc servomechanism

Objectives: Write a model for the system, and simulate for the conditions given.

Solution   Applying Newton's second law to the rotor,

$$J_m \dot{\omega}_m = T_m - T_e$$

where T_e is the torque exerted on the motor shaft by the load, transmitted through the gears. Thus,

$$T_e = T_m - J_m \dot{\omega}_m$$

so the torque at the motor shaft is seen to be the torque generated by the motor, minus the torque required to accelerate the rotor.

The torque exerted by the motor on the load shaft, transmitted through the gears, is NT_e. Newton's second law applied to the load is

$$J\dot{\omega} = NT_e - T_L$$
$$= NT_m - T_L - NJ_m \dot{\omega}_m.$$

Since $\omega_m = N\omega$, this becomes

$$(J + N^2 J_m)\dot{\omega} = NT_m - T_L$$

or

$$J_e \dot{\omega} = NT_m - T_L \tag{2.13}$$

where $J_e = J + N^2 J_m$ is the effective inertia seen at the load shaft. With ϕ constant, let $k_1 \phi = k_2 \phi = K_m$. Then

$$T_m = K_m i$$

so that

$$\dot{\omega} = \frac{NK_m}{J_e}i - \frac{T_L}{J_e}. \tag{2.14}$$

To apply Equation 2.14, we need the current, i. By Kirchhoff's voltage law,

$$L\frac{di}{dt} + Ri = v - K_m\omega_m$$

where $K_m\omega$ is the back emf. This becomes

$$i = -\frac{R}{L}i + \frac{v}{L} - \frac{NK_m}{L}\omega \tag{2.15}$$

where the fact that $\omega_m = N\omega$ has been used. Because the angle θ is of interest, the definition equation

$$\dot{\theta} = \omega \tag{2.16}$$

is added. Equations 2.14, 2.15, and 2.16 are the desired equations. In matrix form,

$$\frac{d}{dt}\begin{bmatrix} \theta \\ \omega \\ i \end{bmatrix} = \begin{bmatrix} 0 & 1 & 0 \\ 0 & 0 & \frac{NK_m}{J_e} \\ 0 & -\frac{NK_m}{L} & -\frac{R}{L} \end{bmatrix}\begin{bmatrix} \theta \\ \omega \\ i \end{bmatrix} + \begin{bmatrix} 0 & 0 \\ 0 & -\frac{1}{J_e} \\ \frac{1}{L} & 0 \end{bmatrix}\begin{bmatrix} v \\ T_L \end{bmatrix}. \tag{2.17}$$

Since θ and ω have been specified as the outputs, the output equation is

$$\begin{bmatrix} \theta \\ \omega \end{bmatrix} = \begin{bmatrix} 1 & 0 & 0 \\ 0 & 1 & 0 \end{bmatrix}\begin{bmatrix} \theta \\ \omega \\ i \end{bmatrix}. \tag{2.18}$$

For the specific values $K_m = .05$ Nm/A, $R = 1.2$ Ω, $L = .05$ H, $J_m = 8 \times 10^{-4}$ kg m^2, $J = 0.020$ kg m^2, and $N = 12$, the state equations are

$$\frac{d}{dt}\begin{bmatrix} \theta \\ \omega \\ i \end{bmatrix} = \begin{bmatrix} 0 & 1 & 0 \\ 0 & 0 & 4.438 \\ 0 & -12 & -24 \end{bmatrix}\begin{bmatrix} \theta \\ \omega \\ i \end{bmatrix} + \begin{bmatrix} 0 & 0 \\ 0 & -7.396 \\ 20 & 0 \end{bmatrix}\begin{bmatrix} v \\ T_L \end{bmatrix}. \tag{2.19}$$

The simulation conditions are as follows: with zero initial conditions and $T_L = 0$, $v(t) = 3$ V for $0 \le t \le 2$ and -3 V for $2 < t \le 4$. The result (MATLAB command lsim) is shown in Figure 2.3.

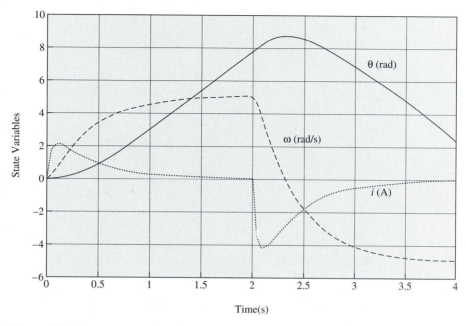

Figure 2.3 Time responses for the dc servo

<hr>

Example 2.2 **(Active Suspension)**

Description: Figure 2.4 shows a simple model of a vehicle suspension. The mass M is the body of the vehicle, and the small mass m is the unsprung part, e.g., the tires, axles, and so forth. Spring K_1 and dash-pot D are passive suspension elements, with a force u, possibly provided by a hydraulic actuator, added for active suspension. Spring K_2 acts between the axle and the road and models the stiffness of the tires. The height y_R of the roadway with respect to an inertial reference varies, and the vehicle is moving at constant speed V. The purpose of control is to ensure ride quality.

Inputs and Outputs: The active suspension force u is the control input, and the roadway height $y_R(t)$ is the disturbance input. The output is the height of the vehicle body, x_1, with respect to the reference.

Basic Principles: The force exerted by a spring is $K(x - x_0)$, where x is the spring length and x_0 is the length when the spring is at rest; this is known as Hooke's law. A dash-pot exerts a force $D\dot{x}$, where x is the distance between the two ends.

Objectives: Write a model for the system, and simulate for the conditions given.

Solution The force exerted by the top spring is $K_1(x_1 - x_2 - x_{10})$, where x_{10} is the rest length. The reference direction is such as to pull the two masses together; physically, if x_2 stays fixed and x_1 increases, the force must act downward on M. The damping

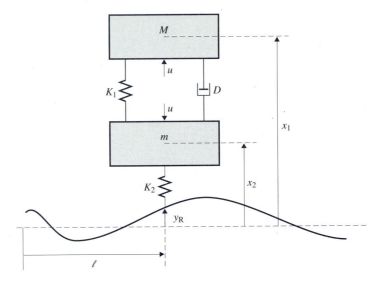

Figure 2.4   An active suspension system

force is $D(\dot{x}_1 - \dot{x}_2)$. If x_2 is fixed and $\dot{x}_1$ is positive, the damper resists motion and hence must pull M downward. With $v_1 = \dot{x}_1$ and $v_2 = \dot{x}_2$, Newton's second law yields

$$M\frac{dv_1}{dt} = -K_1(x_1 - x_2 - x_{10}) - D(v_1 - v_2) - Mg + u$$

$$m\frac{dv_2}{dt} = K_1(x_1 - x_2 - x_{10}) + D(v_1 - v_2) - K_2(x_2 - y_R - x_{20}) - mg - u.$$

Note that, because of Newton's third law, u must act in opposite directions on M and m.

Including the two velocity definitions, the state equations are

$$\dot{x}_1 = v_1$$

$$\dot{x}_2 = v_2$$

$$\dot{v}_1 = -\frac{K_1}{M}(x_1 - x_2 - x_{10}) - \frac{D}{M}(v_1 - v_2) - g + \frac{1}{M}u$$

$$\dot{v}_2 = \frac{K_1}{m}(x_1 - x_2 - x_{10}) + \frac{D}{m}(v_1 - v_2)$$

$$-\frac{K_2}{m}(x_2 - y_R - x_{20}) - g - \frac{1}{m}u. \tag{2.20}$$

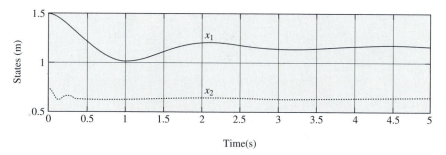

Figure 2.5 Time responses for the active suspension system: Zero-input case

The output equation is

$$x_1 = \begin{bmatrix} 1 & 0 & 0 & 0 \end{bmatrix} \begin{bmatrix} x_1 \\ x_2 \\ v_1 \\ v_2 \end{bmatrix}. \tag{2.21}$$

For the specific values $M = 300$ kg, $m = 50$ kg, $K_1 = 3000$ N/m, $K_2 = 3 \times 10^4$ N/m, $D = 600$ N/ms^{-1}, $x_{10} = 1.5$ m, $x_{20} = .75$ m, and $g = 9.8$ m/s^2, the state equations are

$$\frac{d}{dt} \begin{bmatrix} x_1 \\ x_2 \\ v_1 \\ v_2 \end{bmatrix} = \begin{bmatrix} 0 & 0 & 1 & 0 \\ 0 & 0 & 0 & 1 \\ -10 & 10 & -2 & 2 \\ 60 & -660 & 12 & -12 \end{bmatrix} \begin{bmatrix} x_1 \\ x_2 \\ v_1 \\ v_2 \end{bmatrix}$$
$$+ \begin{bmatrix} 0 \\ 0 \\ .00334 \\ -.02 \end{bmatrix} u + \begin{bmatrix} 0 \\ 0 \\ 0 \\ 600 \end{bmatrix} y_R + \begin{bmatrix} 0 \\ 0 \\ 5.2 \\ 350.2 \end{bmatrix}. \tag{2.22}$$

The simulation conditions are as follows:

1. $x_1(0) = 1.5$ m, $x_2(0) = .75$ m, $v_1(0) = v_2(0) = 0$, $u(t) = y_R(t) = 0$, $0 \leq t \leq 5$ s. This shows the transient behavior of the system. The result (MATLAB command step) is given in Figure 2.5.

2. Initial conditions are the final conditions in 1, i.e., the steady-state equilibrium. The road has a rectangular bump of height 0.15 m and length 2 m. Assume $u(t) = 0$. Simulate for $0 \leq t \leq 5$ s, with $t = 0$ being the time at which the vehicle just hits the bump, for a vehicle speed of 30 km/h. It is easily seen that $y_R(t)$ is a rectangular pulse of height 0.15 m and duration 2.0/V, where V is the speed in meters per second. The results (MATLAB command lsim) are shown in Figure 2.6.

3. Initial conditions are steady-state equilibrium; $0 \leq t \leq 5$ s; the road height is described by the signal $\sum_{n=0}^{4} 0.1 \sin(5 + 4n)t$. This signal is meant to

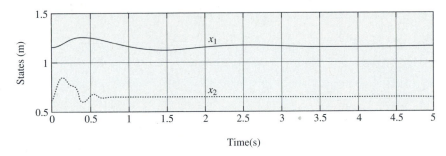

Figure 2.6 Time responses for the active suspension system: Responses to a pulse

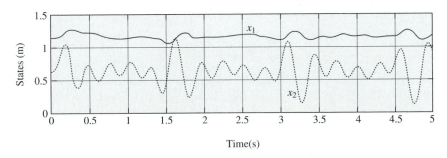

Figure 2.7 Time responses for the active suspension system: Response to a sum of sinusoids

approximate a random process by summing harmonically unrelated sinusoids. The results (MATLAB command lsim) are given in Figure 2.7.

Example 2.3 **(Train)**

Description: A train, consisting of one or two locomotives and many identical cars, is being driven over a horizontal track. One locomotive heads the train, and there may be a second at the rear. The cars and locomotives are joined by identical couplings modeled by a spring-and-damper parallel combination. (See Figure 2.8 for details). The control objectives are to maintain a desired speed and to avoid overstressing the couplings.

Inputs and Outputs: The control inputs are the forces F_1 and F_2 supplied by the locomotives or, more precisely, the forces exerted by the tracks on the locomotives. Velocities are with respect to the ground. The outputs are the velocity of the head locomotive and the spacing between the units of the train.

Basic Principles: The couplers joining the ith and $(i+1)$st units deliver a force $K(x_{i+1} - x_0) + D(v_i - v_{i+1})$ where v_i is the velocity of the ith car with respect to the inertial reference frame.

Objective: Derive a model, and simulate under given conditions.

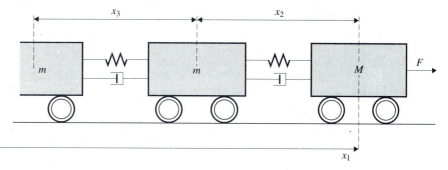

Figure 2.8 Model of a train

Solution For the head locomotive,

$$M\dot{v}_1 = F_1 - K(x_2 - x_0) - D(v_1 - v_2). \tag{2.23}$$

For the ith car,

$$m\dot{v}_i = K(x_i - x_0) + D(v_{i-1} - v_i) - K(x_{i+1} - x_0) - D(v_i - v_{i+1}),$$
$$i = 2, 3, \ldots, N - 1 \tag{2.24}$$

For the Nth car,

$$m\dot{v}_N = K(x_N - x_0) + D(v_{N-1} - v_N). \tag{2.25}$$

If the Nth car is replaced by a second locomotive, m is changed to M and a force F_2 is added to the right-hand side (RHS) of Equation 2.25.

The other state equations are

$$\dot{x}_1 = v_1$$
$$\dot{x}_2 = v_1 - v_2$$
$$\dot{x}_3 = v_2 - v_3$$
$$\vdots$$
$$\dot{x}_N = v_{N-1} - v_N. \tag{2.26}$$

The outputs are v_1, and the state variables $x_2, x_3, \ldots, x_N$.

For the specific values $M = 2 \times 10^5$ kg, $m = 40,000$ kg, $K = 2.5 \times 10^6$ N/m, $D = 1.5 \times 10^5$ N/m/s, and $x_0 = 20$ m, the state equations are Equation 2.26 and

$$\dot{v}_1 = -12.5x_2 - .75v_1 + .75v_2 + 250 + .005F_1$$
$$\dot{v}_i = 62.5x_i - 62.5x_{i+1} + 3.75v_{i-1} - 7.5v_i + 3.75v_{i+1}, \qquad i = 2, 3, \ldots, N - 1$$
$$\dot{v}_N = 62.5x_N + 3.75v_{N-1} - 3.75v_N - 1250 \tag{2.27}$$

if the last unit is a car, and

$$\dot{v}_N = 12.5x_N + .75v_{N-1} - .75v_N - 250 + .005F_2$$

if it is a locomotive. Note that F_1 and F_2 are in kilonewtons.

The simulation conditions are: $N = 5$, with only one locomotive; $x_1(0) = 0, x_2(0) = x_3(0) = x_4(0) = x_5(0) = 20$ m; $F_1 = 750$ kN. The simulation interval is 10 s, and the results (MATLAB command step) are shown in Figures 2.9 and 2.10.

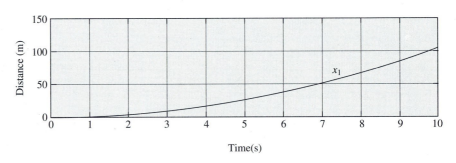

Figure 2.9   Response to a force step of the distance of the locomotive from the origin

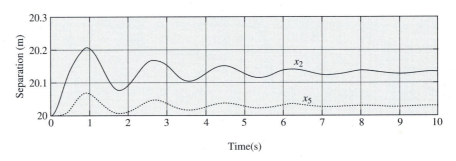

Figure 2.10   Responses to a force step of the coupler extensions, first and last couplers

◆ ◆ ◆ **R E M A R K**

Example 2.3 illustrates the application of engineering common sense to simulation. The locomotive distance coordinate x_1 is defined with respect to a stationary origin, and $x_2, x_3, \ldots$ could have been defined in similar fashion. The forces, expressed as differences $x_i - x_{i-1}$, would then have involved differences of large numbers of almost equal values, leading to numerical inaccuracies.   ◆

Example 2.4 (Level Control)

Description: A tank of uniform cross section is fed at the top by a flow subject to variations. Liquid is withdrawn from the bottom by a control valve. The valve is adjusted to vary the outflow in order to maintain a constant level. (See Fig. 2.11.) The outlet pressure is atmospheric.

Inputs and Outputs: The control input is the valve position u. The disturbance input is the flow F_{in}. The output is the level, ℓ.

Basic Principles: Over a given time interval, the difference between the volumes of liquid flowing in and flowing out of the tank is translated into a change in level. As for the valve, it is an orifice of area proportional to u, the position of the valve spool. The volumetric flow through an orifice is theoretically equal to $S\sqrt{(2\Delta P)/\rho}$, where S is the area of the orifice, ρ is the liquid density, and ΔP is the pressure across the orifice [3]. In actual fact, this expression must be multiplied by a coefficient somewhat less than 1, the *orifice coefficient*. A valve is basically a device with an orifice of variable area. In a *linear* valve, the orifice area is proportional to the displacement of a piston. The displacement is called the *stroke*.

Objectives: Derive a model, and simulate for the following conditions.

Solution

Let A be the cross-sectional area of the tank. Over a small time interval Δt, the volume of liquid flowing in is $F_{in}\Delta t$, and the volume flowing out is $F_{out}\Delta t$. The change in the volume of the liquid is also given by $A\Delta\ell$, where $\Delta\ell$ is the change in level. Thus,

$$A\Delta\ell = F_{in}\Delta t - F_{out}\Delta t.$$

Dividing by Δt and passing to the limit,

$$A\frac{d\ell}{dt} = F_{in} - F_{out}. \tag{2.28}$$

The flow through the valve is given by $c'u\sqrt{(2\Delta P)/\rho}$, where c' is a constant comprising both the constant of proportionality between u and S and the orifice coefficient. The pressure at the bottom of the tank is $\rho g\ell$, where ρ is the density

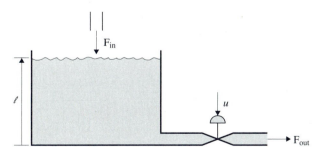

Figure 2.11 Tank with valve pertaining to level control

of the liquid; that is, the pressure is the weight of a column of height ℓ and unit area. Therefore,

$$F_{\text{out}} = c'u\sqrt{\frac{2\rho g \ell}{\rho}}$$

$$= cu\sqrt{\ell} \qquad (2.29)$$

where c is a constant encompassing all constants in the equations.

From Equations 2.28 and 2.29, the state equation is

$$\dot{\ell} = -\frac{c}{A}u\sqrt{\ell} + \frac{F_{\text{in}}}{A}. \qquad (2.30)$$

Specific values are $A = 1$ m^2, $c = 2.0$ m$^{3/2}$/s. With these,

$$\dot{\ell} = -2.0u\sqrt{\ell} + F_{\text{in}}. \qquad (2.31)$$

The simulation conditions are as follows: $\ell(0) = 1$ m, $u(t) = .01$ m, $F_{\text{in}}(t) = 0, 0 \leq t \leq 100$ s. These conditions correspond to the tank being emptied at constant valve opening. Figure 2.12 shows the behavior of the level (MATLAB command ode23). Note that the behavior is *not* exponential: the asymptotic value $\ell = 0$ is reached in finite time.

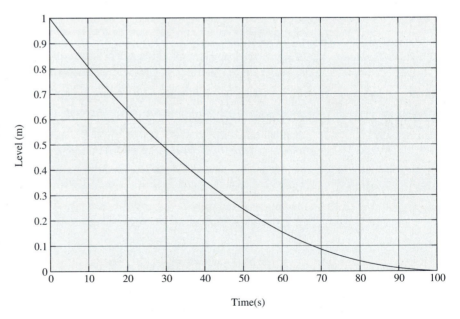

Figure 2.12   Response of level with zero in flow

2.4 MODELING WITH LAGRANGE'S EQUATIONS

Lagrange's equations constitute a well-known and useful technique for the analysis of mechanical systems [4,5]. To use Lagrange's equations, we define a set of *generalized coordinates*, that is, a set of positions and angles that completely describe the motion of the system. These coordinates must be independent; that is, motion obtained by arbitrary specification of coordinate time history must be mechanically possible.

The kinetic energy of the system is a function of the generalized coordinates q_i and their derivatives, and is written as $T(\mathbf{q}, \dot{\mathbf{q}})$. The potential energy is a function of the q_i and is written as $V(\mathbf{q})$.

The *Lagrangian L* is defined as

$$L = T - V. \tag{2.32}$$

To write Lagrange's equations, we need to define the *generalized forces*, F_i. We do this by computing the work done by all nonconservative forces when q_i is changed to $q_i + dq_i$ with all other coordinates held fixed. For infinitesimal dq_i, the work is proportional to dq_i, and the proportionality factor is F_i.

Lagrange's equations are as follows:

$$\frac{d}{dt}\left(\frac{\partial L}{\partial \dot{q}_i}\right) - \frac{\partial L}{\partial q_i} = F_i, \qquad i = 1, 2, \ldots, n. \tag{2.33}$$

Example 2.5   (Pendulum on a Cart)

Description: An inverted pendulum of mass m and length ℓ moves in the vertical plane, about a horizontal axis fixed on a cart. The cart, of mass M, moves horizontally in one dimension, under the influence of a force F. (See Fig. 2.13). The pendulum rod is assumed to have zero mass. There is no friction in the system. The force F is to be manipulated to keep the pendulum vertical.

Inputs and Outputs: The input is the force F, and the outputs are the angle θ and the distance x.

Objective: Write the equations, and simulate under given conditions.

Solution   The generalized coordinates are x and θ. The velocity of m has two components, one due to the motion of the cart and the other due to the angular motion of the pendulum. The velocity of the cart is $\dot{x}$ in the horizontal direction.

The horizontal position of the mass m is $x + \ell \sin \theta$, and its vertical position is $\ell \cos \theta$. Therefore, the total kinetic energy is

$$T = \frac{1}{2}M\dot{x}^2 + \frac{1}{2}m\left[\left\{\frac{d}{dt}(x + \ell \sin\theta)\right\}^2 + \left\{\frac{d}{dt}(\ell \cos\theta)\right\}^2\right]$$

$$= \frac{1}{2}M\dot{x}^2 + \frac{1}{2}m[(\dot{x} + \ell\dot{\theta}\cos\theta)^2 + (-\ell\dot{\theta}\sin\theta)^2].$$

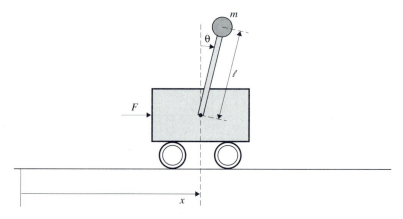

Figure 2.13   Pendulum on a cart

The potential energy of m varies with height. If V_0 is the potential energy of m for $\theta = 90°$, then

$$V = V_0 + mg\ell \cos\theta.$$

Thus,

$$L = \frac{1}{2}M(\dot{x})^2 + \frac{1}{2}m[(\dot{x} + \ell\dot{\theta}\cos\theta)^2 + (\ell\dot{\theta}\sin\theta)^2] - V_0 - mg\ell\cos\theta.$$

The only nonconservative force is F. If x is held fixed and θ is changed to $\theta + d\theta$, F does no work: the generalized force associated with θ is zero. If θ is held fixed and x changes to $x + dx$, the work done is $F\,dx$; therefore, F is the generalized force associated with x.

We may now write Lagrange's equations:

$$\frac{\partial L}{\partial \dot{x}} = M\dot{x} + m(\dot{x} + \ell\dot{\theta}\cos\theta)$$

$$\frac{\partial L}{\partial x} = 0$$

$$\frac{d}{dt}\left(\frac{\partial L}{\partial \dot{x}}\right) = M\ddot{x} + m\ddot{x} + m\ell\ddot{\theta}\cos\theta - m\ell(\dot{\theta})^2\sin\theta.$$

The equation related to x is

$$(M + m)\ddot{x} + m\ell\ddot{\theta}\cos\theta - m\ell(\dot{\theta})^2\sin\theta = F. \qquad (2.34)$$

For the θ equations,

$$\frac{\partial L}{\partial \dot{\theta}} = m(\dot{x} + \ell \dot{\theta} \cos \theta)\ell \cos \theta + m\ell^2 \dot{\theta} \sin^2 \theta$$

$$= m\ell \dot{x} \cos \theta + m\ell^2 \dot{\theta}$$

$$\frac{\partial L}{\partial \theta} = -m(\dot{x} + \ell \dot{\theta} \cos \theta)\ell \dot{\theta} \sin \theta + m\ell^2 (\dot{\theta})^2 \sin \theta \cos \theta + mg\ell \sin \theta$$

$$= -m\ell \dot{\theta}\dot{x} \sin \theta + mg\ell \sin \theta$$

$$\frac{d}{dt}\left(\frac{\partial L}{\partial \dot{\theta}}\right) = m\ell \ddot{x} \cos \theta - m\ell \dot{\theta}\dot{x} \sin \theta + m\ell^2 \ddot{\theta}.$$

The equation pertaining to θ is

$$m\ell \ddot{x} \cos \theta - m\ell \dot{\theta}\dot{x} \sin \theta + m\ell^2 \ddot{\theta} + m\ell \dot{\theta}\dot{x} \sin \theta - mg\ell \sin \theta = 0$$

or

$$\ddot{x} \cos \theta + \ell \ddot{\theta} - g \sin \theta = 0. \qquad (2.35)$$

Equations 2.34 and 2.35 are not state equations.

Define $v = \dot{x}$ and $\omega = \dot{\theta}$, and write Equations 2.34 and 2.35 as

$$\begin{bmatrix} M + m & m\ell \cos \theta \\ \cos \theta & \ell \end{bmatrix} \begin{bmatrix} \dot{v} \\ \dot{\omega} \end{bmatrix} = \begin{bmatrix} F + m\ell \omega^2 \sin \theta \\ g \sin \theta \end{bmatrix}.$$

Solving for $\dot{v}$ and $\dot{\omega}$ yields

$$\dot{v} = \frac{F + m\ell \omega^2 \sin \theta - mg \sin \theta \cos \theta}{M + m(1 - \cos^2 \theta)} \qquad (2.36)$$

$$\dot{\omega} = \frac{-F \cos \theta - m\ell \omega^2 \sin \theta \cos \theta + (M + m)g \sin \theta}{\ell[M + m(1 - \cos^2 \theta)]}. \qquad (2.37)$$

Append the definitions

$$\dot{x} = v \qquad (2.38)$$

$$\dot{\theta} = \omega. \qquad (2.39)$$

Equations 2.36 to 2.39 are the four state equations. Specific values are $\ell = 1$ m and $M = m = 1$ kg. The state equations are

$$\dot{x} = v$$

$$\dot{\theta} = \omega$$

$$\dot{v} = \frac{F + \omega^2 \sin \theta - 9.8 \sin \theta \cos \theta}{2 - \cos^2 \theta}$$

$$\dot{\omega} = \frac{-F \cos \theta - \omega^2 \sin \theta \cos \theta + 19.6 \sin \theta}{2 - \cos^2 \theta}. \qquad (2.40)$$

The simulation conditions are as follows: $x(0) = v(0) = \omega(0) = 0, \theta(0) = 0.1$ rad, $F(t) = 0, 0 \leq t \leq 1$ s. Figure 2.14 shows the results (MATLAB command ode23). This system is seen to be unstable. The pendulum falls to the right ($\theta > 0$) while the cart goes to the left.

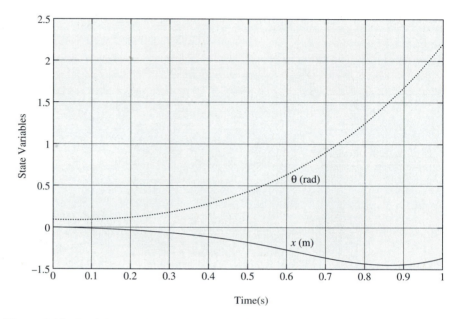

Figure 2.14   Pendulum angle and cart distance from a nonzero initial state

2.5 LINEARIZATION

The task of a control system is often to maintain given constant operating conditions—for example, constant speed, level, position, or basis weight. To achieve this objective, we use a two-step procedure:

1. Select a dc steady state that corresponds to desired constant values of **u** and/or **y**.

2. Design a control strategy to generate increments in the control in response to deviations from the dc steady state.

To do this, we need to study (i) the dc steady state of a system and (ii) the model that relates the deviations from steady state, i.e., the *small-signal*, or *incremental*, model.

We begin with

$$\dot{\mathbf{x}} = \mathbf{f}(\mathbf{x}, \mathbf{u}) \tag{2.41}$$

$$\mathbf{y} = \mathbf{h}(\mathbf{x}, \mathbf{u}). \tag{2.42}$$

Note that the functions $\mathbf{f}$ and $\mathbf{h}$ are not explicitly functions of t, so the system is time-invariant.

For constant $\mathbf{u} = \mathbf{u}^*$, $\mathbf{x}^*$ is an *equilibrium state* if $\mathbf{f}(\mathbf{x}^*, \mathbf{u}^*) = \mathbf{0}$. We shall use the symbol $\mathbf{0}$ to denote a vector whose elements are all 0. If $\mathbf{x} = \mathbf{x}^*$ and $\mathbf{u} = \mathbf{u}^*$, then $\dot{\mathbf{x}} = \mathbf{0}$ and the state remains at $\mathbf{x}^*$; i.e., $\mathbf{x}^*$ is an equilibrium point with $\mathbf{u} = \mathbf{u}^*$.

The output corresponding to an equilibrium state $\mathbf{x}^*$ is $\mathbf{y}^* = \mathbf{h}(\mathbf{x}^*, \mathbf{u}^*)$. Therefore, the dc steady-state quantities satisfy

$$\mathbf{f}(\mathbf{x}^*, \mathbf{u}^*) = \mathbf{0}$$
$$\mathbf{h}(\mathbf{x}^*, \mathbf{u}^*) = \mathbf{y}^*. \tag{2.43}$$

A dc steady state is defined by choosing some of the variables in Equation (2.43) and solving for the others. There is no guarantee that a solution will exist, or that it will be unique. With n states, r inputs, and m outputs, Equation 2.43 represents $n + m$ nonlinear equations with $n + m + r$ variables. In most cases, it will not be possible to predetermine more than r of those variables. For example, it will not usually be possible to set 2 outputs ($m = 2$) at arbitrary values if the system has only one input ($r = 1$).

The next step is to write equations for incremental variables, i.e., for deviations from equilibrium. Let

$$\mathbf{x}(t) = \mathbf{x}^* + \Delta\mathbf{x}(t), \qquad \mathbf{u}(t) = \mathbf{u}^* + \Delta\mathbf{u}(t), \qquad \mathbf{y}(t) = \mathbf{y}^* + \Delta\mathbf{y}(t).$$

Because $\dot{\mathbf{x}}^* = \mathbf{0}$, substitution in Equations 2.41 and 2.42 yields

$$\Delta\dot{\mathbf{x}} = \mathbf{f}(\mathbf{x}^* + \Delta\mathbf{x}, \mathbf{u}^* + \Delta\mathbf{u}) \tag{2.44}$$
$$\Delta\mathbf{y} = \mathbf{h}(\mathbf{x}^* + \Delta\mathbf{x}, \mathbf{u}^* + \Delta\mathbf{u}) - \mathbf{y}^*. \tag{2.45}$$

Expanding the components of $\mathbf{f}$ in a Taylor series, we obtain

$$f_i(\mathbf{x}^* + \Delta\mathbf{x}, \mathbf{u}^* + \Delta\mathbf{u}) = f_i(\mathbf{x}^*, \mathbf{u}^*) + \left.\frac{\partial f_i}{\partial x_1}\right|_* \Delta x_1 + \left.\frac{\partial f_i}{\partial x_2}\right|_* \Delta x_2 + \cdots$$
$$+ \left.\frac{\partial f_i}{\partial x_m}\right|_* \Delta x_n + \left.\frac{\partial f_i}{\partial u_1}\right|_* \Delta u_1 + \cdots + \left.\frac{\partial f_i}{\partial u_r}\right|_* \Delta u_r$$
$$+ \text{ higher-order terms in } \Delta x, \Delta u. \tag{2.46}$$

Here, the notation "$\left.\vphantom{x}\right|_*$" means "evaluated at $\mathbf{x}^*, \mathbf{u}^*$." At this point, it is assumed that the Δx's and Δu's are sufficiently small to justify neglecting the higher-order terms. If the control system to be designed works at all well, that assumption should be satisfied.

Without the higher-order terms, and with $\mathbf{f}(\mathbf{x}^*, \mathbf{u}^*) = \mathbf{0}$, the RHS of Equation 2.46 is the ith member of a set of n equations, written in matrix form as

$$\mathbf{f}(\mathbf{x}^* + \Delta\mathbf{x}, \ \mathbf{u}^* + \Delta\mathbf{u}) = \left.\frac{\partial\mathbf{f}}{\partial\mathbf{x}}\right|_* \Delta\mathbf{x} + \left.\frac{\partial\mathbf{f}}{\partial\mathbf{u}}\right|_* \Delta\mathbf{u} \tag{2.47}$$

where

$$\frac{\partial\mathbf{f}}{\partial\mathbf{x}} = \begin{bmatrix} \dfrac{\partial f_1}{\partial x_1} & \dfrac{\partial f_1}{\partial x_2} & \cdots & \dfrac{\partial f_1}{\partial x_n} \\[2ex] \dfrac{\partial f_2}{\partial x_1} & \dfrac{\partial f_2}{\partial x_2} & \cdots & \dfrac{\partial f_2}{\partial x_n} \\[2ex] \vdots & & & \\[2ex] \dfrac{\partial f_n}{\partial x_1} & \cdots & \cdots & \dfrac{\partial f_n}{\partial x_n} \end{bmatrix}$$

is the Jacobian of $\mathbf{f}$ with respect to $\mathbf{x}$, with a similar definition for $\dfrac{\partial\mathbf{f}}{\partial\mathbf{u}}$, the Jacobian with respect to $\mathbf{u}$. Thus, Equation 2.44 becomes approximately

$$\Delta\dot{\mathbf{x}} = \left.\frac{\partial\mathbf{f}}{\partial\mathbf{x}}\right|_* \Delta\mathbf{x} + \left.\frac{\partial\mathbf{f}}{\partial\mathbf{u}}\right|_* \Delta\mathbf{u}. \tag{2.48}$$

As for Equation 2.45, since $\mathbf{y}^* = \mathbf{h}(\mathbf{x}^*, \mathbf{u}^*)$, we have

$$\Delta\mathbf{y} = \left.\frac{\partial\mathbf{h}}{\partial\mathbf{x}}\right|_* \Delta\mathbf{x} + \left.\frac{\partial\mathbf{h}}{\partial\mathbf{u}}\right|_* \Delta\mathbf{u} \tag{2.49}$$

for small $\Delta\mathbf{x}$, $\Delta\mathbf{u}$.

Note that the Jacobians in Equations 2.48 and 2.49 are constant matrices, because they are evaluated at specific values $\mathbf{x}^*$ and $\mathbf{u}^*$. Note also that the right-hand sides of those equations are linear functions of $\Delta\mathbf{x}$ and $\Delta\mathbf{u}$, so the incremental system is *linear* and *time-invariant*.

It is also possible to linearize about a *trajectory*—a nominal set of time functions, $\mathbf{x}^*(t)$ and $\mathbf{u}^*(t)$, that satisfy the state equations. An example would be a robotic manipulator following a preset path. In such a case, the linearized system is *time-varying* (see Problem 2.21).

If some of the inputs are disturbances, it is often desirable to separate them from the control inputs. The linearized equations become

$$\Delta \dot{\mathbf{x}} = \left. \frac{\partial \mathbf{f}}{\partial \mathbf{x}} \right|_* \Delta \mathbf{x} + \left. \frac{\partial \mathbf{f}}{\partial \mathbf{u}} \right|_* \Delta \mathbf{u} + \left. \frac{\partial \mathbf{f}}{\partial \mathbf{w}} \right|_* \Delta \mathbf{w} \qquad (2.50)$$

$$\Delta \mathbf{y} = \left. \frac{\partial \mathbf{h}}{\partial \mathbf{x}} \right|_* \Delta \mathbf{x} + \left. \frac{\partial \mathbf{h}}{\partial \mathbf{u}} \right|_* \Delta \mathbf{u} + \left. \frac{\partial \mathbf{h}}{\partial \mathbf{w}} \right|_* \Delta \mathbf{w} \qquad (2.51)$$

where $\mathbf{w}$ is the vector of disturbance inputs.

If the original system is linear and time-invariant, it is represented by equations of the form

$$\dot{\mathbf{x}} = A\mathbf{x} + B\mathbf{u} + F\mathbf{w}$$

$$\mathbf{y} = C\mathbf{x} + D\mathbf{u} + G\mathbf{w}. \qquad (2.52)$$

The equilibrium point satisfies

$$\mathbf{0} = A\mathbf{x}^* + B\mathbf{u}^* + F\mathbf{w}^*$$

$$\mathbf{y}^* = C\mathbf{x}^* + D\mathbf{u}^* + G\mathbf{w}^*. \qquad (2.53)$$

If $\mathbf{u}^*$ and $\mathbf{w}^*$ are given, a unique solution $\mathbf{x}^*$ (hence $\mathbf{y}^*$) always exists if A is nonsingular. If A is singular, there are multiple solutions if the vector $B\mathbf{u}^* + F\mathbf{w}^*$ is in the *range space* of A, i.e., can be constructed by a linear combination of the columns of A; if that is not the case, there is no solution.

If $\mathbf{y}^*$ and $\mathbf{w}^*$ are given and we wish to solve for $\mathbf{x}^*$ and $\mathbf{u}^*$, it is useful to write Equation 2.53 as

$$\begin{bmatrix} A & B \\ C & D \end{bmatrix} \begin{bmatrix} \mathbf{x}^* \\ \mathbf{u}^* \end{bmatrix} = \begin{bmatrix} 0 \\ \mathbf{y}^* \end{bmatrix} - \begin{bmatrix} F \\ G \end{bmatrix} \mathbf{w}^*. \qquad (2.54)$$

If $m = r$ (equal number of inputs and outputs), then the matrix on the left-hand side (LHS) of Equation 2.54 is square, and a unique solution exists if that matrix is nonsingular. If $r > m$ (more inputs than outputs), and if the matrix has maximal rank $n + m$, there exist multiple solutions to Equation 2.54. Finally, if $r < m$ (fewer inputs than outputs) and the matrix has maximal rank $n + r$, there is a (unique) solution only in the special case where $\mathbf{y}^*$ and $\mathbf{w}^*$ are such that the RHS of Equation 2.54 in the range space of the matrix $\left[\begin{smallmatrix} A & B \\ C & D \end{smallmatrix} \right]$.

As for the incremental system, with

$$\mathbf{x} = \mathbf{x}^* + \Delta \mathbf{x}, \qquad \mathbf{u} = \mathbf{u}^* + \Delta \mathbf{u}, \qquad \mathbf{y} = \mathbf{y}^* + \Delta \mathbf{y}, \qquad \mathbf{w} = \mathbf{w}^* + \Delta \mathbf{w}$$

Equations 2.52 become

$$\Delta\dot{\mathbf{x}} = A\mathbf{x}^* + B\mathbf{u}^* + F\mathbf{w}^* + A\,\Delta\mathbf{x} + B\,\Delta\mathbf{u} + F\Delta\mathbf{w}$$

$$\mathbf{y}^* + \Delta\mathbf{y} = C\mathbf{x}^* + D\mathbf{u}^* + G\mathbf{w}^* + C\,\Delta\mathbf{x} + D\,\Delta\mathbf{u} + G\,\Delta\mathbf{w}$$

which, in view of Equation 2.53, yields

$$\Delta\dot{\mathbf{x}} = A\,\Delta\mathbf{x} + B\,\Delta\mathbf{u} + F\,\Delta\mathbf{w}$$

$$\Delta\mathbf{y} = C\,\Delta\mathbf{x} + D\,\Delta\mathbf{u} + G\,\Delta\mathbf{w}. \tag{2.55}$$

Equation 2.55 expresses the fact that a linear system is its own incremental system, and therefore no extra work is needed to obtain the incremental system in that case.

Example 2.6   **(dc Servo)**

For the servomechanism of Example 2.1, calculate the constant equilibrium point for $T_L = 0$ and $\theta^* = \theta_d$. Give the incremental model.

Solution   From Equation 2.17 and the first of the two output equations in Equation 2.18, application of Equation 2.53 yields

$$\begin{bmatrix} 0 \\ 0 \\ 0 \end{bmatrix} = \begin{bmatrix} 0 & 1 & 0 \\ 0 & 0 & \dfrac{NK_m}{J_e} \\ 0 & \dfrac{-NK_m}{L} & -\dfrac{R}{L} \end{bmatrix} \begin{bmatrix} \theta^* \\ \omega^* \\ i^* \end{bmatrix} + \begin{bmatrix} 0 \\ 0 \\ \dfrac{1}{L} \end{bmatrix} v^*$$

$$\theta_d = [1 \quad 0 \quad 0] \begin{bmatrix} \theta^* \\ \omega^* \\ i^* \end{bmatrix}.$$

It follows easily that $\omega^* = i^* = v^* = 0$.
    The incremental variables are

$$\Delta\theta = \theta - \theta_d, \qquad \Delta\omega = \omega - \omega^* = \omega, \qquad \Delta i = i - i^* = i, \qquad \Delta v = v - v^* = v.$$

Following Equation 2.55, the incremental model is

$$\frac{d}{dt} \begin{bmatrix} \Delta\theta \\ \omega \\ i \end{bmatrix} = \begin{bmatrix} 0 & 1 & 0 \\ 0 & 0 & \dfrac{NK_m}{J_e} \\ 0 & \dfrac{-NK_m}{L} & \dfrac{-R}{L} \end{bmatrix} \begin{bmatrix} \Delta\theta \\ \omega \\ i \end{bmatrix} + \begin{bmatrix} 0 \\ 0 \\ \dfrac{1}{L} \end{bmatrix} v$$

$$\begin{bmatrix} \Delta\theta \\ \omega \end{bmatrix} = \begin{bmatrix} 1 & 0 & 0 \\ 0 & 1 & 0 \end{bmatrix} \begin{bmatrix} \Delta\theta \\ \omega \\ i \end{bmatrix}.$$

Example 2.7 **(Active Suspension)**

Calculate the dc steady state for the suspension system of Example 2.2, with $y_R = 0, u^* = 0$.

Solution From Equation 2.20,

$$0 = v_1{}^*$$

$$0 = v_2{}^*$$

$$0 = -\frac{K_1}{M}x_1{}^* + \frac{K_1}{M}x_2{}^* - \frac{D}{M}v_1{}^* + \frac{D}{M}v_2{}^* + \frac{K_1}{M}x_{10} - g$$

$$0 = \frac{K_1}{m}x_1{}^* - \frac{K_1+K_2}{m}x_2{}^* + \frac{D}{m}v_1{}^* - \frac{D}{m}v_2{}^* - \frac{K_1}{m}x_{10} + \frac{K_2}{m}x_{20} - g.$$

The solution is

$$v_1{}^* = v_2{}^* = 0$$

$$x_1{}^* = x_{10} + x_{20} - \frac{Mg}{K_1} - \frac{m+M}{K_2}g = 1.156 \text{ m}$$

$$x_2{}^* = x_{20} - \frac{m+M}{K_2}g = 0.636 \text{ m}.$$

Example 2.8 **(Level Control)**

For the level control system of Example 2.4, find the dc steady state with $\ell^* = \ell_d$ and $F_{in}{}^* = F_d$. Derive the incremental model.

Solution From Equation 2.30,

$$-\frac{c}{A}u^*\sqrt{\ell_d} + \frac{F_d}{A} = 0$$

or

$$u^* = \frac{F_d}{c\sqrt{\ell_d}}$$

As for the incremental model, with f being the RHS of Equation (2.30),

$$\frac{\partial f}{\partial \ell} = -\frac{c}{2A}\frac{u}{\sqrt{\ell}}; \qquad \frac{\partial f}{\partial u} \doteq -\frac{c}{A}\sqrt{\ell}; \qquad \frac{\partial f}{\partial F_{in}} = \frac{1}{A}$$

so that

$$\dot{\Delta\ell} = -\frac{cu^*}{2A\sqrt{\ell_d}}\Delta\ell - \frac{c}{A}\sqrt{\ell_d}\Delta u + \frac{1}{A}\Delta F_{in}.$$

⌐┐ **Example 2.9**   **(Pendulum on a Cart)**

For the inverted pendulum of Example 2.5, calculate the incremental system, with $x^* = \theta^* = 0$.

Solution   The equilibrium point is simply the vertical equilibrium, with all quantities equal to zero.

To find the incremental system, it is possible to calculate the Jacobian corresponding to Equations 2.40. An alternative way is to write the Lagrangian for small changes from equilibrium. The Lagrangian is written by throwing out terms of order higher than 2 in the variables. We use the fact that $\sin\theta \approx \theta$ and $\cos\theta \approx 1 - \frac{1}{2}\theta^2$ to write

$$L = \frac{1}{2}M(\dot{x})^2 + \frac{1}{2}m(\dot{x} + \ell\dot{\theta})^2 - V_0 - mg\ell(1 - \frac{1}{2}\theta^2).$$

We write

$$\frac{\partial L}{\partial \dot{x}} = M\dot{x} + m(\dot{x} + \ell\dot{\theta}); \qquad \frac{\partial L}{\partial x} = 0$$

$$\frac{\partial L}{\partial \dot{\theta}} = m\ell(\dot{x} + \ell\dot{\theta}); \qquad \frac{\partial L}{\partial \theta} = mg\ell\theta.$$

Lagrange's equations are

$$(M + m)\ddot{x} + m\ell\ddot{\theta} = F$$

$$m\ell\ddot{x} + m\ell^2\ddot{\theta} - mg\ell\theta = 0.$$

Solving for $\ddot{x}$ and $\ddot{\theta}$ yields

$$\ddot{x} = \dot{v} = -\frac{mg}{M}\theta + \frac{F}{M}$$

$$\ddot{\theta} = \dot{\omega} = \frac{(M + m)g}{M\ell}\theta - \frac{F}{M\ell}.$$

These, plus the two definitions $\dot{x} = v$ and $\dot{\theta} = \omega$, are the incremental state equations (no Δ's are required, because the variables are 0 at equilibrium).

___

This technique can be used to advantage in cases where the equilibrium point is easy to calculate and only the incremental equations are sought. Another, non-Lagrangian method is used in the following example. Basically, since the right-hand sides are linear in the state and control increments, we perturb the states and inputs one at a time to calculate the contribution of each.

Example 2.10   **(Sprung Beam)**

Figure 2.15 shows a bar moving in a vertical plane, resting on a pair of linear spring–damper combinations. The mass of the bar is M, and its moment of inertia about its center of gravity is J. A force u is applied at the center, in the vertical direction.

Write a linear model for small variations from equilibrium of height y and angle θ.

Solution   At equilibrium, the bar rests horizontally at some height y_0. For small motion, Newton's second law states that

$$M \Delta \ddot{y} = \Delta F$$

$$J \Delta \ddot{\theta} = \Delta T$$

where ΔF and ΔT are the incremental force and torque, referred to equilibrium. Invoking linearity, ΔF is expressed as a linear function of the incremental state variables and input. Thus,

$$\Delta F = k_1 \Delta y + k_2 \, \Delta \dot{y} + k_3 \, \Delta \theta + k_4 \, \Delta \dot{\theta} + k_5 \, \Delta u.$$

We obtain the constants by calculating the force, with each term acting alone.

If $\Delta \dot{y} = \Delta \theta = \Delta \dot{\theta} = \Delta u = 0$ and $\Delta y \neq 0$, the bar is at rest in the horizontal position, Δy away from y_0. The force on the bar is $-2K\Delta y$, so $k_1 = -2K$.

If $\Delta y = \Delta \theta = \Delta \dot{\theta} = \Delta u = 0$ and $\Delta \dot{y} \neq 0$, the bar is at its equilibrium position but is moving vertically. The force is $-2D\Delta \dot{y}$, so that $k_2 = -2D$.

If $\Delta y = \Delta \dot{y} = \Delta \dot{\theta} = \Delta u = 0$ and $\Delta \theta \neq 0$, the center of gravity is at its equilibrium position and the bar is at rest, at an angle $\Delta \theta$ to the horizontal. There is no net force, because the incremental forces exerted by the springs are equal and in opposite directions. Thus, $k_3 = 0$. It can be seen that k_4 is also zero. If $\Delta y = \Delta \dot{y} = \Delta \theta = \Delta \dot{\theta} = 0$ and $\Delta u \neq 0$, the net force is Δu, so $k_5 = 1$. Thus,

$$\Delta F = -2K \Delta y - 2D \Delta \dot{y} + \Delta u.$$

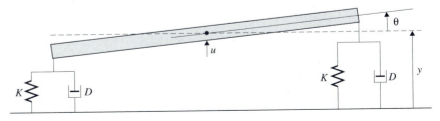

Figure 2.15   Beam moving in a vertical plane

Now we proceed to the torque expression. There are no torque components due to Δy, $\Delta \dot{y}$, and Δu. If $\Delta y = \Delta \dot{y} = \Delta \dot{\theta} = \Delta u = 0$ and $\Delta \theta \neq 0$, the bar is displaced by $L \Delta \theta / 2$ at each end, causing spring forces of magnitude $KL\Delta\theta/2$. The net torque is $-2KL^2 \Delta\theta/4$. Similarly, the net torque due to $\Delta\dot{\theta}$ is $-DL^2 \Delta\dot{\theta}/2$. Therefore,

$$\Delta T = -\frac{K}{2}L^2\Delta\theta - \frac{D}{2}L^2\Delta\dot{\theta}$$

and the state equations of the linearized system are

$$\Delta\dot{y} = \Delta v$$

$$\Delta\dot{\theta} = \Delta\omega$$

$$\Delta\dot{v} = -2\frac{K}{M}\Delta y - 2\frac{D}{M}\Delta v + 1\frac{1}{M}\Delta u$$

$$\Delta\dot{\omega} = -\frac{KL^2}{2J}\Delta\theta - \frac{DL^2}{2J}\Delta\omega$$

or, in vector-matrix form,

$$\frac{d}{dt}\begin{bmatrix} \Delta y \\ \Delta \theta \\ \Delta v \\ \Delta \omega \end{bmatrix} = \begin{bmatrix} 0 & 0 & 1 & 0 \\ 0 & 0 & 0 & 1 \\ -\dfrac{2K}{M} & 0 & -\dfrac{2D}{M} & 0 \\ 0 & -\dfrac{KL^2}{2J} & 0 & -\dfrac{DL^2}{2J} \end{bmatrix}\begin{bmatrix} \Delta y \\ \Delta \theta \\ \Delta v \\ \Delta \omega \end{bmatrix} + \begin{bmatrix} 0 \\ 0 \\ \dfrac{1}{M} \\ 0 \end{bmatrix}\Delta u.$$

2.6 MODEL UNCERTAINTY

A model is an abstract construct used to predict the behavior of output variables, given the time history of the input variables. Whether or not a model is "true" is irrelevant: the only function of a model is to predict the behavior of a physical system. The whole idea of a control system is to generate inputs that will produce desirable outputs; it is not surprising, therefore, that the design process depends on the predictive ability of the model.

We shall see that the achievable performance of a control system depends, in part, on the quality of the model: the more stringent the requirements, the better the model must be. On the other hand, model improvement often goes hand in hand with increased complexity. For example, the simplest model of a physical resistor is a resistance; at high frequencies, that simple model will not be adequate, and the addition of capacitances, inductances, and nonlinearities may be necessary. The development of a complex model requires time and experimental work, both of which have costs. Therefore, the idea is to use a model of no greater complexity than is needed by the performance to be achieved. This implies that the design of the control system needs to take into account the predictive inaccuracy of the model.

Part of the model uncertainty is not truly "uncertain." There is often a *truth model*, the most accurate model we have, and a simplified version, the *design model*, used as the basis for the design. The use of a linearized model of a nonlinear system falls in the latter category, as does the use of a single linear model for all expected operating points, as opposed to a different model for each one.

Some of the uncertainty can be accounted for by changes in the model parameters; this is known as *parametric* uncertainty. All models in our examples contain parameters: masses, spring constants, and so forth. The parameter values are known only to certain tolerances—hence the discrepancy between the model and the physical object if the parameter values are in error.

Some of the uncertainty comes from effects that were left out of the model to keep complexity down. Examples are parasitic capacitances, structural vibration modes in a robot whose links have some compliance, and vibration dynamics due to oil compressibility in a hydraulic system. Since we have no model, we cannot precisely assess the effect of that uncertainty. Engineering judgment and experience are invoked to produce rough estimates of the bounds, in the frequency domain, of the magnitude of the deviations.

Finally, some uncertainties are best described as the effects of extraneous signals, i.e., *disturbances*. If a good case can be made for a time-invariant model, output changes in the presence of constant control inputs are ascribed to unmodeled inputs rather than to time variability of the model.

In some cases, the physical origins of disturbances may be known; examples are the ocean waves that cause a ship to roll and the wind gusts act on an aircraft. In other cases, it is difficult to identify the causes. Disturbances can be modeled as signals in a number of ways—for instance, as random processes with given spectra, as bounded deterministic signals, or as finite sums of sinusoids.

Problems

2.1   Systems are often described by interconnected blocks, each with its own state equations. Figure 2.16 shows a parallel interconnection of two of these blocks, or subsystems, $\mathcal{A}_1$ and $\mathcal{A}_2$, with the following descriptions:

$$\mathcal{A}_1: \quad \dot{\mathbf{x}}1 = A_1\mathbf{x}1 + B_1\mathbf{u}1$$

$$\mathbf{y}1 = C_1\mathbf{x}1 + D_1\mathbf{u}1$$

$$\mathcal{A}_2: \quad \dot{\mathbf{x}}2 = A_2\mathbf{x}2 + B_2\mathbf{u}2$$

$$\mathbf{y}2 = C_2\mathbf{x}2 + D_2\mathbf{u}2.$$

The state of the composite system is $\mathbf{x} = \begin{bmatrix} x1 \\ x2 \end{bmatrix}$, and its input and output are $\mathbf{u}$ and $\mathbf{y}$, respectively. Write state and output equations for the system.

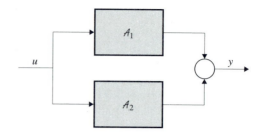

Figure 2.16 Parallel interconnection

2.2 Repeat Problem 2.1 for the system of Figure 2.17.

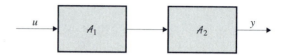

Figure 2.17 Series interconnection

2.3 Repeat Problem 2.1 for the system of Figure 2.18.

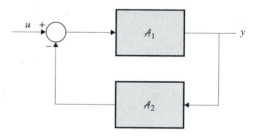

Figure 2.18 Feedback interconnection

M **2.4** ***Servo, simplified model*** Repeat Example 2.1 if the motor inductance L is negligible.

M **2.5** ***Servo with flexible shaft*** The low-velocity side of the gear box in Example 2.1 drives an inertial load through a shaft sufficiently long to exhibit torsional flexibility. The model of Figure 2.19 illustrates the situation: the spring is linear and develops a torque $K(\theta_1 - \theta_2)$.

    a. Model this system. Suggested steps:
       i. With $\omega_2 = \dot{\theta}_2$, $J\dot{\omega}_2 = $ torque from spring.
      ii. $J_m \dot{\omega}_m = T_m - \frac{1}{N}$ (torque exerted by spring).
     iii. Write di/dt as in Example 2.1, and use the fact that $\omega_m = N\dot{\theta}_1$.
     iv. Because $\theta_1 - \theta_2$ is small, the equations for $\dot{\omega}_1$ and $\dot{\omega}_2$ call for differences of almost equal terms. It is numerically preferable to work with $\Delta = \theta_1 - \theta_2$. Define $\dot{\Delta} = \Omega$, and write an equation for $\dot{\Omega}$ in terms of Δ and i. Write the state equations, using the state variables $\theta_2, \Delta, \omega_2, \Omega$, and i, with v as the input.

b. Write the state equations for the specific values of Example 2.1, with $K = 500$ Nm/rad.

c. Simulate under the conditions given in Example 2.1.

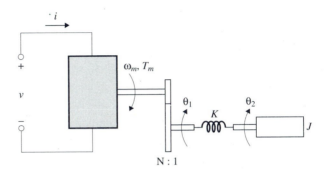

Figure 2.19 Servo with flexible shaft

M **2.6** *Drum speed control* In the system of Figure 2.20, two identical dc motors are working in cooperation to rotate a large drum, which is subject to a load torque. The motors are assumed to have rotor inertia J_m negligible inductance, and their shafts are modeled by rotary springs. (This model is useful to explain why shafts fail from vibration fatigue.)

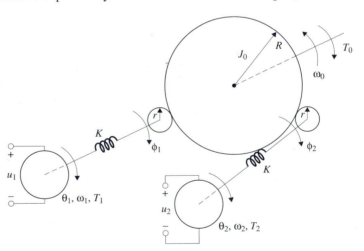

Figure 2.20 Drum driven by two motors

a. Model the system, with the following suggested steps:

  i. Write Newton's law for $\dot{\omega}_1$ and $\dot{\omega}_2$, using the motor torques T_1 and T_2 and the fact that the spring torques are proportional to the differences $\theta_1 - \phi_1$ and $\theta_2 - \phi_2$, respectively.

  ii. Write Newton's law for the large inertia; the torque applied by each motor shaft is multiplied by the ratio R/r.

iii. Write differential equations for $\Delta_1 = \theta_1 - \phi_1$ and $\Delta_2 = \theta_2 - \phi_2$. Note that $\dot{\phi}_1 = \dot{\phi}_2 = -\frac{R}{r}\omega_0$, so $\dot{\Delta}_1$ and $\dot{\Delta}_2$ are expressed in terms of ω_1, ω_2, and ω_0.

iv. Write the motor torques in terms of the applied armature voltages u_1 and u_2, the motor torque constant K_m, and the armature resistance R_a.

You should have five state variables, Δ_1, Δ_2, ω_1, ω_2, and ω_0, and three inputs, u_1, u_2, and T_0.

b. Write the state equations for the following values: $K_m = 6.0$ Nm/A, $R_a = 0.2\ \Omega$, $J_m = 1$ kg m^2, $K = 75{,}000$ Nm/rad, $J_0 = 10{,}000$ kg m^2, $R = 1$ m, $r = 0.07$ m.

c. Simulate for the following conditions: $\Delta_1(0) = \Delta_2(0) = 0$, $\omega_0(0) = 4.67$ rad/s, $\omega_1(0) = \omega_2(0) = 66.68$ rad/s, $u_1 = u_2 = 4$ V, and the load torque T_0 is a step of 1000 Nm.

M   **2.7**   ***Blending tank***   A blending tank has two input flows of products A and B, respectively (gin and vermouth, for example). The flow of B is uncontrolled, and the flows of A and the output are controlled. The flows F_A, F_B, and F_0 are in cubic meters per second. (See Fig. 2.21.) The tank is well mixed, so that uniform composition is assumed. A control system is required to regulate composition and level.

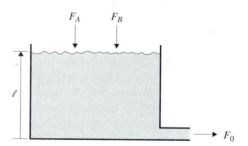

Figure 2.21   Blending tank

a. Derive a model for this system, with the following suggested steps as a guide:

i. Write the total volume V_T in the tank at $t + \Delta t$, in terms of the volume at t and the flows in and out of the tank during the interval t to $t + \Delta t$. Note that a flow F m^3/s carries $F\Delta t$ m^3 of material during Δt. With $V_T = A\ell$ (A = cross-sectional area), write a differential equation for ℓ in terms of the variables F_A, F_B, and F_0.

ii. Proceed as in (i) to write a differential equation for V_A, the volume of substance A in the tank. Note that, if the tank is well mixed, the concentration of A is $C_A = V_A/V_T$, and the rate of outflow of A is $C_A F_0$.

b. Write the state equations for the specific value $A = 0.1 \text{ m}^2$.

c. Simulate for $\ell(0) = 0.2$ m, $C_A(0) = 0.5$, $F_A = F_B = 0.0002 \text{ m}^3/\text{s}$, $F_0 = 0.0003 \text{ m}^3/\text{s}$. The simulation interval is $0 \le t \le 60$ s. (Note: You will need to calculate $V_T(0)$ from $l(0)$ and $C_A(0)$.)

2.8 **Heat exchanger** Figure 2.22 illustrates a counterflow heat exchanger. A hot liquid flows leftward at a rate F_H and transfers heat through a pipe to a cold liquid flowing rightward at a rate F_C. This is actually a distributed system; it is approximated by the system of Figure 2.23, where the fluids are assumed to flow between well-mixed tanks with uniform temperatures. The tanks appear in pairs, and heat is conducted across the separating surfaces.

The basic physical principles are that (i) the heat contained in a mass M of liquid is Mc_vT, where c_v is the heat capacity and T is the temperature, and (ii) the rate of heat conduction across a surface is proportional to the temperature difference across it. All tanks have volume V, and both liquids have density ρ and heat capacity c_v.

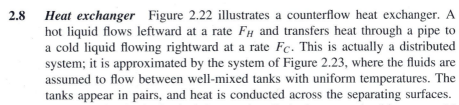

Figure 2.22 Counterflow heat exchanger

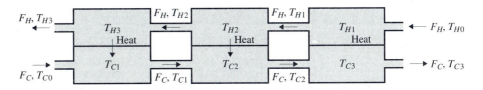

Figure 2.23 Lumped model of the heat exchanger

a. Model the system, with the following suggested steps:

i. Let $Q_{C1} = \rho V c_v T_{C1}$ be the heat contained in the first cold tank. During the interval t to $t + \Delta t$, a mass $\rho F_C \Delta t$ flows into the tank, at temperature T_{C0}. An equal mass flows out, but at temperature T_{C1}. Calculate the net heat conveyed *into* the tank by the two flows. Add $k(T_{H3} - T_{C1}) \Delta t$, the heat conducted across the boundary between the last hot tank and the first cold tank. Let $Q_{C1}(t + \Delta t) = Q_{C1}(t) +$ net heat flow during Δt.

ii. Let $\Delta t \to 0$ and write a differential equation for $Q_{C1}(t)$, then for $T_{C1}(t)$.

iii. Repeat in similar fashion for all tanks. The same flow rate, F_C, applies to all tanks on the cold side, while F_H is the constant flow on the hot side. The six temperatures are state variables, and the inputs are T_{C0}, T_{H0}, F_C, and F_H.

b. Write the state equations for the following specific values: $V = .2 \text{ m}^3$, $\rho = 10^3 \text{ kg/m}^3$, $c_v = 4180 \text{ J/kg}°\text{C}$, $k = 2 \times 10^5 \text{ J}/°\text{C min}$.

c. Simulate for $T_{C1}(0) = T_{C2}(0) = T_{C3}(0) = T_{H1}(0) = T_{H2}(0) = T_{H3}(0) = 20°\text{C}$, $F_C = 0.05 \text{ m}^3/\text{min}$, $F_H = 0.15 \text{ m}^3/\text{min}$, $T_{C0} = 20°\text{C}$, $T_{H0} = 80°\text{C}$.

M **2.9** *Chemical reactor* Figure 2.24 illustrates a so-called continuous stirred tank reactor (CSTR). An aqueous solution of temperature T_0, containing a material A at a concentration c_{A0} kg-moles/m³, flows into the reactor at a rate F m³/s. A controlled quantity of heat is fed to (or taken from) the reactor, at a rate Q watts. At a sufficiently high temperature, a chemical reaction, $A \to B + C$, takes place. The reactor is well mixed, so the concentrations of A, B, and C are uniform over the (constant) volume V of the tank. Density is assumed constant, as is the heat capacity, and the outlet flow is taken to be F. Some relevant principles are as follows: (i) The reaction proceeds at a rate of r kg-moles/s. (ii) In the reaction, one mole of A is converted to one mole each of B and C. (iii) The reaction liberates heat (is exothermic) at a rate proportional to r. The rate of reaction is given by $r = VAc_A e^{-E/RT}$, where V is the volume, c_A is the concentration of A in kilogram-moles per cubic meter, T is the temperature in degrees Kelvin, and A, E, and R are constants. The rate of heat liberated is $\Delta H r$, where ΔH is a positive constant.

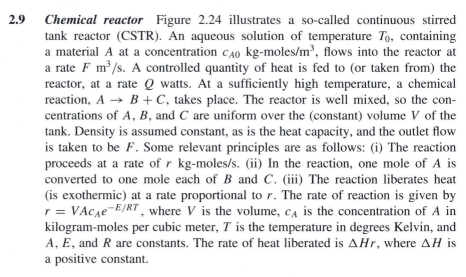

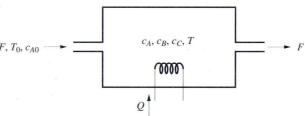

Figure 2.24 Chemical reactor

a. Derive a model for the reactor. Suggested steps are:

   i. *Mass balance.* Let $N_A(t) =$ number of kg-moles of A in the reactor at time t. Between t and $t + \Delta t$, $F c_{A0} \Delta t$ kg-moles of A flow in, $F c_A \Delta t$ kg-moles flow out, and $r \Delta t$ kg-moles are converted to B and C. Write $N_A(t + \Delta t)$ as $N_A(t) +$ net increase of A between t and $t + \Delta t$.

   ii. Obtain a differential equation for N_A. Since $c_A = N_A/V$, obtain a differential equation for c_A.

   iii. Repeat for c_B and c_C. Note that none of B or C flows into the tank; amounts $r \Delta t$ kg-moles of each are created during Δt, and $F c_B \Delta t (F c_C \Delta t)$ kg-moles are removed.

iv. *Energy balance.* The heat energy contained by the tank is $H(t) = \rho V c_v T(t)$. During the interval Δt, the inflow conveys $F \rho c_v T_0 \Delta t$ joules while $F \rho c_v T \Delta t$ joules are taken out. The reaction produces $r \Delta H \Delta t$, and $Q \Delta t$ is added. Write $H(t + \Delta t)$ as $H(t)$ + net energy into the tank during Δt. Let $\Delta t \to 0$, and generate a differential equation for $H(t)$, then for $T(t)$.

The state variables are c_A, c_B, c_C, and T, and the inputs are F, T_0, c_{A0}, and Q.

b. Write the state equations for the following constants: $A = 2.68 \times 10^9$ min^{-1}; $E/R = 7553°$K; $\Delta H = 2.09 \times 10^8$ J/kg-moles; $\rho = 10^3$ kg/m^3; $c_v = 4180$ J/kg°K; $V = 18 \times 10^{-3}$ m^3; $F = 3.6 \times 10^{-3}$ m^3/min; $c_{A0} = 3$ kg-moles/m^3; $T_0 = 293°$K. Use the minute as the time unit.

c. Simulate the reactor for the initial conditions $T(0) = 445°$K, $c_A(0) = .05$ kg-mole/m^3, $c_B(0) = c_C(0) = 0$, $Q = -120,000$ J/min (a constant) and $0 < t < 10$ min.

M **2.10** *Flow control* Figure 2.25 represents a very common element in process control: a valve being used to control liquid flow. The valve stroke u is the input, and the flow F is the output. The flow through a valve is as discussed in Example 2.4.

a. Let the orifice area be $S = Au$, and let c be the orifice coefficient. Write the flow F as a function of u. (This model has no dynamics, since the input–output relationship is algebraic.)

b. For $A = 4 \times 10^{-2}$ m, $c = 0.9$, and $\rho = 10^3$ kg/m^3, plot the valve characteristics—i.e., plot F vs. $\Delta P = P_1 - P_2$—at different constant values of u. Use the range 0 to 8 kpa for ΔP, and 0 to 2×10^{-2} m for u.

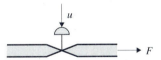

Figure 2.25 Valve for flow control

M **2.11** *Crane* Figure 2.26 shows a crane, used to transport a load of mass M from one point to another. The crane consists of a truck of mass m moving on a linear rail under the influence of a force F. The load swings at the end of a cable of length ℓ, of negligible mass.

a. Using Lagrange's equations, derive a model for this system.

b. Write the state equations for $m = 500$ kg, $M = 2000$ kg, and $\ell = 10$ m.

c. Simulate the system for zero initial conditions and $F = 1000$ N.

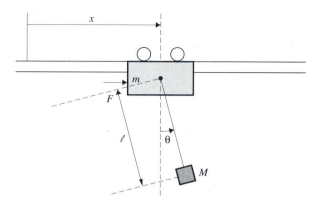

Figure 2.26 Crane

2.12 *High-wire artist* A high-wire artist uses a balancing rod for stability purposes. Figure 2.27 is a simplified representation of the physical situation. The person is represented by a bar of mass M_M pivoting about one of its end points. The moment of inertia of the bar about the pivot is J_M. The rod, of mass M_R, rotates about a point situated at a distance ℓ from the feet of the artist. The person exerts a torque τ on the rod, which has a moment of inertia J_R about its pivoting axis, which goes through the rod's center of gravity.

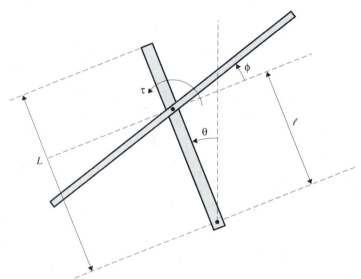

Figure 2.27 Model for a high-wire artist

a. Derive Lagrange's equations.

b. Derive the state equation for the values $M_M = 75$ kg, $J_M = 3.2$ kg m², $M_R = 2$ kg, $J_R = 1.5$ kg m², $L = 1.8$ m, $\ell = 1$ m. Use the symbols ω_θ and ω_ϕ as the derivatives of θ and ϕ, respectively.

Hint   If $\dot{\theta} \neq 0$, the rod rotates even if $\dot{\phi} = 0$. The full rotation of the bar must be used to write the bar's rotational kinetic energy.

◆◆◆ REMARK
The same principles apply to *reaction wheel* stabilization. The rod is replaced by a rotating wheel driven by a motor.   ◆

**2.13 *Blending tank*   For the blending tank of Problem 2.7:

   a. Calculate the equilibrium values F_A^* and F_0^*, given F_B^*, ℓ^*, and C_A^*.

   b. Linearize about the equilibrium point. Consider F_A, F_B, and F_0 as inputs, i.e., as variables.

   c. Give the linearized system for the data of Problem 2.7, with $\ell^* = 0.2$ m, $C_A^* = 0.8$, and $F_B^* = 5.0 \times 10^{-5}$ m^3/s.

**2.14 *Heat exchanger*   For the heat exchanger of Problem 2.8:

   a. Write, but do not solve, the equations for the equilibrium temperatures, for given values of F_C^*, F_H^*, T_{C0}^*, and T_{H0}^*.

   b. Linearize about the steady-state equilibrium, with ΔF_C, ΔF_H ΔT_{C0}, and ΔT_{H0} as inputs.

   c. Solve for the steady-state temperatures, and write the linearized model for the specific values of Problem 2.8c, with F_C^*, F_H^*, T_{C0}^*, and T_{H0}^*.

**2.15 *Chemical reactor*   For the system of Problem 2.9:

   a. Calculate the equilibrium values of Q^*, c_A^*, c_B^*, and c_C^* in terms of c_{A0}^*, T^*, T_0^*, and F^*.

   b. Linearize about the equilibrium point. Consider Q, T_0, c_{A0}, and F as being variable, i.e., as inputs.

   c. Evaluate the linear model for the data of Problem 2.9b, with $c_{A0}^* = 3$ kg-moles/m^3, $T^* = 346°$K, and $F^* = 3.6 \times 10^{-3}$ m^3/min.

**2.16 *Flow control*   For the flow control model of Problem 2.10:

   a. Linearize about a set of values u^*, P_1^*, P_2^*, and F^*, with $P_1^* > P_2^*$. Use ΔF as the output, with Δu, ΔP_1, and ΔP_2 on the inputs.

   b. Evaluate the model for $u^* = 1 \times 10^{-2}$ m, $P_1^* = 3$ kpa, and $P_2^* = 1$ kpa.

**2.17 *Crane*   For the system of Problem 2.11:

   a. Linearize about the equilibrium point $F = 0$, $x = x^*$, $\theta = 0$, $v = \omega = 0$. (Note that the system can be in equilibrium for any x^*.)

   b. Evaluate the linearized system for the parameter values of Problem 2.11.

2.18   *High-wire artist*   For the system of Problem 2.12:

    a. Linearize about the equilibrium point

$$\tau^* = \theta^* = \phi^* = \omega_\theta{}^* = \omega_\psi{}^* = 0.$$

    b. Evaluate the linearized system for the parameter values of Problem 2.12.

2.19   *Two-pendula problem*   Figure 2.28 shows an extension of the inverted-pendulum problem of Example 2.5 to the case of *two* pendula, moving in parallel vertical planes about the same axis. The two bodies at the end of the pendula have the same mass, m. The equilibrium point is easily determined, on physical grounds, to be $\theta_1 = \theta_2 = \dot{\theta}_1 = \dot{\theta}_2 = \dot{x} = 0, F = 0, x = x^*$. (As in Problem 2.17, the system can be in equilibrium for any x^*.) The rods have negligible mass.

    a. Following the method of Example 2.9, write the Lagrangian to second order in the coordinates and their derivatives, and derive the linearized state equations.

    b. Write the equations for $M = m = 1$ kg. Leave ℓ_1 and ℓ_2 as algebraic quantities.

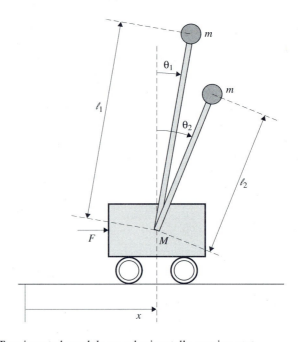

Figure 2.28   Two inverted pendula on a horizontally moving cart

2.20 *Maglev* At speeds exceeding 250 km/h, wheeled vehicles become uneco-
nomical because of wear on the wheels and stringent requirements on the
alignment of the surface or rails. That is why researchers in several countries
are investigating magnetically levitated (Maglev) vehicles [7].

Consider a simplified version of the control problem. Figure 2.29 shows
a rear view of the vehicle, useful for studying lateral motion, i.e., heave
(translation in the z direction), sway (translation in the y direction), and roll
(rotation θ). We will study this planar description, assuming no longitudinal
motion (pitch and yaw).

There are two levitation magnets applying vertical force, and two guidance
magnets for lateral force. Each magnet is assumed to exert a force ki^2/S^2,
where i is the magnet current and S is the gap between the magnet and the
guideway. The guideway (both sides) is at a height $z_G(t)$ with respect to a
fixed datum. The wind generates a force F_W, acting halfway up the side of
the vehicle. We assume also that motions are small. Figure 2.30 shows the
forces and gaps in a simplified manner. Derive a model for this system, with
the following suggested steps:

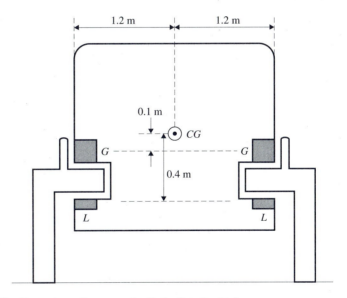

Figure 2.29 Rear view of a magnetically levitated vehicle

a. Calculate the equilibrium levitation forces F_{L1}^* and F_{L2}^*, given a vehicle
mass $M = 4000$ kg, and $F_w^* = 0$.

b. Given magnet forces of $0.0054i^2/S^2$ in *SI* units, calculate the dc steady-
state current i^* for each levitation magnet, given that $S^* = 0.014$ m.

c. Repeat (b) for the guidance magnets, assuming $F_{G1}^* = F_{G2}^* = 5000$ N
and the same S^* as above.

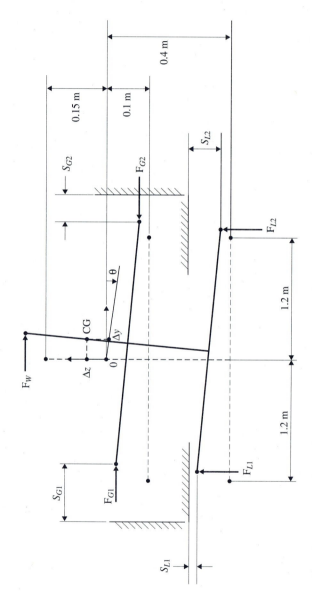

Figure 2.30 Simplified diagram showing gaps and forces. The dotted figure represents the situation at the nominal point.

d. Write Newton's second law for $\ddot{z}$, $\ddot{y}$, and $\ddot{\theta}$ in terms of the incremental forces ΔF_{L1}, ΔF_{L2}, ΔF_{G1}, and ΔF_{G2}; the wind force F_W; and the guideway elevation z_G. (The moment of inertia about the center of gravity is $J = 2500$ kg m^2.)

e. The control inputs are defined to be currents squared; i.e., $u = i^2$. Linearize the magnetic forces about the respective equilibrium values of u and S to obtain linear expressions in Δu and ΔS for each ΔF.

f. Obtain a linear expression for each ΔS in terms of z_G, Δz, Δy, and θ. Do this by calculating ΔS from individual changes in Δz, Δy, and θ (θ small), and summing the contributions. Insert the results in the Newton's law equations, and derive the state equations (six state variables).

 2.21 It is possible to linearize about a trajectory rather than a point. Consider the system

$$\dot{\mathbf{x}} = \mathbf{f}(\mathbf{x}, \mathbf{u})$$

$$\mathbf{y} = \mathbf{h}(\mathbf{x}, \mathbf{u}).$$

Let $\mathbf{x}^*(t)$, $\mathbf{u}^*(t)$, and $\mathbf{y}^*(t)$, the nominal state trajectory, input, and output functions, satisfy the system equations. Define

$$\Delta\mathbf{x}(t) = \mathbf{x}(t) - \mathbf{x}^*(t), \qquad \Delta\mathbf{u}(t) = \mathbf{u}(t) - \mathbf{u}^*(t), \qquad \Delta\mathbf{y}(t) = \mathbf{y}(t) - \mathbf{y}^*(t).$$

a. Derive, under the assumption of small $\Delta\mathbf{x}$, $\Delta\mathbf{u}$, and $\Delta\mathbf{y}$, the equations satisfied by these quantities. Show that this incremental system is time-varying, in general.

b. Given the nonlinear system

$$\dot{x}_1 = x_2$$

$$\dot{x}_2 = -x_1 - x_2^3 + x_2 u$$

$$y = x_1$$

compute the trajectory $x^*(t)$, $0 \le t \le 5$, for $x_1(0) = 1$, $x_2(0) = 0$, and $u(t) = t$. Also compute, as functions of time, the elements of the matrices of the linearized system for $0 \le t \le 5$.

References

[1] Morris, J. L., *Computational Methods in Elementary Numerical Analysis*, Wiley (1983).
[2] Smith, W. A., *Elementary Numerical Analysis*, Prentice-Hall (1986).

[3] Gibson, J. E., and F. B. Tuteur, *Control System Components*, McGraw-Hill (1958).

[4] Goldstein, H., *Classical Mechanics*, Addison-Wesley (1953).

[5] Friedland, B., *Control System Design: An Introduction to State-Space Methods*, McGraw-Hill (1986).

[6] Fortmann, T. E., and K. L. Hitz, *An Introduction to Linear Control Systems*, Marcel Dekker (1977).

[7] Gottzein, E., et al., "Control Aspects of a Tracked Magnetic Levitation High-speed Test Vehicle," *Automatica*, vol. 13, pp. 205–224 (1977).

Chapter 3

Linear Systems Theory

3.1 INTRODUCTION

The subject of this chapter is the theory of linear, time-invariant (LTI) systems. Part of the motivation for the study of this theory is the fact that, in the neighborhood of a constant equilibrium point, systems are adequately modeled by linear, time-invariant systems. The major motivation, however, is simply that LTI systems are the only ones for which we have a nearly complete, practical theory. That is why nonlinearities are often ignored in the design phase, when an LTI approximation is used to obtain a control law; the nonlinearities are restored in the simulation phase, when the control law is tested and tuned. It is perhaps surprising that, in most cases, LTI approximations can yield both meaningful insights and useful control laws.

3.2 LINEARITY PROPERTIES

A linear, time-invariant system is one described by the equations

$$\dot{\mathbf{x}} = A\mathbf{x} + B\mathbf{u}$$
$$\mathbf{y} = C\mathbf{x} + D\mathbf{u}. \tag{3.1}$$

The dimension of the state vector $\mathbf{x}$ is n, the input $\mathbf{u}$ is an r-dimensional vector, and the output $\mathbf{y}$ is of dimension m. The matrices A, B, C, and D are constant matrices of appropriate dimensions. The initial state, $\mathbf{x}(0)$, is $\mathbf{x}_0$.

It can be shown that there exists a unique solution $\mathbf{x}(t)$ to Equation 3.1 such that $\mathbf{x}(0) = \mathbf{x}_0$.

The solution of LTI systems is broken down into two components, whose properties we will now explore. The *zero-input response* satisfies

$$\dot{\mathbf{x}}_{zi} = A\mathbf{x}_{zi}$$
$$\mathbf{y}_{zi} = C\mathbf{x}_{zi} \tag{3.2}$$

with $\quad \mathbf{x}_{zi}(0) = \mathbf{x}_0.$

The *zero-state response* satisfies

$$\dot{\mathbf{x}}_{zs} = A\mathbf{x}_{zs} + B\mathbf{u}$$
$$\mathbf{y}_{zs} = C\mathbf{x}_{zs} + D\mathbf{u} \tag{3.3}$$

with $\mathbf{x}_{zs}(0) = \mathbf{0}$.

LTI systems possess several important *linearity* properties.

a. *Zero-input linearity*: Let $\mathbf{x}_1(t)$ and $\mathbf{x}_2(t)$ be the zero-input responses with $\mathbf{x}_1(0) = \mathbf{x}_{10}$ and $\mathbf{x}_2(0) = \mathbf{x}_{20}$. Then the zero-input response for the initial state $\alpha\mathbf{x}_{10} + \beta\mathbf{x}_{20}$, with α and β constants, is $\mathbf{x}(t) = \alpha\mathbf{x}_1(t) + \beta\mathbf{x}_2(t)$. To show this, insert $\mathbf{x}(t)$ in Equation 3.2:

$$\dot{\mathbf{x}} = \alpha\dot{\mathbf{x}}_1 + \beta\dot{\mathbf{x}}_2 = \alpha A\mathbf{x}_1 + \beta A\mathbf{x}_2 = A(\alpha\mathbf{x}_1 + \beta\mathbf{x}_2) = A\mathbf{x}$$

so that $\mathbf{x}(t)$ is a zero-input solution. Since

$$\mathbf{x}(0) = \alpha\mathbf{x}_1(0) + \beta\mathbf{x}_2(0) = \alpha\mathbf{x}_{10} + \beta\mathbf{x}_{20}$$

it follows that $\mathbf{x}$ also has the required initial value and therefore, by uniqueness, is the desired solution.

b. *Zero-state linearity*: Let $\mathbf{x}_1(t)$ be the zero-state response to an input $\mathbf{u}_1(t)$, and $\mathbf{x}_2(t)$ be the zero-state response to an input $\mathbf{u}_2(t)$. Then the zero-state response to an input $\alpha\mathbf{u}_1(t) + \beta\mathbf{u}_2(t)$, with α and β constants, is $\mathbf{x}(t) = \alpha\mathbf{x}_1(t) + \beta\mathbf{x}_2(t)$. This is established, as above, by inserting $\mathbf{x}(t)$ in Equation 3.3:

$$\dot{\mathbf{x}} = \alpha\dot{\mathbf{x}}_1 + \beta\dot{\mathbf{x}}_2 = \alpha(A\mathbf{x}_1 + B\mathbf{u}_1) + \beta(A\mathbf{x}_2 + B\mathbf{u}_2)$$
$$= A(\alpha\mathbf{x}_1 + \beta\mathbf{x}_2) + B(\alpha\mathbf{u}_1 + \beta\mathbf{u}_2)$$
$$= A\mathbf{x} + B(\alpha\mathbf{u}_1 + \beta\mathbf{u}_2)$$

so that $\mathbf{x}$ is a solution, with $\alpha\mathbf{u}_1 + \beta\mathbf{u}_2$ as the input. Since

$$\mathbf{x}(0) = \alpha\mathbf{x}_1(0) + \beta\mathbf{x}_2(0) = \mathbf{0}$$

$\mathbf{x}$ is also a zero-state solution. Uniqueness is invoked to establish that it is the one and only solution.

c. *Additivity*: The complete solution to Equation 3.1 is the sum of the zero-input and zero-state solutions. This is easily shown by inserting $\mathbf{x}_{zi} + \mathbf{x}_{zs}$ into Equation 3.1. The details are left to the reader.

♦ ♦ ♦ **R E M A R K**

The three linearity properties also apply to time-varying systems, since the fact that A is constant was not invoked in the arguments. The additivity property justifies the separate development of the zero-input and zero-state solutions, as follows.   ♦

3.3 THE ZERO-INPUT SOLUTION

3.3.1 The Matrix Exponential

Consider the zero-input equation

$$\dot{\mathbf{x}} = A\mathbf{x} \tag{3.4}$$

to be solved with a given initial state $\mathbf{x}(0)$.

By differentiation of Equation 3.4, we obtain

$$\ddot{\mathbf{x}} = A\dot{\mathbf{x}} = A^2\mathbf{x}.$$

Repetition k times yields

$$\mathbf{x}^{(3)} = A^2\dot{\mathbf{x}} = A^3\mathbf{x}$$

$$\vdots$$

$$\mathbf{x}^{(k)} = A^k\mathbf{x}. \tag{3.5}$$

The kth derivative at $t = 0$ is

$$\mathbf{x}^{(k)}(0) = A^k\mathbf{x}(0). \tag{3.6}$$

Equation 3.6 shows that $\mathbf{x}^{(k)}(0)$ exists for all finite k. This means that the components of $\mathbf{x}(t)$ are analytic functions, which admit representation by Taylor series. Thus,

$$
\begin{aligned}
\mathbf{x}(t) &= \mathbf{x}(0) + \dot{\mathbf{x}}(0)t + \frac{1}{2!}\ddot{\mathbf{x}}(0)t^2 + \cdots + \frac{1}{k!}\mathbf{x}^{(k)}(0)t^k + \cdots \\
&= \mathbf{x}(0) + A\mathbf{x}(0)t + \frac{1}{2!}A^2\mathbf{x}(0)t^2 + \cdots + \frac{1}{k!}A^k\mathbf{x}(0)t^k + \cdots \\
&= \left(I + At + \frac{1}{2!}A^2t^2 + \cdots + \frac{1}{k!}A^k t^k + \cdots \right)\mathbf{x}(0).
\end{aligned}
\tag{3.7}
$$

The series in parentheses can be shown to converge for all finite t. It is called the *matrix exponential* and given the symbol e^{At}, by analogy with the scalar exponential and its series. Thus,

$$e^{At} = I + At + \frac{1}{2!}A^2t^2 + \cdots + \frac{1}{k!}A^k t^k + \cdots. \tag{3.8}$$

Numerical algorithms to evaluate e^{At} are based on this series, with a few tricks thrown in (see Problem 3.3).

The solution $\mathbf{x}(t) = e^{At}\mathbf{x}(0)$ is seen to be a linear transformation: e^{At} is an $n \times n$ matrix which, by matrix–vector multiplication, transforms vector $\mathbf{x}(0)$ into vector

$\mathbf{x}(t)$. The transformation depends on t, the difference between the initial and final times; the initial time, taken here to be zero, is arbitrary. This fact is used in the numerical solution of $\dot{\mathbf{x}} = A\mathbf{x}$ to generate values of $\mathbf{x}$ at $t = \Delta, 2\Delta, 3\Delta, \ldots$, as follows:

$$\mathbf{x}(\Delta) = e^{A\Delta}\mathbf{x}(0)$$

$$\mathbf{x}(2\Delta) = e^{A\Delta}\mathbf{x}(\Delta)$$

$$\vdots \qquad \vdots$$

$$\mathbf{x}(k\Delta + \Delta) = e^{A\Delta}\mathbf{x}(k\Delta). \tag{3.9}$$

In Equation 3.9, $e^{A\Delta}$ is computed once and used to step forward in time by Δ, iteratively.

The matrix exponential has several important properties, of which six are examined here. The first three are quite analogous to properties of the scalar exponential; the others are specifically matrix properties.

◆ **Property I** $e^{A0} = I$

This is easily seen by inserting $t = 0$ in Equation 3.8.   ◆

◆ **Property II** $e^{A(t_1+t_2)} = e^{At_2}e^{At_1} = e^{At_1}e^{At_2}$

To show this, break the "trip" from 0 to $t_1 + t_2$ into two parts, the first from 0 to t_1 and the second from t_1 to $t_1 + t_2$. Starting with $\mathbf{x}(0)$, we have

$$\mathbf{x}(t_1) = e^{At_1}\mathbf{x}(0)$$

and

$$\mathbf{x}(t_1 + t_2) = e^{At_2}\mathbf{x}(t_1) = e^{At_2}e^{At_1}\mathbf{x}(0).$$

Now,

$$\mathbf{x}(t_1 + t_2) = e^{A(t_1+t_2)}\mathbf{x}(0)$$

so that

$$e^{A(t_1+t_2)}\mathbf{x}(0) = e^{At_2}e^{At_1}\mathbf{x}(0).$$

Since this must hold for any vector $\mathbf{x}(0)$, it follows that

$$e^{A(t_1+t_2)} = e^{At_2}e^{At_1}.$$

The argument is repeated with a first step of t_2 followed by a step of t_1 to establish the second equation of Property II.   ◆

◆ **Property III** $(e^{At})^{-1} = e^{-At}$

To see this, write

$$e^{At}e^{-At} = e^{At}e^{A(-t)} = e^{A(t-t)} = e^{A0} = I.$$

This also shows that e^{At} always has an inverse.   ◆

◆ **Property IV** $e^{A^T t} = (e^{At})^T$

This follows by transposition of each side of Equation 3.8 and the fact that $(A^k)^T = (A^T)^k$.   ◆

◆ **Property V** $Ae^{At} = e^{At}A$

It is seen that

$$A(A^k t^k) = A^{k+1}t^k = (A^k t^k)A$$

so that pre- or postmultiplying Equation 3.8 by A yields the same result.   ◆

◆ **Property VI** $\frac{d}{dt}e^{At} = Ae^{At}$

We write

$$\dot{\mathbf{x}} = \left(\frac{d}{dt}e^{At}\right)\mathbf{x}_0 = A\mathbf{x} = Ae^{At}\mathbf{x}_0.$$

The result follows from the fact that this equality must hold for all $\mathbf{x}_0$.   ◆

3.3.2 Solution by Laplace Transforms

An aphorism credited to A. J. Laub is stated as follows: "If it's a good method for hand calculation, it's a poor method for computer implementation; if it's a good numerical method, it's a poor one for pencil and paper." This holds true for the computation of the matrix exponential: the series evaluation is difficult to do by hand, so we rely on a Laplace transform solution.

Transformation of Equation 3.4 and use of the real differentiation theorem yield

$$s\mathbf{x}(s) - \mathbf{x}(0) = A\mathbf{x}(s)$$

or

$$(sI - A)\mathbf{x}(s) = \mathbf{x}(0)$$

and

$$\mathbf{x}(s) = (sI - A)^{-1}\mathbf{x}(0). \tag{3.10}$$

(Note the introduction of the $n \times n$ identity matrix I in the second step, so as to have a difference of two $n \times n$ matrices in the parentheses.)

Now

$$\mathbf{x}(t) = e^{At}\mathbf{x}(0).$$

Transformation of each side yields

$$\mathbf{x}(s) = \mathcal{L}[e^{At}]\mathbf{x}(0).$$

Comparing this with Equation 3.10, we obtain

$$\mathcal{L}[e^{At}] = (sI - A)^{-1}. \tag{3.11}$$

Example 3.1   Calculate e^{At} for $A = \begin{bmatrix} 0 & 1 \\ -2 & -3 \end{bmatrix}$.

Solution

$$sI - A = \begin{bmatrix} s & 0 \\ 0 & s \end{bmatrix} - \begin{bmatrix} 0 & 1 \\ -2 & -3 \end{bmatrix}$$

$$= \begin{bmatrix} s & -1 \\ 2 & s+3 \end{bmatrix}.$$

Now,

$$\det(sI - A) = s^2 + 3s + 2$$

and

$$(sI - A)^{-1} = \frac{1}{s^2 + 3s + 2} \begin{bmatrix} s+3 & 1 \\ -2 & s \end{bmatrix}$$

$$= \frac{1}{(s+1)(s+2)} \begin{bmatrix} s+3 & 1 \\ -2 & s \end{bmatrix}.$$

To obtain e^{At}, we invert each of the four elements of the matrix by partial fraction expansion.

$$(sI - A)^{-1} = \begin{bmatrix} \dfrac{2}{s+1} - \dfrac{1}{s+2} & \dfrac{1}{s+1} - \dfrac{1}{s+2} \\ \dfrac{-2}{s+1} + \dfrac{2}{s+2} & \dfrac{-1}{s+1} + \dfrac{2}{s+2} \end{bmatrix}$$

and

$$e^{At} = \begin{bmatrix} 2e^{-t} - e^{-2t} & e^{-t} - e^{-2t} \\ -2e^{-t} + 2e^{-2t} & -e^{-t} + 2e^{-2t} \end{bmatrix}.$$

Property I provides an easy check, and it is seen that the identity matrix results if $t = 0$.

3.3.3 Dynamic Modes

We may write

$$(sI - A)^{-1} = \frac{Adj(sI - A)}{\det(sI - A)}. \tag{3.12}$$

The determinant is an nth-order polynomial. As for the adjoint, the cofactors of $sI - A$ are also polynomials, of degree $n - 1$ at most. Therefore, any particular element of $(sI - A)^{-1}$ can be expanded by partial fractions into an expression of the form

$$\frac{A_{11}}{s - s_1} + \frac{A_{12}}{(s - s_1)^2} + \cdots + \frac{A_{21}}{s - s_2} + \frac{A_{22}}{(s - s_2)^2} + \cdots$$

where $s_1, s_2, \ldots$ are the roots of the determinant, some of which may be multiple roots. Upon inverse transforming, there results an expression of the form

$$A_{11}e^{s_1 t} + A_{12}te^{s_1 t} + \cdots + A_{21}e^{s_2 t} + A_{22}te^{s_2 t} + \cdots.$$

Clearly, the elements of e^{At} are sums of exponentials or time-weighted exponentials, whose exponents are the roots of the determinant. These roots satisfy the equation

$$\det(sI - A) = 0.$$

But that is precisely the equation satisfied by the eigenvalues of the matrix A, so the roots $s_1, s_2, \ldots$ are also the eigenvalues of A. In other words, the elements of the matrix exponential are composed of exponentials whose exponents are the eigenvalues of A, with time-weighted exponentials corresponding to repeated eigenvalues. In linear systems theory, we speak of these eigenvalues as the *modes*.

The eigenvectors of A also have special significance. Recall that an eigenvector $\mathbf{v}_i$, corresponding to an eigenvalue s_i, satisfies

$$A\mathbf{v}_i = s_i \mathbf{v}_i.$$

We now show that $\mathbf{v}_i$ is also an eigenvector of e^{At}. Using Equation 3.8,

$$e^{At}\mathbf{v}_i = \mathbf{v}_i + A\mathbf{v}_i t + \frac{1}{2!}A^2\mathbf{v}_i t^2 + \cdots + \frac{1}{k!}A^k\mathbf{v}_i t^k + \cdots.$$

But

$$A^2\mathbf{v}_i = AA\mathbf{v}_i = As_i\mathbf{v}_i = s_i A\mathbf{v}_i = s_i^2\mathbf{v}_i$$
$$A^3\mathbf{v}_i = AA^2\mathbf{v}_i = As_i^2\mathbf{v}_i = s_i^3\mathbf{v}_i$$

$$\vdots$$

$$A^k\mathbf{v}_i = s_i^k\mathbf{v}_i.$$

Therefore,

$$
\begin{aligned}
e^{At}\mathbf{v}_i &= \mathbf{v}_i + s_i\mathbf{v}_i t + \frac{1}{2!}s_i^2\mathbf{v}_i t^2 + \cdots + \frac{1}{k!}s_i^k\mathbf{v}_i t^k + \cdots \\
&= \left(1 + s_i t + \frac{1}{2!}s_i^2 t^2 + \cdots + \frac{1}{k!}s_i^k t^k + \cdots\right)\mathbf{v}_i \\
&= e^{s_i t}\mathbf{v}_i
\end{aligned}
\tag{3.13}
$$

so that $\mathbf{v}_i$ is an eigenvector of the matrix e^{At}, with eigenvalue $e^{s_i t}$.

The significance of this mathematical fact is this: the zero-input response to an initial state $\mathbf{x}_0 = \mathbf{v}_i$ is $\mathbf{x}(t) = e^{s_i t}\mathbf{v}_i$. Only the mode s_i is excited. The solution $\mathbf{x}(t)$ is a vector whose direction, defined by $\mathbf{v}_i$, is constant in time, and whose magnitude is $|e^{s_i t}|$ times that of $\mathbf{x}_0$. The eigenvectors are called the *modal vectors*.

Example 3.2 Derive the state equations for the circuit of Figure 3.1. Calculate the system modes and modal vectors for $R = C = 1$, and the zero-input responses for the two initial states equal to the two eigenvectors.

Solution In a circuit, capacitance voltages and inductance currents are generally valid choices for the state variables, since they describe the circuit initial conditions. (The exceptions are the case in which capacitive loops or inductive cut sets are present.) Since $dv/dt = i/C$ for a capacitance and $di/dt = v/L$ for an inductance, calculating the derivatives of the state variables amounts to obtaining capacitance currents and inductance voltages, as functions of the voltage or current source inputs and the state variables.

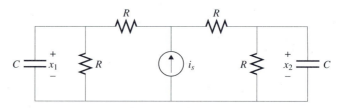

Figure 3.1 A symmetric circuit

Since the state variable derivatives are to be calculated as functions of the state variables, the latter may be assumed known, along with the source signals. This suggests replacing capacitances with voltage sources and inductances with current sources, as in Figure 3.2. Superposition is used to calculate the separate contributions of each source to the two capacitance currents, as shown in Figure 3.3. The result is

$$i_{c1} = -\frac{3x_1}{2R} + \frac{x_2}{2R} + \frac{i_s}{2}$$

$$i_{c2} = \frac{x_1}{2R} - \frac{3x_2}{2R} + \frac{i_s}{2}$$

and therefore the state equations are

$$\dot{x}_1 = \frac{1}{C}i_{c1} = -\frac{3}{2RC}x_1 + \frac{1}{2RC}x_2 + \frac{1}{2C}i_s$$

$$\dot{x}_2 = \frac{1}{C}i_{c2} = \frac{1}{2RC}x_1 - \frac{3}{2RC}x_2 + \frac{1}{2C}i_s.$$

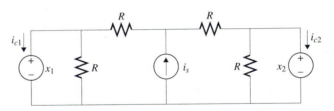

Figure 3.2   Pertaining to the state equations of the circuit

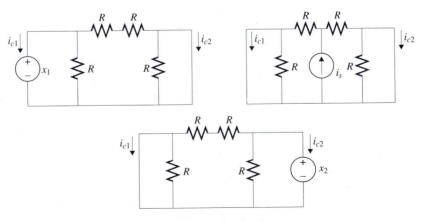

Figure 3.3   Application of a superposition to the circuit

For $R = C = 1$, they become

$$\dot{\mathbf{x}} = \begin{bmatrix} -\dfrac{3}{2} & \dfrac{1}{2} \\ \dfrac{1}{2} & -\dfrac{3}{2} \end{bmatrix} \mathbf{x} + \begin{bmatrix} \dfrac{1}{2} \\ \dfrac{1}{2} \end{bmatrix} i_s.$$

To calculate the eigenvalues, write

$$\det(sI - A) = \det \begin{bmatrix} s + \dfrac{3}{2} & -\dfrac{1}{2} \\ -\dfrac{1}{2} & s + \dfrac{3}{2} \end{bmatrix}$$

$$= s^2 + 3s + 2$$

$$= (s + 1)(s + 2).$$

The eigenvalues are -1 and -2.

The eigenvectors satisfy

$$A\mathbf{v}_i = s_i \mathbf{v}_i.$$

For the eigenvalue $s_1 = -1$,

$$\begin{bmatrix} -\dfrac{3}{2} & \dfrac{1}{2} \\ \dfrac{1}{2} & -\dfrac{3}{2} \end{bmatrix} \begin{bmatrix} v_{11} \\ v_{12} \end{bmatrix} = -\begin{bmatrix} v_{11} \\ v_{12} \end{bmatrix}$$

or

$$-\frac{3}{2}v_{11} + \frac{1}{2}v_{12} = -v_{11}$$

$$\frac{1}{2}v_{11} - \frac{3}{2}v_{12} = -v_{12}.$$

The first equation yields $v_{12} = v_{11}$. So does the second equation; when we solve for an eigenvector, one equation is always redundant. The reason is that an eigenvector is not unique: if $\mathbf{v}_i$ is an eigenvector, so is $\alpha\mathbf{v}_i$, for any nonzero constant α. (A numerical algorithm will return a single vector, to be scaled if desired.)

The eigenvector corresponding to $s_i = -1$ is thus

$$\mathbf{v}_1 = \alpha \begin{bmatrix} 1 \\ 1 \end{bmatrix}$$

where α is any nonzero constant.

Corresponding to $s_2 = -2$, we have

$$\begin{bmatrix} -\dfrac{3}{2} & \dfrac{1}{2} \\ \dfrac{1}{2} & -\dfrac{3}{2} \end{bmatrix} \begin{bmatrix} v_{21} \\ v_{22} \end{bmatrix} = -2 \begin{bmatrix} v_{21} \\ v_{22} \end{bmatrix}$$

which yields $v_{22} = -v_{21}$. The eigenvector corresponding to $s_2 = -2$ is

$$\mathbf{v}_2 = \beta \begin{bmatrix} 1 \\ -1 \end{bmatrix}$$

where β is an arbitrary constant.

For $\mathbf{x}(0) = \alpha \begin{bmatrix} 1 \\ 1 \end{bmatrix}$, the solution follows immediately from Equation 3.13; it is

$$\mathbf{x}(t) = e^{-t} \alpha \begin{bmatrix} 1 \\ 1 \end{bmatrix}.$$

For $\mathbf{x}(0) = \beta \begin{bmatrix} 1 \\ -1 \end{bmatrix}$, the solution is

$$\mathbf{x}(t) = e^{-2t} \beta \begin{bmatrix} 1 \\ -1 \end{bmatrix}.$$

(The reader is invited to check this by calculating $\mathbf{x}(t)$ as $e^{At}\mathbf{x}_0$.)

In terms of the circuit, the first modal vector corresponds to equal voltages on the capacitances. With $i_s = 0$, the voltages remain equal and there is no current through the two top resistances, so both voltages decay according to the $RC = 1$ time constant.

The second modal vector corresponds to an initial state where the two voltages are equal but of opposite signs. The symmetry of the network forces them to remain so. Since $x_2(t) = -x_1(t)$, the current leaving node 1 through the two top resistances is $(2x_1)/2R = (x_1)/R$. The total current drawn from the capacitance at node 1 is thus $(2x_1)/R$, for an effective time constant of $\frac{R}{2}C = \frac{1}{2}$; this explains the mode -2.

Example 3.3   **(Pendulum on a Cart)**

Calculate the modes and the corresponding eigenvectors for the linearized inverted pendulum system of Example 2.9, using the parameter values of Example 2.5.

Solution   From Example 2.9, the state equations with zero input are

$$\frac{d}{dt} \begin{bmatrix} x \\ v \\ \theta \\ \omega \end{bmatrix} = \begin{bmatrix} 0 & 1 & 0 & 0 \\ 0 & 0 & -9.8 & 0 \\ 0 & 0 & 0 & 1 \\ 0 & 0 & +19.6 & 0 \end{bmatrix} \begin{bmatrix} x \\ v \\ \theta \\ \omega \end{bmatrix}.$$

To solve for the eigenvalues, solve

$$\det(sI - A) = s^2(s^2 - 19.6) = 0.$$

The eigenvalues are 0, 0, and ± 4.43.
The eigenvectors are as follows:

$$\text{For } s = 0, \quad \alpha \begin{bmatrix} 1 \\ 0 \\ 0 \\ 0 \end{bmatrix}; \qquad \text{for } s = 4.43, \quad \beta \begin{bmatrix} -.5 \\ -2.21 \\ 1 \\ 4.43 \end{bmatrix};$$

$$\text{for } s = -4.43, \quad \gamma \begin{bmatrix} -.5 \\ 2.21 \\ 1 \\ -4.43 \end{bmatrix}.$$

Note that there are only three eigenvectors. In cases of repeated eigenvalues, there may be fewer than n eigenvectors, in which case it is also possible to define a generalized eigenvector. (This shall not be pursued here.)

Corresponding to $s = 0$, the modal vector calls for an initial cart position $x = \alpha$ (arbitrary), with a motionless cart and a motionless vertical pendulum. Since $e^{0t} = 1$, the system remains in this state, which is a state of equilibrium (albeit unstable).

Corresponding to $s = -4.43$ is an initial state where the angle is γ, the angular velocity is -4.43γ (i.e., toward the vertical), and the cart velocity is 2.21γ (to the right for $\gamma > 0$). The cart position is $-.5\gamma$ (to the left of the origin). The motion is described by multiplying this state by $e^{-4.43t}$, so the system eventually comes to rest with the cart at the origin and a vertical pendulum. It is seen that, with just the right initial conditions, the system is sent to equilibrium.

The other eigenvector produces a time evolution governed by $e^{4.43t}$. The angle is β, the angular velocity is such as to increase it, and the cart is moving leftward.

The foregoing also applies to complex frequencies, the difference being that it is impossible to have a complex vector as an initial state. It is known that, for real matrices, both eigenvalues and eigenvectors occur in complex conjugate pairs. A real initial state, $\mathbf{x}_0$, is formed from pairs of complex conjugate eigenvectors $\mathbf{v}_i$ and $\mathbf{v}_i^*$ corresponding to complex conjugate modes s_i and s_i^*, as follows:

$$\mathbf{x}_0 = \alpha \mathbf{v}_i + \alpha^* \mathbf{v}_i^* = 2 Re(\alpha \mathbf{v}_i) \tag{3.14}$$

with α a complex constant. By zero-input linearity, the response is

$$\begin{aligned} \mathbf{x}(t) &= e^{At} \alpha \mathbf{v}_i + e^{At} \alpha^* \mathbf{v}_i^* \\ &= \alpha e^{s_i t} \mathbf{v}_i + \alpha^* e^{s_i^* t} \mathbf{v}_i^* \\ &= 2 Re(\alpha e^{s_i t} \mathbf{v}_i). \end{aligned} \tag{3.15}$$

For $s_i = \sigma_i + j\omega_i$, $e^{s_i t}$ yields sinusoids of frequency ω_i, multiplied by the exponential $e^{\sigma_i t}$. Since $\mathbf{v}_i = Re(\mathbf{v}_i) + jIm(\mathbf{v}_i)$, $\mathbf{x}(t)$ is a linear combination of the real and imaginary parts; $\mathbf{x}(t)$ does not move along one vector but lies in the plane described by the two vectors $Re(\mathbf{v}_i)$ and $Im(\mathbf{v}_i)$.

Example 3.4   **(Sprung Beam)**

Calculate eigenvalues and eigenvectors for the beam on springs of Example 2.10 in Chapter 2 (see Fig. 3.4) with $D = 0$, $L = 1$ m, $M = 2$ kg, $J = .167$ kg m^2, $K = 50$ N/m. Study the motion that results when the initial state is a linear combination of the real and imaginary parts of a complex eigenvector.

Solution   From Example 2.10, the state equations, with zero input, are

$$\frac{d}{dt}\begin{bmatrix} y \\ v \\ \theta \\ \omega \end{bmatrix} = \begin{bmatrix} 0 & 1 & 0 & 0 \\ -50 & 0 & 0 & 0 \\ 0 & 0 & 0 & 1 \\ 0 & 0 & -150 & 0 \end{bmatrix}\begin{bmatrix} y \\ v \\ \theta \\ \omega \end{bmatrix}.$$

To calculate eigenvalues, compute

$$\det\begin{bmatrix} s & -1 & 0 & 0 \\ 50 & s & 0 & 0 \\ 0 & 0 & s & -1 \\ 0 & 0 & 150 & s \end{bmatrix} = (s^2 + 50)(s^2 + 150).$$

The eigenvalues are:

$$s = \pm j\sqrt{50}$$
$$s = \pm j\sqrt{150}.$$

Now we compute the eigenvectors. For $s = j\sqrt{50}$, write

$$\begin{bmatrix} 0 & 1 & 0 & 0 \\ -50 & 0 & 0 & 0 \\ 0 & 0 & 0 & 1 \\ 0 & 0 & -150 & 0 \end{bmatrix}\begin{bmatrix} v_{11} \\ v_{12} \\ v_{13} \\ v_{14} \end{bmatrix} = j\sqrt{50}\begin{bmatrix} v_{11} \\ v_{12} \\ v_{13} \\ v_{14} \end{bmatrix}.$$

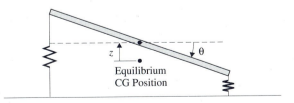

z θ

Equilibrium
CG Position

Figure 3.4   A sprung beam

This generates the equations

$$v_{12} = j\sqrt{50}\,v_{11}$$
$$-50v_{11} = j\sqrt{50}\,v_{12}$$
$$v_{14} = j\sqrt{50}\,v_{13}$$
$$-150v_{13} = j\sqrt{50}\,v_{14}.$$

The second equation is simply the first, times $j\sqrt{50}$; therefore, $v_{12} = j\sqrt{50}\,v_{11}$. Insertion of v_{14} from the third equation into the fourth yields

$$-150v_{13} = -50v_{13}$$

which can hold only if $v_{13} = 0$. So, $v_{13} = v_{14} = 0$ and the eigenvector is

$$\mathbf{v}_1 = \alpha \begin{bmatrix} 1 \\ j\sqrt{50} \\ 0 \\ 0 \end{bmatrix}$$

with α any real or complex constant.

It is easy to show that the eigenvector corresponding to $-j\sqrt{50}$ is $\mathbf{v}_1^*$. The eigenvector corresponding to $j\sqrt{150}$ is

$$\mathbf{v}_2 = \beta \begin{bmatrix} 0 \\ 0 \\ 1 \\ j\sqrt{150} \end{bmatrix}$$

with $\mathbf{v}_2^*$ corresponding to $-j\sqrt{150}$.

To study the motion resulting from the excitation of the mode $\pm j\sqrt{50}$, consider Equation 3.14, with $\alpha = Ve^{j\phi}$ and with V real and positive. Then,

$$\mathbf{x}(0) = 2Re\left\{ Ve^{j\phi} \begin{bmatrix} 1 \\ j\sqrt{50} \\ 0 \\ 0 \end{bmatrix} \right\} = 2Re\left\{ V(\cos\phi + j\sin\phi) \begin{bmatrix} 1 \\ j\sqrt{50} \\ 0 \\ 0 \end{bmatrix} \right\}$$

$$= 2V\cos\phi \begin{bmatrix} 1 \\ 0 \\ 0 \\ 0 \end{bmatrix} - 2V\sin\phi \begin{bmatrix} 0 \\ \sqrt{50} \\ 0 \\ 0 \end{bmatrix} = \begin{bmatrix} 2V\cos\phi \\ -2\sqrt{50}V\sin\phi \\ 0 \\ 0 \end{bmatrix}$$

This vector describes an initial state with nonzero vertical position and/or velocity, but zero angular position and velocity. By appropriate choices of V and ϕ, the vertical position and velocity can be selected arbitrarily; i.e., $\mathbf{x}(0)$ may be any linear combination of the real and imaginary parts of the eigenvector.

From Equation 3.15, the subsequent motion is

$$\mathbf{x}(t) = 2Re\left[Ve^{j\phi}e^{j\sqrt{50}t}\mathbf{v}_1 \right]$$

$$= 2V\cos(\sqrt{50}t + \phi)\begin{bmatrix} 1 \\ 0 \\ 0 \\ 0 \end{bmatrix} - 2V\sin(\sqrt{50}t + \phi)\begin{bmatrix} 0 \\ \sqrt{50} \\ 0 \\ 0 \end{bmatrix}$$

$$= \begin{bmatrix} 2V\cos(\sqrt{50}t + \phi) \\ -2V\sqrt{50}\sin(\sqrt{50}t + \phi) \\ 0 \\ 0 \end{bmatrix}.$$

The physical interpretation is this: given an initial state with vertical motion only, the subsequent motion is also purely vertical, and is sinusoidal with a frequency of $\sqrt{50}$ rad/s. Mechanical engineers call this the *heave mode*.

The modes with frequencies $\pm j\sqrt{150}$ rad/s are analyzed in the same manner. The result is sinusoidal motion at $\sqrt{150}$ rad/s in the last two coordinates, θ and ω, with no motion of the center of gravity, provided that the initial state has only angular displacement and velocity. That mode is called the *pitch mode*.

This example has a simple structure, which we did not exploit. The state equations are seen to be composed of two independent sets of two equations:

$$\begin{bmatrix} y \\ \dot{v} \end{bmatrix} = \begin{bmatrix} 0 & 1 \\ -50 & 0 \end{bmatrix}\begin{bmatrix} y \\ v \end{bmatrix}$$

and

$$\begin{bmatrix} \dot{\theta} \\ \dot{\omega} \end{bmatrix} = \begin{bmatrix} 0 & 1 \\ -150 & 0 \end{bmatrix}\begin{bmatrix} \theta \\ \omega \end{bmatrix}$$

which explains why initial conditions in y and v do not affect θ and ω, and vice versa. Actually, many LTI systems can be written in this decoupled way, if one is willing to accept transformed state coordinates; the topic will be pursued later in this chapter.

In the examples just presented, eigenvalues and eigenvectors were calculated "by hand." For more complex problems, powerful algorithms quite unlike those used in pen-and-paper calculations can be brought to bear.

3.4 THE ZERO-STATE RESPONSE

3.4.1 The Matrix Transfer Function

The zero-state response is best studied by means of the Laplace transform. The system is represented by

$$\dot{\mathbf{x}} = A\mathbf{x} + B\mathbf{u}$$

$$\mathbf{y} = C\mathbf{x} + D\mathbf{u}. \tag{3.16}$$

Taking transforms of the state equation [and remembering that $\mathbf{x}(0) = 0$],

$$s\mathbf{x}(s) = A\mathbf{x}(s) + B\mathbf{u}(s)$$

$$\mathbf{x}(s) = (sI - A)^{-1}B\mathbf{u}(s). \tag{3.17}$$

Use of the output equation yields

$$\mathbf{y}(s) = [C(sI - A)^{-1}B + D]\mathbf{u}(s). \tag{3.18}$$

The matrix

$$H(s) = C(sI - A)^{-1}B + D \tag{3.19}$$

has m rows (the number of outputs) and r columns (the number of inputs). It is the *matrix transfer function*. For a single-input, single-output (SISO) system, $m = r = 1$ and $H(s)$ is just a scalar valued function.

A time-domain expression is obtained by writing Equation 3.18 as

$$\mathbf{y}(s) = C\mathcal{L}[e^{At}]B\mathbf{u}(s) + D\mathbf{u}(s).$$

By the real convolution theorem, a product of transforms goes into the time domain as a convolution integral, so that

$$\mathbf{y}(t) = \int_0^t Ce^{A(t-\tau)}B\mathbf{u}(\tau)d\tau + D\mathbf{u}(t). \tag{3.20}$$

Example 3.5   Calculate the transfer function for

$$A = \begin{bmatrix} 0 & 1 \\ -2 & -3 \end{bmatrix}, \qquad B = \mathbf{b} = \begin{bmatrix} 1 \\ 1 \end{bmatrix}, \qquad C = \mathbf{c}^T = \begin{bmatrix} 1 & 0 \end{bmatrix}, \qquad D = 0.$$

Solution By Equation 3.19,

$$H(s) = \mathbf{c}^T (sI - A)^{-1}\mathbf{b}.$$

Now, $(sI - A)^{-1}$ has already been calculated in Example 3.2. Using that result,

$$H(s) = \begin{bmatrix} 1 & 0 \end{bmatrix} \frac{1}{s^2 + 3s + 2} \begin{bmatrix} s+3 & 1 \\ -2 & s \end{bmatrix} \begin{bmatrix} 1 \\ 1 \end{bmatrix}$$

$$= \begin{bmatrix} 1 & 0 \end{bmatrix} \frac{1}{s^2 + 3s + 2} \begin{bmatrix} s+4 \\ s-2 \end{bmatrix}$$

$$= \frac{s+4}{s^2 + 3s + 2}.$$

Example 3.6 **(dc Servo)**

Calculate the transfer function θ/v and θ/T_L for the system of Equation 2.19, Example 2.1 (Chapter 2).

Solution In this case,

$$A = \begin{bmatrix} 0 & 1 & 0 \\ 0 & 0 & 4.438 \\ 0 & -12 & -24 \end{bmatrix}, \qquad B = \begin{bmatrix} 0 & 0 \\ 0 & -7.396 \\ 20 & 0 \end{bmatrix}, \qquad C = \begin{bmatrix} 1 & 0 & 0 \end{bmatrix}.$$

The first column of B is used to compute θ/v, the second to obtain θ/T_L. The results are

$$\frac{\theta}{v} = \frac{88.76}{s(s+21.526)(s+2.474)} \tag{3.21}$$

$$\frac{\theta}{T_L} = \frac{-7.396(s+24)}{s(s+21.526)(s+2.474)}. \tag{3.22}$$

For future reference, other transfer functions are computed (MATLAB ss2zp) for the examples of Chapter 2. They are as follows:
For Example 2.2 (active suspension), with $y = x_1 - x_2$ ($C = \begin{bmatrix} 1 & -1 & 0 & 0 \end{bmatrix}$),

$$\frac{y}{u} = \frac{.02334(s^2 + 85.861)}{(s^2 + 12.305s + 639.76)(s^2 + 1.6954s + 9.3787)} \tag{3.23}$$

$$\frac{y}{y_R} = \frac{-600s^2}{(s^2 + 12.305s + 639.76)(s^2 + 1.6954s + 9.3787)}. \tag{3.24}$$

For Example 2.8 (level control), with $F_d = .01$ m^3/s, $\ell_d = 1$ m,

$$\frac{\Delta \ell}{\Delta u} = \frac{-2.0}{s + .005} \tag{3.25}$$

$$\frac{\Delta \ell}{\Delta F_{\text{in}}} = \frac{1}{s + .005}. \tag{3.26}$$

For Example 2.9 (pendulum on a cart), with $y = x$,

$$\frac{x}{F} = \frac{(s + 3.1305)(s - 3.1305)}{s^2(s - 4.4272)(s + 4.4272)}. \tag{3.27}$$

3.4.2 Poles and Zeros

From Equation 3.19, the transfer function is written as

$$H(s) = C(sI - A)^{-1}B + D.$$

This is a matrix of dimensions $m \times r$. To obtain the element $H_{ij}(s)$, we write C in terms of its rows and B in terms of its columns:

$$H(s) = \begin{bmatrix} \mathbf{c}_1^T \\ \mathbf{c}_2^T \\ \vdots \\ \mathbf{c}_m^T \end{bmatrix} (sI - A)^{-1} \begin{bmatrix} \mathbf{b}_1 & \mathbf{b}_2 & \cdots & \mathbf{b}_r \end{bmatrix} + D.$$

By the rules of matrix multiplication, we see that

$$H_{ij}(s) = \mathbf{c}_i^T(sI - A)^{-1}\mathbf{b}_j + d_{ij}. \tag{3.28}$$

This is written as

$$H_{ij}(s) = \frac{\mathbf{c}_i^T Adj(sI - A)\mathbf{b}_j + d_{ij}\det(sI - A)}{\det(sI - A)}.$$

The matrix $Adj(sI - A)$ is a matrix of polynomials of order $n - 1$ at most. Pre- or postmultiplication by a vector results in weighted sums of such polynomials, so that $\mathbf{c}_i^T Adj(sI - A)\mathbf{b}_i$ is a polynomial of degree $n - 1$ at most. The denominator, $\det(sI - A)$, is a polynomial of degree n.

The numerator of $H_{ij}(s)$ is of degree n because of the term $d_{ij}\det(sI - A)$; $H_{ij}(s)$ [and $H(s)$] is said to be *proper* because its numerator and denominator have the same degree. If $D = 0$, $H_{ij}(s)$ [and $H(s)$] is *strictly proper* because the degree of the numerator is less than that of the denominator. It is not possible for the matrix transfer function of an LTI system represented by state equations to have an excess of zeros over poles.

The denominator of H_{ij} is $\det(sI - A)$. Since the roots of the denominator are the eigenvalues of A, it follows that *all poles of $H(s)$ are eigenvalues of A.* The converse does not necessarily hold. If $\det(sI - A)$ has a factor $(s - s_i)^k$, where k is the multiplicity of s_i, it is possible that all $H_{ij}(s)$ also contain this factor in their numerators, in which case cancellation takes place and s_i is not a pole.

Note that B, C, and D do not influence the pole locations at all. They do, however, influence the zeros, because they enter in the numerators. The reader will recall that, for a scalar-valued $H(s)$, s_0 is a zero if $H(s_0) = 0$. For the multi-input, multi-output (MIMO) case, we define a *transmission zero* as a complex number s_0 that satisfies

$$H(s_0)\mathbf{w} = \mathbf{0} \tag{3.29}$$

for some $\mathbf{w} \neq \mathbf{0}$.

Solving Equation 3.29 for s_0 and $\mathbf{w}$ involves polynomial manipulations and root finding. Whereas this is easily done by hand for low-order cases, it is unsuitable for computer calculation. Rewrite Equation 3.29 as

$$[C(s_0 I - A)^{-1}B + D]\mathbf{w} = \mathbf{0} \tag{3.30}$$

and let

$$\boldsymbol{\theta} = (s_0 I - A)^{-1}B\mathbf{w}. \tag{3.31}$$

Then Equations 3.30 and 3.31 become

$$C\boldsymbol{\theta} + D\mathbf{w} = \mathbf{0}$$

and

$$-(s_0 I - A)\boldsymbol{\theta} + B\mathbf{w} = \mathbf{0}.$$

We write this as one matrix-vector equation,

$$\begin{bmatrix} -s_0 I + A & B \\ C & D \end{bmatrix} \begin{bmatrix} \boldsymbol{\theta} \\ \mathbf{w} \end{bmatrix} = \mathbf{0}. \tag{3.32}$$

The matrix in Equation 3.32 has dimensions $(n+m) \times (n+r)$. For the case $m = r$ (equal numbers of inputs and outputs), then,

$$\det \begin{bmatrix} -s_0 I + A & B \\ C & D \end{bmatrix} = 0 \tag{3.33}$$

which is also written as

$$\det \left\{ s_0 \begin{bmatrix} -I & 0 \\ 0 & 0 \end{bmatrix} + \begin{bmatrix} A & B \\ C & D \end{bmatrix} \right\} = 0. \tag{3.34}$$

This latter equation defines a *generalized eigenvalue problem*, for which good algorithms are available.

Example 3.7   Calculate the poles and zeros for the system of Example 3.5 by computing the eigenvalues of A for the poles and using Equation 3.33 for the zeros.

Solution   The eigenvalues of the A matrix have already been calculated and are equal to -1 and -2. To calculate the zeros, apply Equation 3.33:

$$\det \left[\begin{array}{cc:c} -s_0 & 1 & 1 \\ -2 & -3-s_0 & 1 \\ \hdashline 1 & 0 & 0 \end{array} \right] = 0$$

or

$$1 - (-3 - s_0) = 0$$

$$s_0 = -4$$

which tallies with the result of Example 3.5.

◆ ◆ ◆ **R E M A R K**

Numerical computations often return extraneous zeros of large magnitudes. To see why, replace the two zeros in the bottom row of the matrix with ϵ_1 and ϵ_2, respectively, to represent the effect of finite arithmetic. The determinant is then equal to

$$(4 + s) + \epsilon_1 (s - 2) + \epsilon_2 (s^2 + 3s + 2).$$

For small ϵ_1 and ϵ_2, this polynomial has one root near -4 and the other at some large value of s.   ◆

3.5 OBSERVABILITY

3.5.1 Introduction

To introduce the concept of observability, consider the system of Example 3.4 (with zero input), and suppose there is a sensor that measures y, the height of the center of gravity referred to equilibrium. Let there be an initial angular displacement but no vertical displacement or velocity. As was shown in Example 3.4, the motion for

$t > 0$ will have a pitch (angular) component but no heave (vertical) component. The sensor will read zero and will provide no information at all about the motion in the system.

Example 3.2 provides another case of an uninformative sensor, if the sensor is a voltmeter across the two top resistances of value R. For $i_s = 0$ and an initial state $x_1 = x_2$, the two voltages x_1 and x_2 remain equal for $t > 0$, so that the voltmeter constantly reads zero.

The concept of observability addresses the issue of sensing and the ability of the sensors to capture the dynamical behavior of the system. The C matrix describes how the sensed outputs are generated from the states, i.e., the way in which the sensors "couple into" the states. The relationship between the C and A matrices will turn out to be crucial.

DEFINITION
An LTI system is *observable* if the initial state $\mathbf{x}(0) = \mathbf{x}_0$ can be uniquely deduced from knowledge of the input $\mathbf{u}(t)$ and output $\mathbf{y}(t)$ for all t between 0 and any $T > 0$.

If $\mathbf{x}_0$ can be deduced, then so can $\mathbf{x}(t)$, for $0 < t \leq T$. To show this, we use additivity to write

$$\mathbf{x}(t) = e^{At}\mathbf{x}_0 + \int_0^t e^{A(t-\tau)}B\mathbf{u}(\tau)d\tau \tag{3.35}$$

Once $\mathbf{x}_0$ is known, $\mathbf{u}$ can be inserted in Equation 3.35 to compute $\mathbf{x}(t)$ for all t between 0 and T. We conclude, then, that observability allows us to deduce from measurements the state of the system.

To study observability, it is necessary only to consider the zero-input solution. The complete output is

$$\mathbf{y}(t) = Ce^{At}\mathbf{x}_0 + \int_0^t Ce^{A(t-\tau)}B\mathbf{u}(\tau)d\tau + D\mathbf{u}(t).$$

Since $\mathbf{u}(t)$ is given, the zero-state part can be calculated and subtracted from $\mathbf{y}(t)$, leaving only

$$\mathbf{y}_{zi}(t) = Ce^{At}\mathbf{x}_0$$

which is the only component that depends on $\mathbf{x}_0$. We shall dispense with the subscript zi in the remainder of this section.

3.5.2 Unobservable States and Observability

DEFINITION

A state $\mathbf{x}^* \neq \mathbf{0}$ is said to be *unobservable* if the zero-input solution $\mathbf{y}(t)$, with $\mathbf{x}(0) = \mathbf{x}^*$, is zero for all $t \geq 0$.

Mathematically, $\mathbf{x}^*$ is an unobservable state if $Ce^{At}\mathbf{x}^* = \mathbf{0}$ for all $t \geq 0$. For the sprung bar of Example 3.4, with a sensor measuring the height of the center of gravity referred to equilibrium, any state of the form

$$\mathbf{x}^* = \begin{bmatrix} 0 \\ 0 \\ \alpha \\ \beta \end{bmatrix}$$

with α and β constants is unobservable. Such a state excites only pitch motion and causes no motion of the center of gravity. It can be formally verified that $Ce^{At}\mathbf{x}^* = \mathbf{0}$ for $C = \begin{bmatrix} 1 & 0 & 0 & 0 \end{bmatrix}$, corresponding to a height sensor.

For a system with one output, the equation $y(t) = \mathbf{c}^T\mathbf{x}(t)$ expresses the geometric fact that $y(t)$ is the scalar product of the vectors $\mathbf{c}$ and $\mathbf{x}(t)$. A state $\mathbf{x}^*$ is unobservable if, for $\mathbf{x}(0) = \mathbf{x}^*$, the zero-input solution $\mathbf{x}(t)$ is orthogonal to $\mathbf{c}$ for all t.

If, in the circuit of Example 3.2, $i_s = 0$ and the output is the voltage across the two top resistances of value R, then $y = x_1 - x_2 = \begin{bmatrix} 1 & -1 \end{bmatrix}\mathbf{x}$. Figure 3.5 illustrates the vector $\mathbf{c} = \begin{bmatrix} 1 \\ -1 \end{bmatrix}$. The output for a state $\mathbf{x}(t)$ is the length (with a sign) of the projection of $\mathbf{x}(t)$ upon $\mathbf{c}$, as shown. For any $\mathbf{x}(t)$ where $x_1(t) = x_2(t)$, the output is zero.

In this particular case, if the initial state satisfies $x_1(0) = x_2(0)$, it follows that $x_1(t) = x_2(t)$—if the two voltages are initially equal, they remain so for all $t > 0$.

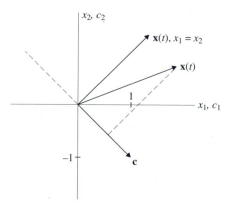

Figure 3.5   Illustration of observability for the circuit of Example 3.2

Therefore, the state space trajectory, or path, follows the line orthogonal to **c**, right to the origin, and $y(t)$ is always zero.

In the case of several outputs, the geometric interpretation is that it is possible to find an initial state such that $\mathbf{x}(t)$ is orthogonal to all rows of C, for all $t > 0$.

To establish the relationship between system observability and the existence of unobservable states, we shall need these results from the theory of quadratic forms:

- A real, symmetric matrix M is *positive definite* if $\mathbf{x}^T M \mathbf{x} > 0$ for every $\mathbf{x} \neq \mathbf{0}$.

- A positive definite matrix is always nonsingular; i.e., M^{-1} exists.

We now present our theorem.

■ **Theorem 3.1**   An LTI system is observable [for short, the pair (C, A) is observable] if, and only if, it has no unobservable states.

Proof: To show necessity ("only if"), we show that the existence of an unobservable state is enough to destroy observability. Let $\mathbf{x}^*$ be an unobservable state and let $\mathbf{x}_1$ be some n-vector. For $\mathbf{x}(0) = \mathbf{x}_1$, the output response is

$$\mathbf{y}_1(t) = C e^{At} \mathbf{x}_1.$$

By zero-input linearity, the response for $\mathbf{x}(0) = \mathbf{x}_1 + \alpha \mathbf{x}^*$ is

$$\mathbf{y}_2(t) = C e^{At} \mathbf{x}_1 + \alpha C e^{At} \mathbf{x}^* = C e^{At} \mathbf{x}_1 = \mathbf{y}_1(t)$$

because $C e^{At} \mathbf{x}^* = \mathbf{0}$.

Since $\mathbf{y}_2(t) = \mathbf{y}_1(t)$, different initial states produce identical outputs. It is thus impossible to determine uniquely the initial state by observing the output; i.e., the system is not observable.

To show sufficiency ("if"), we show that the absence of unobservable states implies observability. Let $\mathbf{x}(0) = \mathbf{x}_0 \neq \mathbf{0}$ and write

$$\mathbf{y}(t) = C e^{At} \mathbf{x}_0. \tag{3.36}$$

Premultiply both sides by $(C e^{At})^T$ to get

$$(C e^{At})^T \mathbf{y}(t) = (C e^{At})^T C e^{At} \mathbf{x}_0.$$

Now integrate:

$$\int_0^T (C e^{At})^T \mathbf{y}(t) dt = M(T) \mathbf{x}_0 \tag{3.37}$$

where

$$M(T) = \int_0^T (C e^{At})^T C e^{At} dt.$$

Note that $M(T)$ is symmetric, since the integrand is a product of a matrix and its transpose.

We now show that $M(T)$ is positive definite for every $T > 0$. Write

$$\mathbf{x}_0^T M(T)\mathbf{x}_0 = \int_0^T \mathbf{x}_0^T (Ce^{At})^T Ce^{At}\mathbf{x}_0 dt$$

$$= \int_0^T (Ce^{At}\mathbf{x}_0)^T (Ce^{At}\mathbf{x}_0)dt$$

$$= \int_0^T \|Ce^{At}\mathbf{x}_0\|^2 dt$$

where $\|\mathbf{v}\|$ is the Euclidean norm, i.e., the length of the vector $\mathbf{v}$.

Because the integrand is the magnitude squared of a vector, $\mathbf{x}_0^T M(T)\mathbf{x}_0 \geq 0$. For the integral to be zero, the integrand must be zero everywhere, except possibly at isolated time points, in the interval 0 to T. If that is so, there is at least one interval of nonzero duration over which $\|Ce^{At}\mathbf{x}_0\|^2 = 0$. But $Ce^{At}\mathbf{x}_0$ is an analytic function (a sum of exponentials). If it is zero over a nonzero time interval, all its derivatives are zero at some interior point of the interval. By Taylor's theorem, $Ce^{At}\mathbf{x}_0$ must be zero for all t. Therefore, $\mathbf{x}_0^T M(T)\mathbf{x}_0 = 0$ only if $Ce^{At}\mathbf{x}_0 = \mathbf{0}$ for all t. Since $\mathbf{x}_0 \neq \mathbf{0}$ and since there are no unobservable states, that is not possible, so

$$\mathbf{x}_0^T M(T)\mathbf{x}_0 > 0 \qquad \text{for any} \quad \mathbf{x}_0 \neq \mathbf{0}.$$

This shows that $M(T)$ is positive definite and hence nonsingular. From Equation 3.37,

$$\mathbf{x}_0 = M^{-1}(T) \int_0^T (Ce^{At})^T \mathbf{y}(t)dt \tag{3.38}$$

and $\mathbf{x}_0$ is uniquely determined, so the system is observable. ∎

◆ ◆ ◆ **R E M A R K**
Equation 3.38 proves the desired result but is not a practical method of obtaining $\mathbf{x}_0$. ◆

3.5.3 Observability Tests

Our results so far suggest that system observability be tested by searching for a vector $\mathbf{x}_0$ such that $Ce^{At}\mathbf{x}_0 = 0$ for all $t \geq 0$. That is feasible, but difficult. Better tests are available, and shall be developed presently.

We shall need the *Cayley–Hamilton theorem*, which states that a matrix satisfies its own characteristic equation. If the characteristic equation of a matrix A is

$$s^n + a_{n-1}s^{n-1} + \cdots + a_1 s + a_0 = 0$$

then the Cayley–Hamilton theorem states that

$$A^n + a_{n-1}A^{n-1} + \cdots + a_1 A + a_0 I = 0.$$

We now derive a result that will lead to our first observability test.

■ **Theorem 3.2**   The vector $\mathbf{x}^*$ is an unobservable state if, and only if,

$$\begin{bmatrix} C \\ CA \\ CA^2 \\ \vdots \\ CA^{n-1} \end{bmatrix} \mathbf{x}^* = \mathbf{0}. \tag{3.39}$$

Proof: To show necessity ("only if"), let $\mathbf{x}^*$ be an unobservable state. Then $Ce^{At}\mathbf{x}^* = 0, t \geq 0$. Since $Ce^{At}\mathbf{x}^*$ is analytic, all its derivatives exist and must be zero at $t = 0$. Using Property VI of the matrix exponential,

$$Ce^{At}\mathbf{x}^* \Big|_{t=0} = C\mathbf{x}^* = 0$$

$$\frac{d}{dt}Ce^{At}\mathbf{x}^* \Big|_{t=0} = CAe^{At}\mathbf{x}^* \Big|_{t=0} = CA\mathbf{x}^* = 0$$

$$\frac{d^2}{dt^2}Ce^{At}\mathbf{x}^* \Big|_{t=0} = CA^2 e^{At}\mathbf{x}^* \Big|_{t=0} = CA^2\mathbf{x}^* = 0$$

$$\vdots \qquad\qquad \vdots \qquad\qquad \vdots$$

$$\frac{d^k}{dt^k}Cd^{At}\mathbf{x}^* \Big|_{t=0} = CA^k e^{At}\mathbf{x}^* \Big|_{t=0} = CA^k\mathbf{x}^* = 0 \tag{3.40}$$

which is precisely what Equation 3.39 expresses in matrix form.

To show sufficiency ("if"), assume that Equation 3.39 holds for some $\mathbf{x}^*$. We show that $Ce^{At}\mathbf{x}^* = \mathbf{0}$ for all t; i.e., $\mathbf{x}^*$ is an unobservable state. From Equation 3.39,

$$C\mathbf{x}^* = CA\mathbf{x}^* = CA^2\mathbf{x}^* = \cdots = CA^{n-1}\mathbf{x}^* = \mathbf{0}. \tag{3.41}$$

From the Cayley–Hamilton theorem,

$$A^n = -a_{n-1}A^{n-1} - a_{n-2}A^{n-2} - \cdots - a_1 A - a_0 I \tag{3.42}$$

and

$$CA^n\mathbf{x}^* = -a_{n-1}CA^{n-1}\mathbf{x}^* - \cdots - a_1 CA\mathbf{x}^* - a_0 C\mathbf{x}^*$$

$$= \mathbf{0}$$

by Equation 3.41. Multiplying each side of Equation 3.42 by A yields

$$A^{n+1} = -a_{n-1}A^n - \cdots - a_1A^2 - a_0A$$

and therefore

$$CA^{n+1}\mathbf{x}^* = -a_{n-1}CA^n\mathbf{x}^* - \cdots - a_1CA^2\mathbf{x}^* - a_0CA\mathbf{x}^*$$
$$= \mathbf{0}.$$

We proceed in the same manner to show that $CA^k\mathbf{x}^* = \mathbf{0}$ for all $k > n - 1$. Using Equation 3.40, we see that $Ce^{At}\mathbf{x}^*$ and all its derivatives are zero at $t = 0$. It follows that $Ce^{At}\mathbf{x}^* = \mathbf{0}$ for all t; i.e., $\mathbf{x}^*$ is unobservable.   ∎

This theorem leads to a test of system observability, by the requirement that there exist no observable states. That will be true if Equation 3.39 cannot be satisfied for any nonzero $\mathbf{x}^*$. Equation 3.39 has a geometric interpretation. The vector $\mathbf{x}^*$ is orthogonal to all rows of the matrix it multiplies; for $\mathbf{x}^* \neq \mathbf{0}$, that is possible only if the rows, considered as n-vectors, do not span the full n-dimensional space. In such a case, the rank of the matrix is less than its maximum possible value, n. Conversely, if the rank is less than n, it is always possible to find a vector that is orthogonal to all rows. It follows that a necessary and sufficient condition for observability is

$$\text{rank } \mathcal{O} = \text{rank} \begin{bmatrix} C \\ CA \\ \vdots \\ CA^{n-1} \end{bmatrix} = n. \tag{3.43}$$

The matrix $\mathcal{O}$, called the *observability matrix*, is the matrix on the left-hand side (LHS) of Equation 3.39.

Example 3.8   Verify that the system of Example 3.2 is unobservable for $y = x_1 - x_2$.

Solution   The relevant matrices are

$$A = \begin{bmatrix} -\dfrac{3}{2} & \dfrac{1}{2} \\ \dfrac{1}{2} & -\dfrac{3}{2} \end{bmatrix} \qquad C = \begin{bmatrix} 1 & -1 \end{bmatrix}.$$

Here, $n = 2$, so the observability matrix is

$$\mathcal{O} = \begin{bmatrix} C \\ CA \end{bmatrix} = \begin{bmatrix} 1 & -1 \\ -2 & 2 \end{bmatrix}.$$

Because this matrix is square, the determinant is calculated to check for full rank. The determinant is zero, so the rank of the matrix is less than 2, and the system is unobservable.

Since

$$\begin{bmatrix} 1 & -1 \\ -2 & 2 \end{bmatrix} \begin{bmatrix} a \\ a \end{bmatrix} = \mathbf{0}$$

the state $\begin{bmatrix} a \\ a \end{bmatrix}$ is unobservable, as expected from previous discussion.

Example 3.9   (dc Servo)

Assess the observability of the dc servo of Example 2.1 (Chapter 2) under the two separate measurement conditions $y = \theta$ and $y = \omega$.

Solution   In this example,

$$A = \begin{bmatrix} 0 & 1 & 0 \\ 0 & 0 & \dfrac{NK_m}{J_e} \\ 0 & \dfrac{-NK_m}{L} & \dfrac{-R}{L} \end{bmatrix} \qquad C = \begin{bmatrix} 1 & 0 & 0 \end{bmatrix}.$$

Here, $n = 3$, so the observability matrix is

$$\mathcal{O} = \begin{bmatrix} C \\ CA \\ CA^2 \end{bmatrix} = \begin{bmatrix} 1 & 0 & 0 \\ 0 & 1 & 0 \\ 0 & 0 & \dfrac{NK_m}{J_e} \end{bmatrix}$$

which is clearly of rank 3: the system is observable with angular position as output. For $y = \omega$, $C = \begin{bmatrix} 0 & 1 & 0 \end{bmatrix}$ and the observability matrix is

$$\mathcal{O} = \begin{bmatrix} 0 & 1 & 0 \\ 0 & 0 & \dfrac{NK_m}{J_e} \\ 0 & \dfrac{-N^2 K_m{}^2}{LJ_e} & \dfrac{-NK_m R}{LJ_e} \end{bmatrix}.$$

The column of zeros establishes that this square matrix is singular and hence of rank less than 3. The state vector $\begin{bmatrix} a \\ 0 \\ 0 \end{bmatrix}$ is orthogonal to all rows and hence unobservable. This makes physical sense: if the velocity is observed but not the angular position, it is impossible to tell what the initial angle was, and hence what the angle is at any t.

⌐ Example 3.10   **(Pendulum on a Cart)**

Assess the observability of the inverted pendulum-and-cart system, using the linearized equations of Example 2.9. Study two sensor configurations: x, v, and θ in the first, and only x and v in the second.

Solution   The state equations are

$$
\begin{bmatrix} \dot{x} \\ \dot{v} \\ \dot{\theta} \\ \dot{\omega} \end{bmatrix} = \begin{bmatrix} 0 & 1 & 0 & 0 \\ 0 & 0 & \dfrac{-mg}{M} & 0 \\ 0 & 0 & 0 & 1 \\ 0 & 0 & \dfrac{(M+m)g}{m\ell} & 0 \end{bmatrix} \begin{bmatrix} x \\ v \\ \theta \\ \omega \end{bmatrix} + \begin{bmatrix} 0 \\ \dfrac{1}{M} \\ 0 \\ \dfrac{-1}{m\ell} \end{bmatrix} F.
$$

For the first configuration,

$$
C = \begin{bmatrix} 1 & 0 & 0 & 0 \\ 0 & 1 & 0 & 0 \\ 0 & 0 & 1 & 0 \end{bmatrix}.
$$

In this case, we need not write the whole observability matrix; we can stop the procedure of constructing $\mathcal{O}$ as soon as we have four independent rows. Since C by itself already has three, we need only find a fourth. We write

$$
\left.\begin{bmatrix} 1 & 0 & 0 & 0 \\ 0 & 1 & 0 & 0 \\ 0 & 0 & 1 & 0 \\ 0 & 1 & 0 & 0 \\ 0 & 0 & \dfrac{-mg}{M} & 0 \\ 0 & 0 & 0 & 1 \end{bmatrix}\right\} \begin{array}{l} \\ C \\ \\ \\ CA \\ \end{array}
$$

and stop, because the last row and the first three span all four dimensions.
With the second configuration,

$$
C = \begin{bmatrix} 1 & 0 & 0 & 0 \\ 0 & 1 & 0 & 0 \end{bmatrix}.
$$

We write

$$
\begin{bmatrix}
1 & 0 & 0 & 0 \\
0 & 1 & 0 & 0 \\
0 & 1 & 0 & 0 \\
0 & 0 & \dfrac{-mg}{M} & 0 \\
0 & 0 & \dfrac{-mg}{M} & 0 \\
0 & 0 & 0 & \dfrac{-mg}{M}
\end{bmatrix}
\begin{array}{l}
\left.\rule{0pt}{20pt}\right\} C \\[6pt]
\left.\rule{0pt}{20pt}\right\} CA \\[6pt]
\left.\rule{0pt}{20pt}\right\} CA^2.
\end{array}
$$

We may stop, since rows 1, 2, 4, and 6 are independent. This system is thus seen to be observable without the angle sensor.

Despite its usefulness, the rank test is not completely satisfactory from a numerical point of view because of the difficulty of establishing the rank of a matrix. The *eigenvector test* is somewhat easier to implement, and also gives a modal interpretation of observability.

Before introducing the result, we recall the concept of *linearly independent* functions. The functions $f_1(t), f_2(t), \ldots, f_n(t)$ form a linearly independent set over the interval 0 to T, $T > 0$, if the relation

$$a_1 f_1(t) + a_2 f_2(t) + \cdots + a_n f_n(t) = 0, \qquad 0 \le t \le T$$

implies $a_1 = a_2 = \cdots = a_n = 0$, i.e., is satisfied only if all coefficients are zero. For example, the functions 1, t, t^2 are linearly independent over any interval of nonzero duration. A linear combination $a_1 + a_2 t + a_3 t^2$ may be zero at isolated points (two at most), but not over the whole interval, unless $a_1 = a_2 = a_3 = 0$.

■ **Theorem 3.3**   The system (C, A) is unobservable if, and only if, there exists an eigenvector $\mathbf{v}$ of the matrix A, such that $C\mathbf{v} = \mathbf{0}$.

Proof: To show sufficiency ("if"), let $\mathbf{v}_1, \mathbf{v}_2, \ldots, \mathbf{v}_k$ be eigenvectors of A. We show that, if $C\mathbf{v}_i = \mathbf{0}$ for some i, the system is not observable. To show this, let $\mathbf{x}(0) = \mathbf{v}_i$ and assume $C\mathbf{v}_i = \mathbf{0}$. Then, by Equation 3.13,

$$\mathbf{x}(t) = e^{s_i t} \mathbf{v}_i$$

where s_i is the eigenvalue corresponding to $\mathbf{v}_i$. It follows that

$$C\mathbf{x}(t) = e^{s_i t} C\mathbf{v}_i = \mathbf{0}, \qquad \text{for all } t$$

and that $\mathbf{v}_i$ is therefore an unobservable state and the system is not observable.

We show necessity ("only if") by proving that there can be no unobservable states if $C\mathbf{v}_i \neq \mathbf{0}$ for all i. The result will be proved for the special case where A has n distinct eigenvalues, but it is true in general.

If the eigenvalues of A are distinct, it is known that its eigenvectors are linearly independent. This implies that the eigenvectors $\mathbf{v}_1, \mathbf{v}_2, \ldots, \mathbf{v}_n$ span the whole state space, so that any n-vector can be expressed as a unique linear combination of $\mathbf{v}_1, \mathbf{v}_2, \ldots, \mathbf{v}_n$. Let $\mathbf{x}(0)$ be so expressed; i.e., let

$$\mathbf{x}(0) = a_1\mathbf{v}_1 + a_2\mathbf{v}_2 + \cdots + a_n\mathbf{v}_n$$

where the a_i are constants. Then, by zero-input linearity and Equation 3.13,

$$\mathbf{x}(t) = a_1 e^{s_1 t}\mathbf{v}_1 + a_2 e^{s_2 t}\mathbf{v}_2 + \cdots + a_n e^{s_n t}\mathbf{v}_n$$

and

$$\mathbf{y}(t) = C\mathbf{x}(t) = a_1 C\mathbf{v}_1 e^{s_1 t} + a_2 C\mathbf{v}_2 e^{s_2 t} + \cdots + a_n C\mathbf{v}_n e^{s_n t}.$$

We now show that an unobservable state cannot exist if $C\mathbf{v}_i \neq \mathbf{0}$ for all i. If one does exist, then $\mathbf{y}(t) = \mathbf{0}$ for all t, for some $\mathbf{x}(0) \neq \mathbf{0}$. Because of the linear independence of the exponentials, this could happen only if the coefficient of each exponential vanishes, i.e., if

$$a_1 C\mathbf{v}_1 = a_2 C\mathbf{v}_2 = \cdots = a_n C\mathbf{v}_n = \mathbf{0}. \tag{3.44}$$

For $\mathbf{x}(0)$ to be nonzero, at least one of the a_i must be nonzero; let that be a_k. To satisfy Equation 3.44, we must have $C\mathbf{v}_k = \mathbf{0}$, which is contrary to the assumption. Therefore, there are no unobservable states and the system is observable. ∎

If an eigenvector $\mathbf{v}_i$ satisfies $C\mathbf{v}_i = \mathbf{0}$, mode s_i is called an *unobservable mode*. What the theorem says is that there is always at least one such mode when the system is unobservable. The first part of the theorem demonstrates that such an eigenvector is itself an unobservable state (there may be others).

To implement the eigenvector test, we compute the eigenvectors $\mathbf{v}_i$ of the A matrix and calculate $C\mathbf{v}_i$. Because of finite arithmetic, the result will never be exactly zero, even if the system is in fact unobservable: it is usually necessary to give a certain tolerance, i.e., to decide how small a number has to be before we call it zero.

If to each eigenvalue there corresponds but a single linearly independent eigenvector, the eigenvector test is applied by computing a full set of eigenvectors of A and calculating $C\mathbf{v}_i$.

If there are repeated eigenvalues, it is possible to have several independent eigenvectors associated with one eigenvalue—say $\mathbf{v}_{i1}^0, \mathbf{v}_{i2}^0, \ldots, \mathbf{v}_{iK}^0$ corresponding to s_i. Since a linear combination of eigenvectors is also an eigenvector, the test consists of ascertaining whether constants $a_1, a_2, \ldots, a_k$, not all zero, can be found such that $C(a_1\mathbf{v}_{i1}^0 + a_2\mathbf{v}_{i2}^0 + \cdots + a_k\mathbf{v}_{iK}^0) = 0$ or $a_1 C\mathbf{v}_{i1}^0 + a_2 C\mathbf{v}_{i2} + \cdots + a_k C\mathbf{v}_{iK}^0 = 0$. Such a set of constants will exist if

$$\text{rank } [C\mathbf{v}_{i1}^0 C\mathbf{v}_{i2}^0 \cdots C\mathbf{v}_{iK}^0] < K. \tag{3.45}$$

If Equation 3.45 holds, then there exists an eigenvector corresponding to the eigenvalue s_i that meets the condition $C\mathbf{v} = \mathbf{0}$, and the system is unobservable. On the other hand, if

$$\text{rank } [C\mathbf{v}_{i1}{}^0, C\mathbf{v}_{i2}{}^0 \cdots C\mathbf{v}_{iK}{}^0] = K$$

then no eigenvector of s_i is orthogonal to all rows of C.

For eigenvalues with only a single eigenvector, Equation 3.45 reduces to rank $C\mathbf{v}_i < 1$, which is the same as the condition $C\mathbf{v}_i = 0$. In effect, Equation 3.45 is just a more specific version of the condition of Theorem 3.3.

◆ ◆ ◆ **REMARK**

If there are more independent eigenvectors associated with some repeated eigenvalue than there are outputs ($K > m$), the matrix in Equation 3.45 has fewer rows (m) than columns (K) and, since rank cannot exceed the number of rows, meets the condition for unobservability. ◆

Example 3.11   Identify the unobservable mode for the network of Example 3.2, with $y = x_1 - x_2$.

Solution   The eigenvectors are

$$\begin{bmatrix} 1 \\ 1 \end{bmatrix} \qquad \text{corresponding to the mode } -1$$

$$\begin{bmatrix} 1 \\ -1 \end{bmatrix} \qquad \text{corresponding to the mode } -2.$$

For $y = x_1 - x_2$, the C matrix is $\begin{bmatrix} 1 & -1 \end{bmatrix}$. Clearly,

$$\begin{bmatrix} 1 & -1 \end{bmatrix} \begin{bmatrix} 1 \\ 1 \end{bmatrix} = 0$$

so the mode -1 is not observable.

Example 3.12   **(Active Suspension)**

Assess the observability of the suspension system in Example 2.2, with the two output measurements $y_1 = x_1 - x_2$ and $y_2 = v_1 - v_2$.

Solution   The eigenvalues and eigenvectors are computed (MATLAB command eig) to be

$$-6.1523 \pm j24.5339 \quad \text{and} \quad \begin{bmatrix} -2.551 \times 10^{-3} \\ -9.617 \times 10^{-3} \\ -2.885 \times 10^{-2} \\ 1 \end{bmatrix} \pm j \begin{bmatrix} 1.816 \times 10^{-3} \\ -3.835 \times 10^{-2} \\ -7.376 \times 10^{-2} \\ 0 \end{bmatrix}$$

$$-0.8477 \pm j2.9428 \quad \text{and} \quad \begin{bmatrix} -9.038 \times 10^{-2} \\ 8.841 \times 10^{-3} \\ 1 \\ 8.005 \times 10^{-2} \end{bmatrix} \pm j \begin{bmatrix} -3.138 \times 10^{-1} \\ -2.975 \times 10^{-2} \\ 0 \\ 5.123 \times 10^{-2} \end{bmatrix}.$$

The C matrix is

$$C = \begin{bmatrix} 1 & -1 & 0 & 0 \\ 0 & 0 & 1 & -1 \end{bmatrix}$$

and none of the eigenvectors is orthogonal to C.

This shows that measurements of position and velocity differences between the vehicle body and the unsprung mass are sufficient to guarantee observability. That is a welcome conclusion, because such measurements are much easier—and less expensive—than measurements of, say, x_1, x_2, v_1, and v_2, all of which are referred to an inertial coordinate system.

This last example might lead one to think that unobservability is a fluke. Indeed, matrices A and C formed at random will yield an observable system with probability 1. In most cases, it is the locations of the zeros and ones that cause unobservability to occur; those are determined by the structure of the system, not by parameters taking on particular values.

3.6 CONTROLLABILITY

3.6.1 Introduction

To introduce the concept of controllability, let us consider once more the sprung beam of Example 3.4. Suppose the input is a vertical force u applied at the center of gravity. If the beam is initially at equilibrium, u will impart to the beam a vertical translational motion, but the beam will remain horizontal. The force u cannot produce angular motion, because it has zero moment about the center of gravity. Conversely, if an angular oscillation exists, it cannot be altered by the input u.

Controllability addresses the issue of actuation and the ability of the actuators to control the state of the system. The B matrix describes the manner in which the inputs are "coupled into" the states; the relationship between the A and B matrices will turn out to be crucial.

Controllability is to actuators what observability is to sensors. A close connection exists between the two in terms of theory and tests; that connection will be exploited.

DEFINITION
An LTI system is *controllable* if, for every $\mathbf{x}_1$ and every $T > 0$, there exists an input function $\mathbf{u}(t), 0 < t \leq T$, such that the system state is taken from $\mathbf{0}$ at $t = 0$ to $\mathbf{x}_1$ at $t = T$.

The definition may appear restrictive due to the special choice of $\mathbf{0}$ as initial state. That is not the case: if the system satisfies the definition, the state can be taken from any initial state to any other state in arbitrary nonzero time. This follows from the fact that

$$\mathbf{x}(t_0 + T) = e^{AT}\mathbf{x}(t_0) + \int_{t_0}^{t_0+T} e^{A(t_0+T-\tau)}B\mathbf{u}(\tau)d\tau. \tag{3.46}$$

With $\tau' = \tau - t_0$, the integral becomes

$$\int_{t_0}^{t_0+T} e^{A(t_0+T-\tau)}B\mathbf{u}(\tau)d\tau = \int_0^T e^{A(T-\tau')}B\mathbf{u}(t_0 + \tau')d\tau'.$$

If we let $\mathbf{v}(\tau') = \mathbf{u}(t_0 + \tau')$, the integral is just the zero-state response at $t = T$ to an input $\mathbf{v}(t'), 0 < t' \leq T$. If the system is controllable, $\mathbf{u}(t')$ can be chosen to give the integral any desired value; this implies that $\mathbf{x}(t_0 + T)$ in Equation 3.46 can also have any desired value.

What this means is that, to study controllability, we need only focus on the zero-state response; recall that only the zero-input response was necessary in the study of observability.

3.6.2 Uncontrollable States and Controllability

DEFINITION
A state $\mathbf{x}^* \neq \mathbf{0}$ is said to be *uncontrollable* if the zero-state response $\mathbf{x}(t)$ is orthogonal to $\mathbf{x}^*$ for all $t > 0$ and all input functions.

In the circuit of Example 3.2, it is clear that, due to symmetry, the zero-state response is such that $x_1(t) = x_2(t)$, regardless of the form of the input $i_s(t)$. In Figure 3.6, the only state values reachable if $\mathbf{x}(0) = \mathbf{0}$ are those on the line $x_1 = x_2$. The zero-state response is orthogonal to $\mathbf{x}^*$, i.e., has no component in that direction; $\mathbf{x}^*$ is an uncontrollable state.

Mathematically, this condition is

$$\mathbf{x}^{*T} \int_0^t e^{A\tau}B\mathbf{u}(t - \tau)d\tau = \int_0^t \mathbf{x}^{*T}e^{A\tau}B\mathbf{u}(t - \tau)d\tau = 0 \tag{3.47}$$

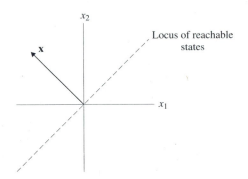

Figure 3.6 Illustration of controllability for the circuit of Example 3.2

for all t. Equation 3.47 must hold for all possible functions $\mathbf{u}(t - \tau)$. This can be only if $\mathbf{x}^{*^T} e^{A\tau} B = \mathbf{0}$ for all $\tau, 0 < \tau \leq t$; otherwise, $\mathbf{u}(t - \tau)$ could always be chosen to make the integrand positive for all values of τ where $\mathbf{x}^{*^T} e^{A\tau} B \neq \mathbf{0}$, and the integral would not be zero. Thus, an uncontrollable state $\mathbf{x}^*$ satisfies the equation

$$\mathbf{x}^{*^T} e^{At} B = \mathbf{0} \quad \text{for all } t \geq 0. \tag{3.48}$$

We can now relate the existence of uncontrollable states to controllability through the following theorem.

■ Theorem 3.4   An LTI system is controllable [for short, the pair (A, B) is controllable] if, and only if, it has no uncontrollable states.

Proof: To show necessity ("only if"), we show that the existence of a single uncontrollable state precludes controllability. This follows immediately from the definition of an uncontrollable state: no state $\mathbf{x}(T)$ that is not orthogonal to $\mathbf{x}^*$ is reachable.

To show sufficiency ("if"), we show that controllability follows if there is no uncontrollable state. We try a control function $\mathbf{u}(T - \tau) = B^T e^{A^T \tau} \mathbf{v}$, where $\mathbf{v}$ is an n-vector to be chosen. We have

$$\mathbf{x}(T) = \int_0^T e^{A\tau} B \mathbf{u}(T - \tau) d\tau$$

$$= \left[\int_0^T e^{A\tau} B B^T e^{A^T \tau} d\tau \right] \mathbf{v}$$

$$= N(T)\mathbf{v}. \tag{3.49}$$

To exploit a result already derived for observability, $N(T)$ is written as

$$N(T) = \int_0^T (B^T e^{A^T \tau})^T B^T e^{A^T \tau} d\tau.$$

By the same reasoning as in Theorem 3.2, $N(T)$ is shown to be positive definite if there exists no uncontrollable state. This means that $N(T)$ can be inverted and Equation 3.49 can be solved uniquely for $\mathbf{v}$, for arbitrary $\mathbf{x}(T)$, and proves that a control of the form specified (among others) can take the state from the origin to any desired $\mathbf{x}(T)$. ∎

3.6.3 Controllability Tests

In parallel with the case of observability, controllability can be ascertained by showing that a system has no uncontrollable states. Duality between controllability and observability is now established to apply the observability tests to controllability.

The transpose of Equation 3.48 yields

$$B^T e^{A^T t} \mathbf{x}^* = \mathbf{0} \tag{3.50}$$

which is precisely the condition for $\mathbf{x}^*$ to be an *unobservable* state of the *dual system*

$$\dot{\mathbf{x}} = A^T \mathbf{x}$$
$$\mathbf{y} = B^T \mathbf{x}. \tag{3.51}$$

The problem of looking for uncontrollable states of the system $\dot{\mathbf{x}} = A\mathbf{x} + B\mathbf{u}$ is reduced to that of seeking unobservable states of the dual system of Equation 3.51. The tests are thus directly transferable. From Equation 3.39, an uncontrollable state satisfies

$$\begin{bmatrix} B^T \\ B^T A^T \\ B^T (A^T)^2 \\ \vdots \\ B^T (A^T)^{n-1} \end{bmatrix} \mathbf{x}^* = \mathbf{0}$$

or, transposing,

$$\mathbf{x}^{*^T} [B \quad AB \quad A^2 B \quad \cdots \quad A^{n-1} B] = 0. \tag{3.52}$$

The system is controllable if no $\mathbf{x}^* \neq \mathbf{0}$ satisfies Equation 3.56, i.e., if

$$\text{rank } \mathcal{C} = \text{rank} \begin{bmatrix} B & AB & A^2 B & \cdots & A^{n-1} B \end{bmatrix} = n \tag{3.53}$$

where $\mathcal{C}$ is called the *controllability matrix*.

The eigenvector test is also applicable. The system (A, B) is uncontrollable if, and only if, there exists an eigenvector $\mathbf{w}$ of the matrix A^T such that $B^T \mathbf{w} = \mathbf{0}$. If $B^T \mathbf{w}_i = \mathbf{0}$, the mode corresponding to $\mathbf{w}_i$ is called an *uncontrollable mode*.

This last sentence deserves some elaboration. Starting from

$$\dot{\mathbf{x}} = A\mathbf{x} + B\mathbf{u}$$

we premultiply each side by $\mathbf{w}_i^T$ to get

$$\mathbf{w}_i^T \dot{\mathbf{x}} = \mathbf{w}_i^T A\mathbf{x} + \mathbf{w}_i^T B\mathbf{u}.$$

Now, $\mathbf{w}_i^T A = (A^T \mathbf{w}_i)^T = (s_i \mathbf{w}_i)^T = \mathbf{w}_i^T s_i$. Thus,

$$\mathbf{w}_i^T \dot{\mathbf{x}} = \frac{d}{dt}(\mathbf{w}_i^T \mathbf{x}) = s_i \mathbf{w}_i^T \mathbf{x} + \mathbf{w}_i^T B\mathbf{u}.$$

If the mode s_i is uncontrollable, $\mathbf{w}_i^T B = \mathbf{0}$ and

$$\frac{d}{dt}(\mathbf{w}_i^T \mathbf{x}) = s_i(\mathbf{w}_i^T \mathbf{x})$$

which is a differential equation for the signed length of the projection of the state $\mathbf{x}$ upon the vector $\mathbf{w}_i$. The solution is $\mathbf{w}_i^T \mathbf{x}(t) = e^{s_i t}\mathbf{w}_i \mathbf{x}(0)$, which is completely unaffected by the input. Hence, the mode is seen to be impervious to control, i.e., uncontrollable.

Example 3.13   Verify that the network of Example 3.2 is uncontrollable. Find the uncontrollable mode.

Solution   The relevant matrices are

$$A = \begin{bmatrix} -\dfrac{3}{2} & \dfrac{1}{2} \\ \dfrac{1}{2} & -\dfrac{3}{2} \end{bmatrix} \qquad B = \begin{bmatrix} \dfrac{1}{2} \\ \dfrac{1}{2} \end{bmatrix}.$$

Since $n = 2$, the controllability matrix is

$$C = \begin{bmatrix} B & AB \end{bmatrix} = \begin{bmatrix} \dfrac{1}{2} & -\dfrac{1}{2} \\ \dfrac{1}{2} & -\dfrac{1}{2} \end{bmatrix}.$$

Since det$\mathcal{C} = 0$, its rank is less than 2 and the system is not controllable. Since

$$\begin{bmatrix} a & -a \end{bmatrix} \begin{bmatrix} \dfrac{1}{2} & -\dfrac{1}{2} \\[2mm] \dfrac{1}{2} & -\dfrac{1}{2} \end{bmatrix} = \mathbf{0}$$

the state $\begin{bmatrix} a \\ -a \end{bmatrix}$ is uncontrollable. It should be an eigenvector of A^T; indeed, $A^T \begin{bmatrix} a \\ -a \end{bmatrix} = -2 \begin{bmatrix} a \\ -a \end{bmatrix}$, so the mode $s = -2$ is not controllable.

Example 3.14   **(Pendulum on a Cart)**

Assess controllability for the inverted pendulum-and-cart system, using the linearized model of Example 2.9 in Chapter 2.

Solution   The equations are given in Example 3.10. The controllability matrix is

$$\mathcal{C} = \begin{bmatrix} 0 & \dfrac{1}{M} & 0 & \dfrac{g}{M\ell} \\[3mm] \dfrac{1}{M} & 0 & \dfrac{g}{M\ell} & 0 \\[3mm] 0 & -\dfrac{1}{m\ell} & 0 & -\dfrac{(M+m)g}{M^2\ell^2} \\[3mm] -\dfrac{1}{m\ell} & 0 & -\dfrac{(M+m)g}{M^2\ell^2} & 0 \end{bmatrix}.$$

$$\underbrace{}_{B}\quad \underbrace{}_{AB}\quad \underbrace{}_{A^2B}\quad \underbrace{}_{A^3B}$$

Because $\mathcal{C}$ is a square matrix (which is always the case with a single input), fullness of rank is assessed by verifying that $\det \mathcal{C} \neq 0$. In fact,

$$\det \mathcal{C} = \frac{g^2}{M^4 \ell^4}$$

so the system is controllable.

Example 3.15   **(Active Suspension)**

Examine the controllability of the (linearized) suspension system in Example 2.2 (Chapter 2).

Solution   We use the eigenvector method, for no special reason except to illustrate its use.

The eigenvalues and eigenvectors of the A matrix were calculated in Example 3.11; we need them for A^T. The eigenvalues of A^T are the same as those of A, but the eigenvectors are not, and must be calculated. The eigenvalues and eigenvectors of A^T are as follows:

$$-6.1523 \pm j24.534 \quad \text{and} \quad \begin{bmatrix} -9.332 \times 10^{-2} \\ 1 \\ -1.800 \times 10^{-2} \\ 9.049 \times 10^{-3} \end{bmatrix} \pm j \begin{bmatrix} -6.063 \times 10^{-3} \\ 0 \\ 2.425 \times 10^{-3} \\ -3.714 \times 10^{-2} \end{bmatrix}$$

$$-0.8477 \pm j2.9428 \quad \text{and} \quad \begin{bmatrix} 1 \\ -1.635 \times 10^{-1} \\ 1.096 \times 10^{-1} \\ 4.142 \times 10^{-3} \end{bmatrix} \pm j \begin{bmatrix} 0 \\ 6.035 \times 10^{-1} \\ -3.138 \times 10^{-1} \\ -3.250 \times 10^{-3} \end{bmatrix}$$

The B matrix is

$$B = \begin{bmatrix} 0 \\ 0 \\ .00334 \\ .02 \end{bmatrix}.$$

Clearly, neither vector is orthogonal to B, and the system is controllable.

3.7 REALIZATIONS

3.7.1 Introduction

The transfer function corresponding to a given set of state equations is unique; however, a given transfer function (scalar or matrix) can be generated by an infinite number of state equations, each of which is called a *realization* of the transfer function. For example, the state equations

$$\dot{x} = -x + u$$

$$y = x$$

and

$$\dot{x} = -x + au$$

$$y = \frac{1}{a}x, \quad a \text{ real}$$

have the same transfer function, $1/(s + 1)$; both are realizations of that transfer function.

DEFINITION
A state description $\dot{\mathbf{x}} = A\mathbf{x} + B\mathbf{u}$, $\mathbf{y} = C\mathbf{x} + D\mathbf{u}$ is a realization of $H(s)$ if

$$C(sI - A)^{-1}B + D = H(s). \qquad (3.54)$$

3.7.2 Similarity Transformations

Given a realization of $H(s)$, it is possible to generate others through *similarity transformations*. A similarity transformation is really just a change in coordinates. Given

$$\dot{\mathbf{x}} = A\mathbf{x} + B\mathbf{u}$$

$$\mathbf{y} = C\mathbf{x} + D\mathbf{u}$$

let

$$\mathbf{x} = T\mathbf{z}$$

or

$$\mathbf{z} = T^{-1}\mathbf{x}$$

where T is a constant $n \times n$ nonsingular matrix. The vector $\mathbf{z}$ is used as the new state vector, i.e., the new set of coordinates. The state equations for $\mathbf{z}$ are easily derived, as

$$\dot{\mathbf{z}} = T^{-1}\dot{\mathbf{x}} = T^{-1}A\mathbf{x} + T^{-1}B\mathbf{u}$$

or

$$\dot{\mathbf{z}} = T^{-1}AT\mathbf{z} + T^{-1}B\mathbf{u} \qquad (3.55)$$

and

$$\mathbf{y} = CT\mathbf{z} + D\mathbf{u}. \qquad (3.56)$$

Equations 3.55 and 3.56 are the state equations of the system in terms of the vector $\mathbf{z}$. To show that the transfer function has not changed, note that the transfer function corresponding to Equations 3.55 and 3.56 is

$$H'(s) = CT(sI - T^{-1}AT)^{-1}T^{-1}B + D. \qquad (3.57)$$

Now,

$$sI - T^{-1}AT = sT^{-1}T - T^{-1}AT$$
$$= T^{-1}(sI - A)T$$

so that

$$(sI - T^{-1}AT)^{-1} = [T^{-1}(sI - A)T]^{-1}$$
$$= T^{-1}(sI - A)^{-1}T.$$

Insertion of this in Equation 3.57 yields

$$H'(s) = C(sI - A)^{-1}B + D = H(s).$$

This result is not surprising. The transfer function relates the input and output of the system, and should not be affected if we use a different coordinate system to describe the "inside of the box."

It should also come as no surprise that controllability and observability are not affected by a similarity transformation. To see this, we show that, if $\mathbf{v}_i$ is an eigenvector of A, corresponding to the eigenvalue s_i, then $T^{-1}\mathbf{v}_i$ is an eigenvector of $T^{-1}AT$. That follows from

$$(T^{-1}AT)T^{-1}\mathbf{v}_i = T^{-1}A\mathbf{v}_i = s_i T^{-1}\mathbf{v}_i.$$

(Note that the eigenvalues of A and $T^{-1}AT$ are the same.)

Application of the eigenvector test for observability on the realization of Equations 3.55 and 3.56 yields

$$(CT)(T^{-1}\mathbf{v}_i) = C\mathbf{v}_i$$

which shows that the new realization is unobservable if, and only if, the original one is. A similar argument is applicable to controllability.

3.7.3 The Jordan Form

Diagonal Form
A special similarity transformation generates the so-called *Jordan form*. A special case of the Jordan form, the *diagonal* form, is considered first. It can be generated in cases where the A matrix has n independent eigenvectors, as when its eigenvalues are distinct. Let these eigenvectors be $\mathbf{v}_1, \mathbf{v}_2, \ldots, \mathbf{v}_n$, and let

$$T = [\mathbf{v}_1 \quad \mathbf{v}_2 \quad \cdots \quad \mathbf{v}_n]. \tag{3.58}$$

Then,

$$\mathbf{x} = T\mathbf{z} = z_1\mathbf{v}_1 + z_2\mathbf{v}_2 + \cdots + z_n\mathbf{v}_n. \tag{3.59}$$

The state vector $\mathbf{x}$ is seen to be expressed in terms of its components along each eigenvector.

The inverse of T is written in terms of rows, as

$$T^{-1} = \begin{bmatrix} \mathbf{w}_1{}^T \\ \mathbf{w}_2{}^T \\ \vdots \\ \mathbf{w}_n{}^T \end{bmatrix}. \tag{3.60}$$

Since $T^{-1}T = I$, it follows that

$$T^{-1}T = \begin{bmatrix} \mathbf{w}_1{}^T\mathbf{v}_1 & \mathbf{w}_1{}^T\mathbf{v}_2 & \cdots & \mathbf{w}_1{}^T\mathbf{v}_n \\ \mathbf{w}_2{}^T\mathbf{v}_1 & \mathbf{w}_2{}^T\mathbf{v}_2 & \cdots & \mathbf{w}_2{}^T\mathbf{v}_n \\ \vdots & \vdots & \vdots & \vdots \\ \mathbf{w}_n{}^T\mathbf{v}_1 & \mathbf{w}_n{}^T\mathbf{v}_2 & \cdots & \mathbf{w}_n^T\mathbf{v}_n \end{bmatrix} = I$$

and hence

$$\mathbf{w}_i{}^T\mathbf{v}_j = \begin{cases} 1 & \text{if } j = i \\ 0 & \text{if } j \neq i. \end{cases} \tag{3.61}$$

It turns out that $\mathbf{w}_i$ is an eigenvector of A^T, corresponding to the eigenvalue s_i. We may now calculate $T^{-1}AT$, as needed by Equation 3.55. We have

$$AT = A[\mathbf{v}_1 \quad \mathbf{v}_2 \quad \cdots \quad \mathbf{v}_n] = [A\mathbf{v}_1 \quad A\mathbf{v}_2 \quad \cdots \quad A\mathbf{v}_n]$$

$$= [s_1\mathbf{v}_1 \quad s_2\mathbf{v}_2 \quad \cdots \quad s_n\mathbf{v}_n]$$

and, using Equation 3.60,

$$T^{-1}AT = \begin{bmatrix} \mathbf{w}_1{}^T \\ \mathbf{w}_2{}^T \\ \vdots \\ \mathbf{w}_n{}^T \end{bmatrix} [s_1\mathbf{v}_1 \quad s_2\mathbf{v}_2 \quad \cdots \quad s_n\mathbf{v}_n] = \begin{bmatrix} s_1 & 0 & \cdots & 0 \\ 0 & s_2 & \cdots & 0 \\ \vdots & \vdots & \vdots & \vdots \\ 0 & 0 & \cdots & s_n \end{bmatrix}. \tag{3.62}$$

The transformed A matrix is diagonal, and its entries are the eigenvalues. The other two quantities required by Equations 3.55 and 3.56 are

$$T^{-1}B = \begin{bmatrix} \mathbf{w}_1{}^T B \\ \mathbf{w}_2{}^T B \\ \vdots \\ \mathbf{w}_n{}^T B \end{bmatrix} \tag{3.63}$$

and

$$CT = [C\mathbf{v}_1 \quad C\mathbf{v}_2 \quad \cdots \quad C\mathbf{v}_n]. \tag{3.64}$$

Use of these results in Equation 3.55 yields

$$\dot{z}_i = s_i z_i + \mathbf{w}_i^T B \mathbf{u}, \qquad i = 1, 2, \ldots, n$$
$$\mathbf{y} = C\mathbf{v}_1 z_1 + C\mathbf{v}_2 z_2 + \cdots + C\mathbf{v}_n z_n + D\mathbf{u}. \tag{3.65}$$

This realization in diagonal form has the very simple block-diagram form of Figure 3.7. The state variables are decoupled from each other. They have direct meaning in terms of modes, because they represent the eigenvector content of the original state vector $\mathbf{x}$. The diagram provides a neat interpretation of controllability and observability. The condition for the lack of controllability of the ith mode, $\mathbf{w}_i^T B = \mathbf{0}$, shows up as a simple decoupling of z_i from $\mathbf{u}$. If the ith mode is unobservable, $C\mathbf{v}_i = \mathbf{0}$ and z_i is cut off from $\mathbf{y}$.

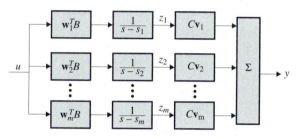

Figure 3.7 A system in diagonal Jordan form

Example 3.16   **(dc Servo)**

For the dc servo of Example 2.1, let $y = \theta$. Transform the given realization to diagonal Jordan form.

Solution   Starting from the state equation, Equation 2.19 (Chapter 2), we compute the eigenvalues and the matrix T of eigenvectors (MATLAB command eig). They are $s_1 = 0, s_2 = -2.474, s_3 = -21.526$, and

$$T = \begin{bmatrix} 1 & -.4042 & .0096 \\ 0 & 1 & -.2062 \\ 0 & -.5575 & 1 \end{bmatrix}.$$

The inverse of T (MATLAB command inv) is

$$T^{-1} = \begin{bmatrix} 1 & .4507 & .0833 \\ 0 & 1.1299 & .2329 \\ 0 & .6299 & 1.1299 \end{bmatrix}.$$

It is verified that

$$T^{-1}AT = \text{diag}[0, -2.474, -21.526].$$

We compute

$$T^{-1}B = \begin{bmatrix} 1.6667 & -3.3330 \\ 4.6588 & -8.3564 \\ 22.5971 & -4.6584 \end{bmatrix}$$

and, with $C = \begin{bmatrix} 1 & 0 & 0 \end{bmatrix}$,

$$CT = \begin{bmatrix} 1 & -.4042 & .0096 \end{bmatrix}.$$

The diagonal Jordan form realization is

$$\dot{\mathbf{z}} = \begin{bmatrix} 0 & 0 & 0 \\ 0 & -2.474 & 0 \\ 0 & 0 & -21.526 \end{bmatrix} \dot{\mathbf{z}} + \begin{bmatrix} 1.6667 & -3.3330 \\ 4.6588 & -8.3564 \\ 22.5971 & -4.6584 \end{bmatrix} \begin{bmatrix} v \\ T_L \end{bmatrix}$$

$$\theta = \begin{bmatrix} 1 & -.4042 & .0096 \end{bmatrix} \mathbf{z}.$$

All three modes are observable and controllable. They are, in fact, controllable from each input taken separately, which means that both the control input v and the disturbance input T_L are capable of exciting all modes.

Note that the state variables z_1, z_2, and z_3 do not have the direct physical significance of the state variables θ, ω, and i in the original realization. Of course, θ, ω, and i can be calculated from $\mathbf{z}$ by use of the relation $\mathbf{x} = T^{-1}\mathbf{z}$.

If some of the eigenvalues are complex, so are the eigenvectors and so is the Jordan form. That is inconvenient, and requires some modification. Since complex eigenvalues and eigenvectors come in conjugate pairs, let us focus our attention on the two equations corresponding to complex conjugate eigenvalues:

$$\dot{z}_i = s_i z_i + (\mathbf{w}_i^T B)\mathbf{u} \tag{3.66}$$

$$\dot{z}_{i+1} = s_i^* z_{i+1} + (\mathbf{w}_i^{*T} B)\mathbf{u} \tag{3.67}$$

$$\mathbf{y} = C\mathbf{v}_i z_i + C\mathbf{v}_i^* z_{i+1}. \tag{3.68}$$

The contribution to the state at $t = 0$ is

$$z_i(0)\mathbf{v}_i + z_{i+1}(0)\mathbf{v}_i^*$$

which must be real, so

$$Re z_i(0)I_m\mathbf{v}_i + I_m z_i(0)Re\mathbf{v}_i - Re z_{i+1}(0)I_m\mathbf{v}_i + I_m z_{i+1}(0)Re\mathbf{v}_i = 0$$

or

$$[Re z_i(0) - Re z_{i+1}(0)]I_m\mathbf{v}_i + [I_m z_i(0) + I_m z_{i+1}(0)]Re\mathbf{v}_i = 0.$$

Since the real and imaginary parts of a complex eigenvector are linearly independent, the coefficients of both $I_m\mathbf{v}_i$ and $Re\mathbf{v}_i$ are zero; i.e.,

$$Rez_{i+1}(0) = Rez_i(0)$$

$$I_mz_{i+1}(0) = -I_mz_i(0)$$

or $z_{i+1}(0) = z_i^*(0)$.

Conjugating each side of Equation 3.66,

$$\dot{z}_i^* = s_i^*z_i^* + (\mathbf{w}_i^{*T}B)u$$

so that z_i^* and z_{i+1} satisfy the same differential equation. Since $z_{i+1}(0) = z_i^*(0)$, it follows that $z_{i+1}(t) = z_i^*(t)$.

We transform Equations 3.66 and 3.67 by (1) addition and (2) subtraction and multiplication by $-j$. This amounts to taking the real and imaginary parts of Equation 3.66:

$$\frac{d}{dt}(Rez_i) = (Res_i)(Rez_i) - (I_ms_i)(I_mz_i) + (Re\mathbf{w}_i^TB)\mathbf{u}$$

$$\frac{d}{dt}(I_mz_i) = (I_ms_i)(Rez_i) + (Res_i)(I_mz_i) + (I_m\mathbf{w}_i^TB)\mathbf{u} \tag{3.69}$$

$$y = 2C(Re\mathbf{v}_i)(Rez_i) - 2C(I_m\mathbf{v}_i)(I_mz_i). \tag{3.70}$$

The two decoupled, complex differential equations are replaced by a pair of coupled, real equations. In the A matrix, a complex pair generates a 2×2 diagonal block.

Example 3.17   (Active Suspension)

In the active suspension of Example 2.2 (Chapter 2), $y = x_1 - x_2$. Compute the diagonal Jordan form, and transform it to real, block-diagonal form.

Solution   Proceeding as in Example 3.16 (MATLAB command eig), we obtain

$$\mathbf{z} = \begin{bmatrix} -6.15 + j24.5 & 0 & 0 & 0 \\ 0 & -6.15 - j24.5 & 0 & 0 \\ 0 & 0 & -.848 + j2.94 & 0 \\ 0 & 0 & 0 & -.848 - j2.94 \end{bmatrix} \mathbf{z}$$

$$+ \begin{bmatrix} -9.11e - 3 + j5.51e - 3 & 252 + j178 \\ -9.11e - 3 - j5.51e - 3 & 252 - j178 \\ 4.50e - 4 + j1.69e - 3 & -1.92 + j5.04 \\ 4.50e - 4 - j1.69e - 3 & -1.92 - j5.04 \end{bmatrix} \begin{bmatrix} u \\ y_R \end{bmatrix}$$

$$y = [-.0254 + j.0317 \quad -.0254 - j.0317 \quad -.282 - j.0414 \quad -.282 + j.0414]\mathbf{z}.$$

Application of Equations 3.69 and 3.70 yields

$$\frac{d}{dt}\begin{bmatrix} Rez_1 \\ I_m z_1 \\ Rez_2 \\ I_m z_2 \end{bmatrix} = \begin{bmatrix} -6.15 & -24.5 & 0 & 0 \\ 24.5 & -6.15 & 0 & 0 \\ 0 & 0 & -.848 & -2.94 \\ 0 & 0 & 2.94 & -.848 \end{bmatrix} \begin{bmatrix} Rez_1 \\ I_m z_1 \\ Rez_2 \\ I_m z_2 \end{bmatrix}$$

$$+ \begin{bmatrix} -9.11e-3 & 253 \\ 5.50e-3 & -179 \\ 4.50e-4 & -1.93 \\ 1.69e-3 & 5.04 \end{bmatrix} \begin{bmatrix} u \\ y_r \end{bmatrix}$$

$$y = [-5.08e-2 \quad -6.34e-2 \quad -.563 \quad 8.29e-2] \begin{bmatrix} Rez_1 \\ I_m z_1 \\ Rez_2 \\ I_m z_2 \end{bmatrix}.$$

General Case[1]

The eigenstructure can be more complex if A has repeated eigenvalues. Corresponding to a repeated root of multiplicity k, there will be at least one eigenvector and, at most k independent eigenvectors; a symmetric matrix, for example, always has k independent eigenvectors for such a root. If there are fewer than k eigenvectors, some (or all) of the eigenvectors will serve as starting points for chains of *generalized eigenvectors*. Let $\mathbf{v}_i^0$ be an eigenvector of A, corresponding to the eigenvalue s_i, and let $\mathbf{v}_i^1, \mathbf{v}_i^2, \ldots$ be the chain of generalized eigenvectors engendered by $\mathbf{v}_i^0$. The vectors satisfy

$$(A - s_i I)\mathbf{v}_i^1 = \mathbf{v}_i^0$$

$$(A - s_i I)\mathbf{v}_i^2 = \mathbf{v}_i^1$$

$$\vdots$$

$$(A - s_i I)\mathbf{v}_i^{j+1} = \mathbf{v}_i^j. \tag{3.71}$$

The chain continues as long as a nontrivial solution can be found for the next generalized eigenvector.

If several independent eigenvectors correspond to s_i, each of those eigenvectors generates a chain of generalized eigenvectors. It can be shown that the eigenvectors and generalized eigenvectors corresponding to the eigenvalue s_i are independent of each other and of the eigenvectors and generalized eigenvectors corresponding to other eigenvalues.

To generate the Jordan form of a matrix with repeated eigenvalues, we define the transformation matrix T whose columns are the eigenvectors and generalized

---

[1]This section may be skipped without loss of continuity.

eigenvectors of the matrix A. The general development requires unwieldy notation, so we present it with the special case

$$T = [\mathbf{v}_{11}{}^0 \ \mathbf{v}_{11}{}^1 \ \mathbf{v}_{11}{}^2 \ \mathbf{v}_{12}{}^0 \ \mathbf{v}_{12}{}^1 \ \mathbf{v}_{21}{}^0 \ \mathbf{v}_{21}{}^1]. \tag{3.72}$$

Here, $\mathbf{v}_{11}{}^0$ and $\mathbf{v}_{12}{}^0$ are eigenvectors corresponding to the eigenvalue s_1; $\mathbf{v}_{21}{}^0$ corresponds to s_2. The other columns of T are generalized eigenvectors.

The matrix T^{-1} is defined by its rows, i.e.,

$$T^{-1} = \begin{bmatrix} \mathbf{w}_{11}{}^{0^T} \\ \mathbf{w}_{11}{}^{1^T} \\ \mathbf{w}_{11}{}^{2^T} \\ \mathbf{w}_{12}{}^{0^T} \\ \mathbf{w}_{12}{}^{1^T} \\ \mathbf{w}_{21}{}^{0^T} \\ \mathbf{w}_{21}{}^{1^T} \end{bmatrix} \tag{3.73}$$

where the ith row of T^{-1} is orthogonal to all columns of T except the ith column, whose scalar product with the ith row of T^{-1} is unity. It turns out that the rows of T^{-1} are left eigenvectors and generalized eigenvectors of A. The order is the reverse of that of the right eigenvectors, in that the last generalized eigenvector in the chain comes first, down to the eigenvector. Thus, $\mathbf{w}_{11}{}^{2^T}$, $\mathbf{w}_{12}{}^{1^T}$, and $\mathbf{w}_{21}{}^{1^T}$ are left eigenvectors of A.

We now form the product $T^{-1}AT$, starting with

$$AT =$$

$$[s_1\mathbf{v}_{11}{}^0 \quad \mathbf{v}_{11}{}^0 + s_1\mathbf{v}_{11}{}^1 \quad \mathbf{v}_{11}{}^1 + s_1\mathbf{v}_{11}{}^2 \quad s_1\mathbf{v}_{12}{}^0\mathbf{v}_{12}{}^0 + s_1\mathbf{v}_{12}{}^1 \quad s_2\mathbf{v}_{21}{}^0 \quad \mathbf{v}_{21}{}^0 + s_2\mathbf{v}_{21}{}^1]$$

where Equation 3.71 was used in the form

$$A\mathbf{v}_i{}^{j+1} = \mathbf{v}_i{}^j + s_i\mathbf{v}_i{}^{j+1}.$$

The product $T^{-1}AT$ is formed, generating the ith column by multiplying each row of T^{-1} by the ith column of AT. The result is

$$T^{-1}AT = \begin{bmatrix} s_1 & 1 & 0 & 0 & 0 & 0 & 0 \\ 0 & s_1 & 1 & 0 & 0 & 0 & 0 \\ 0 & 0 & s_1 & 0 & 0 & 0 & 0 \\ 0 & 0 & 0 & s_1 & 1 & 0 & 0 \\ 0 & 0 & 0 & 0 & s_1 & 0 & 0 \\ 0 & 0 & 0 & 0 & 0 & s_2 & 1 \\ 0 & 0 & 0 & 0 & 0 & 0 & s_2 \end{bmatrix}. \tag{3.74}$$

This matrix is block diagonal. The blocks are called *Jordan blocks*, and there are as many of them as there are generalized eigenvector chains. The independent-eigenvector case generates 1×1 Jordan blocks, as a special case.

Equations for $T^{-1}B$ and CT are easily written, as were Equations 3.63 and 3.64. The following state equations result:

$$\dot{\mathbf{z}}_1 = \begin{bmatrix} s_1 & 1 & 0 \\ 0 & s_1 & 1 \\ 0 & 0 & s_1 \end{bmatrix} \dot{\mathbf{z}}_1 + \begin{bmatrix} \mathbf{w}_{11}^{0^T} B \\ \mathbf{w}_{11}^{1^T} B \\ \mathbf{w}_{11}^{2^T} B \end{bmatrix} \mathbf{u}$$

$$\dot{\mathbf{z}}_2 = \begin{bmatrix} s_1 & 1 \\ 0 & s_1 \end{bmatrix} \mathbf{z}_2 + \begin{bmatrix} \mathbf{w}_{12}^{0^T} B \\ \mathbf{w}_{12}^{1^T} B \end{bmatrix} \mathbf{u}$$

$$\dot{\mathbf{z}}_3 = \begin{bmatrix} s_2 & 1 \\ 0 & s_2 \end{bmatrix} \mathbf{z}_3 + \begin{bmatrix} \mathbf{w}_{21}^{0^T} B \\ \mathbf{w}_{21}^{1^T} B \end{bmatrix} \mathbf{u}$$

$$\mathbf{y} = C[\mathbf{v}_{11}^{0} \mathbf{v}_{11}^{1} \mathbf{v}_{11}^{2}]\mathbf{z}_1 + C[\mathbf{v}_{12}^{0} \mathbf{v}_{12}^{1}]\mathbf{z}_2 + C[\mathbf{v}_{21}^{0} \mathbf{v}_{21}^{1}]\mathbf{z}_3.$$

The structure is that of parallel blocks, each corresponding to a Jordan block.

The first Jordan block can be used as an example to explore the structure of those parallel blocks. Using Laplace transforms,

$$(\mathbf{z}_1)_1 = \frac{1}{s - s_1}[(\mathbf{z}_1)_2 + \mathbf{w}_{11}^{0^T} B\mathbf{u}]$$

$$(\mathbf{z}_1)_2 = \frac{1}{s - s_1}[(\mathbf{z}_1)_3 + \mathbf{w}_{11}^{1^T} B\mathbf{u}]$$

$$(\mathbf{z}_1)_3 = \frac{1}{s - s_1}\mathbf{w}_{11}^{2^T} B\mathbf{u}$$

$$\mathbf{y}_1 = C\mathbf{v}_{11}^{0}(\mathbf{z}_1)_1 + C\mathbf{v}_{11}^{1}(\mathbf{z}_1)_2 + C\mathbf{v}_{11}^{2}(\mathbf{z}_1)_3.$$

Figure 3.8 shows the relevant block diagram. It is immediately obvious that this block is not controllable if $\mathbf{w}_{11}^{2^T} B = \mathbf{0}$, because the state variable $(\mathbf{z}_1)_3$ is unaffected by the input. It is also clear that this block is not observable if $C\mathbf{v}_{11}^{0} = \mathbf{0}$, because the output is not influenced by the state variable $(\mathbf{z}_1)_1$.

Since $\mathbf{v}_{11}^{0}$ is an eigenvector of A and $\mathbf{w}_{11}^{2}$ is an eigenvector of A^T, the relation $C\mathbf{v}_{11}^{0} = \mathbf{0}(\mathbf{w}_{11}^{2^T} B = \mathbf{0})$ shows, by the eigenvector tests, that the subsystem under study is unobservable (uncontrollable), thus confirming the intuitive conclusion.

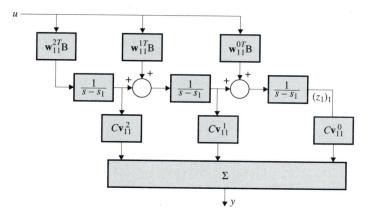

Figure 3.8 Illustration of a Jordan block for the case of repeated eigenvalues

---

Example 3.18 (Pendulum on a Cart)

For the values $M = m = \ell = 1, g = 9.8$, the system of Example 3.10 has the matrices

$$A = \begin{bmatrix} 0 & 1 & 0 & 0 \\ 0 & 0 & -9.8 & 0 \\ 0 & 0 & 0 & 1 \\ 0 & 0 & 19.6 & 0 \end{bmatrix} \qquad B = \begin{bmatrix} 0 \\ 1 \\ 0 \\ -1 \end{bmatrix}.$$

Assuming, for example, that the position x and the angle θ are measured,

$$C = \begin{bmatrix} 1 & 0 & 0 & 0 \\ 0 & 0 & 1 & 0 \end{bmatrix}.$$

Transform this realization to Jordan form.

Solution   The eigenvalues of A are $0, 0, \pm 4.4272$. The vector $\begin{bmatrix} 1 & 0 & 0 & 0 \end{bmatrix}^T$ is the only eigenvector corresponding to the double eigenvalue $s = 0$. We calculate the generalized eigenvector through Equation 3.75:

$$(A - 0I)\mathbf{v}_1{}^1 = \begin{bmatrix} 1 \\ 0 \\ 0 \\ 0 \end{bmatrix}.$$

The solution is

$$\mathbf{v}_1{}^1 = \begin{bmatrix} x \\ 1 \\ 0 \\ 0 \end{bmatrix}$$

where x is arbitrary. Choose $x = 0$.

The eigenvectors corresponding to ± 4.4272 are computed in the normal manner. From Equation 3.72,

$$T = \begin{bmatrix} 1 & 0 & 1 & 1 \\ 0 & 1 & 4.4272 & -4.4272 \\ 0 & 0 & -2 & -2 \\ 0 & 0 & -8.8544 & 8.8544 \end{bmatrix}$$

whose inverse is

$$T^{-1} = \begin{bmatrix} 1 & 0 & .5 & 0 \\ 0 & 1 & 0 & .5 \\ 0 & 0 & -.25 & -.0565 \\ 0 & 0 & -.25 & .0565 \end{bmatrix}.$$

As expected,

$$T^{-1}AT = \begin{bmatrix} 0 & 1 & 0 & 0 \\ 0 & 0 & 0 & 0 \\ 0 & 0 & 4.4272 & 0 \\ 0 & 0 & 0 & -4.4272 \end{bmatrix}.$$

We obtain

$$T^{-1}B = \begin{bmatrix} 0 \\ .5 \\ .0565 \\ -.0565 \end{bmatrix}$$

$$CT = \begin{bmatrix} 1 & 0 & 1 & 1 \\ 0 & 0 & -2 & -2 \end{bmatrix}.$$

Figure 3.9 shows a block diagram of this system. The system is seen to be controllable and observable. Note, however, that the system would not be observable from θ alone.

3.7.4 The Canonical Decomposition

In the diagonal form, the state variables can be divided into four categories:

- Controllable and observable ($\mathbf{w}_i^T B \neq \mathbf{0}, C\mathbf{v}_i \neq \mathbf{0}$)
- Uncontrollable and observable ($\mathbf{w}_i^T B = \mathbf{0}, C\mathbf{v}_i \neq \mathbf{0}$)
- Controllable and unobservable ($\mathbf{w}_i^T B \neq \mathbf{0}, C\mathbf{v}_i = \mathbf{0}$)
- Uncontrollable and unobservable ($\mathbf{w}_i^T B = \mathbf{0}, C\mathbf{v}_i = \mathbf{0}$)

This decomposition, called the *canonical decomposition*, is illustrated in Figure 3.10. Figures 3.11, 3.12, and 3.13 show the more general structure of this decomposition, which is applicable in the general case of multiple eigenvalues. Figure 3.11 illus-

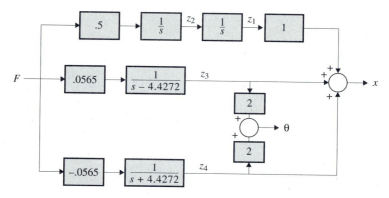

Figure 3.9 Block diagram for the Jordan form of the pendulum-and-cart system

trates the decomposition into controllable and uncontrollable parts. In Figure 3.12 the system is split into observable and unobservable parts. Figure 3.13 combines Figures 3.11 and 3.12 in the sense that the controllable and uncontrollable blocks are both further divided into observable and unobservable parts. Note that there is no path, direct or through a block, from the input to either of the uncontrollable blocks. Similarly, the unobservable blocks have no path to the output.

The transfer function of the system is that of the controllable and observable block; the other three blocks have no influence. To see this, recall that the transfer function is the transform of the zero-state response. If the initial state is zero, the two uncontrollable blocks remain in the zero state and have zero effect on **y**. The state of the controllable, unobservable block does move away from zero under the influence of the input, but that has no effect on the output because that block has no connection to **y**. That leaves only the controllable, observable block.

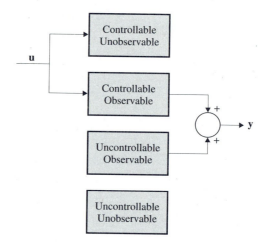

Figure 3.10 The canonical decomposition for the case of independent eigenvectors

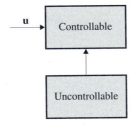

Figure 3.11 Decomposition into controllable and uncontrollable blocks

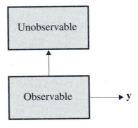

Figure 3.12 Decomposition into observable and unobservable blocks

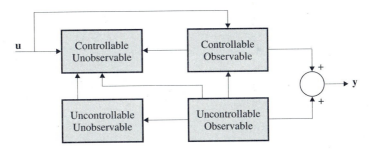

Figure 3.13 The canonical decomposition for the general case

Example 3.19   **(dc Servo)**

Repeat Example 3.16 but with the output $y = \omega$ $\left(C = \begin{bmatrix} 0 & 1 & 0 \end{bmatrix}\right)$. Display the canonical decomposition. (For simplicity, use only the input v.)

Solution   The steps are as in Example 3.16, up to

$$CT = \begin{bmatrix} 0 & 1 & -.2062 \end{bmatrix}.$$

Figure 3.14 displays the canonical decomposition.

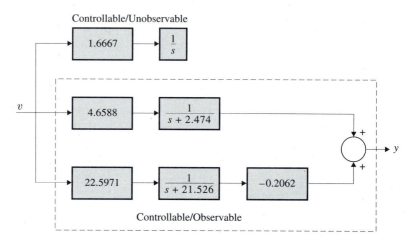

Figure 3.14   Canonical decomposition of the dc servo

Example 3.20   **(Pendulum on a Cart)**

Give a canonical decomposition for the pendulum-and-cart system of Example 3.10, with the output $y = \theta$.

Solution   For the state equations used in Example 3.10, with $C = \begin{bmatrix} 0 & 0 & 1 & 0 \end{bmatrix}$, the system breaks down by inspection into observable and unobservable parts, both controllable, as shown in Figure 3.15. Since the linear position x and velocity v do not couple into the output, they cannot be observed.

3.7.5 Minimal Realizations

The equations

$$\dot{x}_1 = -x_1 + u$$

$$y = x_1 \tag{3.75}$$

describe a realization of the transfer function $1/(s + 1)$. A different realization of the same transfer function is given by

$$\dot{x}_1 = -x_1 + x_2 + u$$

$$\dot{x}_2 = -2x_2 \tag{3.76}$$

$$y = x_1 + x_2.$$

To see this, recall that the transfer function is equal to the transform of the zero-state output, divided by $u(s)$. The zero-state solution for x_2 is clearly $x_2(t) = 0$, which reduces Equation 3.76 to Equation 3.75.

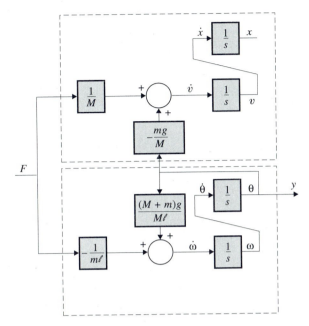

Figure 3.15   Canonical decomposition of the pendulum-and-cart system

The realization of Equation 3.76 has two states. It is termed *nonminimal* because it is obviously possible to realize the same transfer function with fewer states. The realization of Equation 3.75 is *minimal*, because there needs to be at least one state equation to generate one pole, and it is not possible to do this with zero differential equations.

DEFINITION
A realization of a transfer function $H(s)$ is *minimal* if there exists no realization of lesser order whose transfer function is $H(s)$.

The concept of minimality is directly related to controllability and observability through the following theorem.

■ **Theorem 3.5**   A realization is minimal if, and only if, it is controllable and observable.

Proof: To show necessity ("only if"), we show that minimality is precluded if the realization is either uncontrollable or unobservable. This follows directly from observation of the canonical decomposition: the uncontrollable or unobservable blocks can be removed without affecting the transfer function. If a realization is uncontrollable or unobservable, it can be transformed into its canonical decomposition and the uncontrollable and unobservable blocks removed, to generate a realization of lower order with the same transfer function.

We show sufficiency for the special case where the A matrix has distinct eigenvalues. In such a case, transformation into diagonal form is possible, leading to equations of the form of Equation 3.65. From that equation,

$$z_i(s) = \frac{1}{s - s_i} \mathbf{w_i}^{\mathsf{T}} B \mathbf{u}(s)$$

and

$$\mathbf{y}(s) = \left[\sum_i = 1^n \frac{1}{s - s_i} (C\mathbf{v}_i)(\mathbf{w}_i^T B) + D \right] \mathbf{u}(s).$$

If the system is controllable and observable, $C\mathbf{v}_i \neq \mathbf{0}$ and $\mathbf{w}_i^T B \neq \mathbf{0}$ for $i = 1, 2, \ldots, n$; hence, the coefficient of each term $\frac{1}{s-s_i}$ is nonzero and the transfer function has n poles, $s_1, s_2, \ldots, s_n$. Since all transfer function poles must be eigenvalues of A, then A must have at least n eigenvalues, so a realization with fewer than n states is not possible. ∎

The method of proof used in the second part of Theorem 3.5 uncovers an interesting fact. For the case of distinct eigenvalues, the coefficient of $1/(s - s_i)$ is zero if, and only if, the ith mode is either uncontrollable ($\mathbf{w}_i^T B = \mathbf{0}$) or unobservable ($C\mathbf{v}_i = \mathbf{0}$). In such a case, s_i is not a pole of the transfer function and the mode is called a *hidden mode*. The factor $(s - s_i)$ appears in $\det(sI - A)$, of course, but is cancelled out by the numerator, $C Adj(sI - A)B$. The absence of an eigenvalue from the set of poles indicates either uncontrollability or unobservability (or both) of the corresponding mode. That turns out to hold for the repeated eigenvalue case, also. In the single-input, single-output case, the conclusion is also true if an eigenvalue of multiplicity r appears as a pole of multiplicity less than r; that, however, is not true for multi-input, multi-output systems, in which a decrease in multiplicity does not necessarily signify loss of controllability or observability.

3.7.6 Realization of Transfer Functions

This section addresses the problem of deriving a realization for a given transfer function. There are several situations where this is useful. In filter or controller design, the outcome is sometimes a specification in terms of a transfer function to be implemented by analog elements. An implementation in terms of "op-amp" integrators is essentially the same as a realization. There are also cases in control where the plant is described by its transfer function, which must be converted to state form to permit the use of a state-based design method.

The starting point is a strictly proper SISO transfer function,

$$\frac{y(s)}{u(s)} = H(s) = \frac{b_{n-1}s^{n-1} + b_{n-2}s^{n-2} + \cdots + b_0}{s^n + a_{n-1}s^{n-1} + \cdots + a_0}. \tag{3.77}$$

Define an intermediate quantity $z(s)$:

$$z(s) = \frac{1}{s^n + a_{n-1}s^{n-1} + \cdots + a_0} u(s). \tag{3.78}$$

The differential equation corresponding to Equation 3.78 is

$$z^{(n)} = -a_{n-1}z^{(n-1)} - a_{n-1}z^{(n-2)} - \cdots - a_0 z + u. \tag{3.79}$$

Figure 3.16 is a simulation diagram representing Equation 3.79. The variable z and its derivatives up to the $(n-1)$st are the outputs of integrators, and the nth derivative is constructed as prescribed by Equation 3.79.

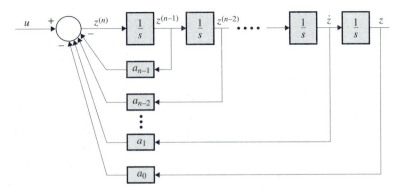

Figure 3.16   Block diagram used in the construction of the controllable canonical form

From Equations 3.77 and 3.78, it follows that

$$y(s) = (b_{n-1}s^{n-1} + b_{n-2}s^{n-2} + \cdots + b_0)z(s)$$

so that

$$y = b_{n-1}z^{(n-1)} + b_{n-2}z^{(n-2)} + \cdots + b_0 z. \tag{3.80}$$

Figure 3.17 generates y from z and its first $(n-1)$ derivatives, according to Equation 3.80; this completes the simulation.

Given a simulation diagram, it is easy to write state equations. The integrator outputs are taken to be the state variables; the derivative of any state variable is simply the input to the corresponding integrator. With

$$x_1 = z, \qquad x_2 = \dot{z}, \qquad \ldots, \qquad x_n = z^{(n-1)}$$

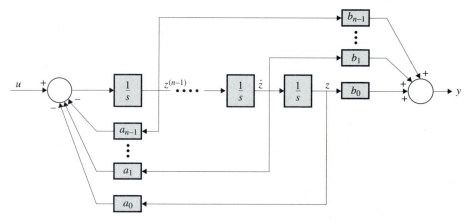

Figure 3.17 Block diagram for the controllable canonical form

we have

$$\dot{x}_1 = x_2$$
$$\dot{x}_2 = x_3$$
$$\vdots$$
$$\dot{x}_{n-1} = x_n$$
$$\dot{x}_n = -a_0 x_1 - a_1 x_2 - \cdots - a_{n-1} x_n + u$$

and

$$y = b_0 x_1 + b_1 x_2 + \cdots + b_{n-1} x_n.$$

In matrix form,

$$\dot{\mathbf{x}} = \begin{bmatrix} 0 & 1 & 0 & 0 & \ldots & 0 \\ 0 & 0 & 1 & 0 & \ldots & 0 \\ \vdots & \vdots & \vdots & \vdots & \vdots & \vdots \\ 0 & 0 & \ldots & \ldots & 0 & 1 \\ -a_0 & -a_1 & \ldots & \ldots & \ldots & -a_{n-1} \end{bmatrix} \mathbf{x} + \begin{bmatrix} 0 \\ 0 \\ \vdots \\ \vdots \\ 1 \end{bmatrix} u$$

$$y = \begin{bmatrix} b_0 & b_1 & \ldots & b_{n-1} \end{bmatrix} \mathbf{x}. \tag{3.81}$$

Equation 3.81 is the *controllable canonical* form. It is easily written by inspection, if the transfer function is expressed as a ratio of polynomials.

Example 3.21

Give the realization in controllable canonical form and the associated simulation diagram for $H(s) = (2s + 1)/(s^3 + 2s^2 + 3s + 4)$.

Solution

By inspection, the realization is

$$\dot{\mathbf{x}} = \begin{bmatrix} 0 & 1 & 0 \\ 0 & 0 & 1 \\ -4 & -3 & -2 \end{bmatrix} \mathbf{x} + \begin{bmatrix} 0 \\ 0 \\ 1 \end{bmatrix} u$$

$$y = \begin{bmatrix} 1 & 2 & 0 \end{bmatrix} \mathbf{x}.$$

The diagram is shown in Figure 3.18.

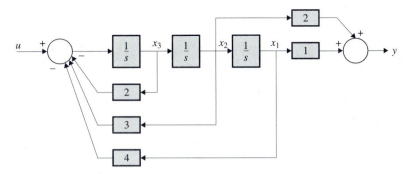

Figure 3.18 Controllable canonical form

Another realization is generated by writing Equation 3.77 as

$$s^n y = b_{n-1}s^{n-1}u + b_{n-2}s^{n-2}u + \cdots + b_0 u$$
$$-a_{n-1}s^{n-1}y - a_{n-2}s^{n-2}y - \cdots - a_0 y$$

or

$$y = \frac{1}{s}(b_{n-1}u - a_{n-1}y) + \frac{1}{s^2}(b_{n-2}u - a_{n-2}y) + \cdots + \frac{1}{s^n}(b_0 u - a_0 y). \quad \textbf{(3.82)}$$

The simulation diagram is shown in Figure 3.19. Taking the integrator outputs as state variables, we obtain

$$\dot{x}_1 = x_2 + b_{n-1}u - a_{n-1}x_1$$
$$\dot{x}_2 = x_3 + b_{n-2}u - a_{n-2}x_1$$
$$\vdots$$
$$\dot{x}_n = b_0 u - a_0 x_1$$
$$y = x_1$$

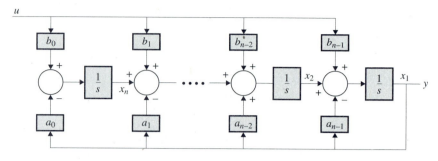

Figure 3.19   Block diagram for the observable canonical form

or, in matrix form,

$$\dot{\mathbf{x}} = \begin{bmatrix} -a_{n-1} & 1 & 0 & \cdots & 0 \\ -a_{n-2} & 0 & 1 & \cdots & 0 \\ \vdots & \vdots & \vdots & \vdots & \vdots \\ -a_1 & 0 & 0 & \cdots & 1 \\ -a_0 & 0 & 0 & \cdots & 0 \end{bmatrix} \mathbf{x} + \begin{bmatrix} b_{n-1} \\ b_{n-2} \\ \vdots \\ b_1 \\ b_0 \end{bmatrix} u$$

$$y = \begin{bmatrix} 1 & 0 & 0 & \cdots & 0 & 0 \end{bmatrix} \mathbf{x}.$$

(3.83)

Equation 3.83 is known as the *observable canonical form.*

Example 3.22   Repeat Example 3.21, but use the observable canonical form.

Solution   By inspection, the realization is

$$\dot{\mathbf{x}} = \begin{bmatrix} -2 & 1 & 0 \\ -3 & 0 & 1 \\ -4 & 0 & 0 \end{bmatrix} \mathbf{x} + \begin{bmatrix} 0 \\ 2 \\ 1 \end{bmatrix} u$$

$$y = \begin{bmatrix} 1 & 0 & 0 \end{bmatrix} \mathbf{x}.$$

The simulation diagram is shown in Figure 3.20.

It is possible to extend the two canonical forms to the MIMO case, but the result is of somewhat limited usefulness because the realization is usually not minimal. There are better ways to derive MIMO realizations, but they are beyond the scope of this text.

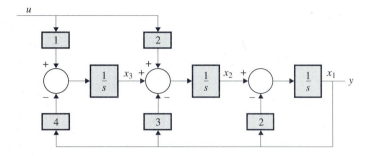

Figure 3.20   Observable canonical form

3.8 STABILITY

There are two different definitions of stability: internal stability and input–output stability. The concept of internal stability is applicable to the zero-input solution of an LTI system; input–output stability is concerned with the zero-state solution.

DEFINITION
An LTI system is *internally stable* if the zero-input solution $\mathbf{x}_{zi}(t)$ converges to zero, for any initial state.

Because the definition is in terms of the state (rather than the output) and because the input is zero, the matrices C and B are clearly irrelevant to internal stability. The conditions for internal stability will therefore be imposed on the A matrix only.

■ **Theorem 3.6**   An LTI system is internally stable if, and only if, all eigenvalues of A lie in the open left-half plane.

Proof: The words "open left-half plane" exclude the imaginary axis, i.e., refer to $Re(s) < 0$.

The terms of the zero-input solution $\mathbf{x}_{zi}(t)$ are of the form $t^k e^{s_i t}$, where the s_i are the eigenvalues of A. If all eigenvalues have negative real parts, then all terms of $\mathbf{x}_{zi}$ tend to zero; this shows sufficiency.

If one eigenvalue of A, say s_1, is such that $Re s_1 \geq 0$, we show that the system is not internally stable with the following argument. Let $\mathbf{v}_1$ be an eigenvector corresponding to s_1. If the initial state is $\mathbf{v}_1$, then $\mathbf{x}_{zi}(t) = \mathbf{v}_1 e^{s_1 t}$. Since $Re s_1 \geq 0$, $\mathbf{x}_{zi}$ does not tend to zero; this establishes necessity.   ■

We call eigenvalues of A with negative real parts *stable modes*, and those with real parts that are positive or zero, *unstable modes*. The test of internal stability is simply to check that all modes are stable.

The concept of input–output stability is concerned with the boundedness of the zero-state output in response to bounded inputs.

> **DEFINITION**
> An LTI system is *input–output stable* if the zero-state output is bounded for all bounded inputs.

Conditions for this type of stability are expressed in terms of the impulse response or its transform, the transfer function.

■ **Theorem 3.7**   A single-input, single-output LTI system is input–output stable if, and only if, its impulse response $h(t)$ satisfies

$$\int_0^\infty |h(t)|\,dt < \infty. \tag{3.84}$$

Proof: To show sufficiency, assume the condition is satisfied. The output is given by the convolution integral,

$$y(t) = \int_0^t h(\tau)u(t - \tau)\,d\tau.$$

Let the input be bounded; i.e., let $|u(t - \tau)| \le M$. Then

$$|y(t)| \le \int_0^t |h(\tau)||u(t - \tau)|\,d\tau \le M \int_0^t |h(\tau)|\,d\tau < \infty$$

and $|y(t)|$ is bounded.

To show necessity, we show that, if the condition is violated, it is always possible to construct a bounded input that makes y "blow up." With t fixed, choose

$$u(t - \tau) = \begin{cases} +1 & \text{if } h(\tau) \ge 0 \\ -1 & \text{if } h(\tau) < 0. \end{cases}$$

Then $h(\tau)u(t - \tau) = |h(\tau)|$, and

$$y(t) = \int_0^t |h(\tau)|\,d\tau.$$

Since $\lim_{t\to\infty} \int_0^t |h(\tau)|\,d\tau = \infty$, we can make $y(t)$ as large as desired simply by making t sufficiently large. Therefore, $y(t)$ is not bounded for all bounded inputs, and the system is not input–output stable. ■

In the multi-input, multi-output case, all elements of the matrix of impulse responses must satisfy the absolute integrability condition; otherwise, one of the inputs could be excited by a bounded function that would make at least one output grow unbounded.

The condition of Equation 3.84 holds for all LTI systems, lumped or distributed. For the lumped systems of this text, a condition in terms of the transfer function is more easily applicable.

■ **Theorem 3.8**   A single-input, single-output LTI system is input–output stable if, and only if, its transfer function has all its poles in the open left-half plane.

Proof: First, let us demonstrate sufficiency. If $p_1, p_2, \ldots, p_K$ are the poles of the transfer function, the impulse response is a sum of terms of the form $t^k e^{p_i t}$. If $Re(p_i) < 0$ for $i = 1, 2, \ldots, K$, then $h(t)$ decreases exponentially and is absolutely integrable, so the system is input–output stable.

To show necessity, assume that at least one pole has a nonnegative real part. Let p_1 be the pole with the largest (i.e., "most positive") real part. For t sufficiently large, the terms $t^k e^{p_1 t}$ will dominate the others in $h(t)$; i.e.,

$$h(t) \approx a t^k e^{p_1 t}$$

for large t. Since the integral of $t^k |e^{p_1 t}|$ diverges if $Re p_1 \geq 0$, the integral of $h(t)$ blows up and the system is not input–output stable.   ■

In the case of multi-input, multi-output systems, the necessary and sufficient condition is that *all* individual elements of the matrix transfer function have all their poles in the open left-half plane.

The two types of stability are related. If an LTI system is internally stable, it is also input–output stable, because all transfer function poles are eigenvalues of the A matrix: if all eigenvalues of A are in the open left-half plane, so are the poles.

The converse is not always true: if all transfer function poles lie in the open left-half plane, it does not follow that all eigenvalues of A do. The A matrix could have unstable hidden modes that do not appear in the transfer function.

This leads to the concepts of detectability and stabilizability, defined as follows.

DEFINITION
A system is *detectable* (*stabilizable*) if all unstable modes are observable (controllable).

Detectability is weaker than observability. An observable system is clearly detectable, since *all* its modes are observable; in general, the converse is not true. Similarly, stabilizability is a weaker property than controllability.

Internal stability (which implies input–output stability) is required in practice. This cannot be achieved unless the plant is both detectable and stabilizable. The presence of an unstable, unobservable mode means that instability exists but goes undetected by the available sensors. An unstable, uncontrollable mode is almost sure to be excited by some unmodeled disturbance input, while the control inputs are powerless to suppress it.

If the realization is minimal (i.e., controllable and observable), all eigenvalues of A appear as transfer function poles. In that case, if all poles are stable, so are the eigenvalues. Therefore, if a realization is known to be minimal, input–output stability guarantees internal stability. A formal proof may be found in Fortmann and Hitz.

To summarize, internal stability implies input–output stability; input–output stability implies internal stability if the realization is controllable and observable, i.e., minimal.

Problems

3.1 An LTI system with two states and one output has the following zero-input responses:

$$y(t) = e^{-t} - .5e^{-2t} \quad \text{if} \quad \mathbf{x}(0) = \begin{bmatrix} 1 \\ .5 \end{bmatrix}$$

$$y(t) = -.5e^{-t} - e^{-2t} \quad \text{if} \quad \mathbf{x}(0) = \begin{bmatrix} -1 \\ 1 \end{bmatrix}.$$

Using linearity properties, calculate $y(t)$ if $\mathbf{x}(0) = \begin{bmatrix} 2 \\ .5 \end{bmatrix}$.

3.2 The response of an LTI second-order system to a unit step is

$$y(t) = .5 - .5e^{-t} + e^{-2t} \quad \text{if} \quad \mathbf{x}(0) = \begin{bmatrix} 1 \\ 1 \end{bmatrix}$$

and

$$y(t) = .5 - e^{-t} + 1.5e^{-2t} \quad \text{if} \quad \mathbf{x}(0) = \begin{bmatrix} 2 \\ 2 \end{bmatrix}.$$

Calculate the zero-state response to a unit step.

3.3 One algorithm for the computation of e^A is the Padé formula, where e^A is approximated by

$$\phi = \left(I - \frac{1}{2}A + \frac{1}{12}A^2 \right)^{-1} \left(I + \frac{1}{2}A + \frac{1}{12}A^2 \right).$$

Assume a series expansion for ϕ, i.e.,

$$\phi = I + c_1 A + c_2 A^2 + \cdots$$

and calculate the coefficients of the series by matching coefficients on both sides of

$$\left(I - \frac{1}{2}A + \frac{1}{12}A^2 \right) \phi = I + \frac{1}{2}A + \frac{1}{12}A^2.$$

For what power of A does the series for ϕ begin to deviate from the exponential series?

3.4 Using Laplace transforms, compute e^{At} for $A = [\begin{smallmatrix} 0 & 1 \\ -1 & -2 \end{smallmatrix}]$.

3.5 Repeat Problem 3.4 for $A = [\begin{smallmatrix} -1 & -2 \\ 2 & -1 \end{smallmatrix}]$.

M **3.6** **a.** Compute e^{At} for

$$
A = \begin{bmatrix} -1 & 1 & 0 \\ 0 & -1 & -1 \\ 0 & 1 & -1 \end{bmatrix}
$$

by Laplace transforms.

b. Evaluate the result of part (a) for $t = 1, 2, 4$.

c. Verify your answers to part (b) by numerical computation of e^{At} (MATLAB expm).

d. Verify numerically that $e^{A2} = (e^A)^2$ and $e^{A.4} = (e^{A.2})^2$.

M **3.7** For the system

$$
\dot{\mathbf{x}} = \begin{bmatrix} 0 & 1 \\ -3 & -4 \end{bmatrix} \mathbf{x}
$$

a. Plot $x_1(t)$ and $x_2(t)$ vs. t, if $\mathbf{x}(0) = [\begin{smallmatrix} 1 \\ 1 \end{smallmatrix}]$.

b. Plot $x_2(t)$ vs. $x_1(t)$ for the same $\mathbf{x}(0)$.

c. Calculate the eigenvalues and eigenvectors of the matrix, and repeat parts (a) and (b) using each eigenvector as an initial state.

M **3.8** Repeat Problem 3.7 for the system

$$
\dot{\mathbf{x}} = \begin{bmatrix} 0 & 1 \\ 1 & 0 \end{bmatrix} \mathbf{x}.
$$

M **3.9** For the system

$$
\dot{\mathbf{x}} = \begin{bmatrix} 0 & 1 \\ -1 & 0 \end{bmatrix} \mathbf{x}
$$

a. Calculate e^{At}.

b. Plot $x_2(t)$ vs. $x_1(t)$ for $\mathbf{x}(0) = [\begin{smallmatrix} 1 \\ 1 \end{smallmatrix}]$.

Interpret the result.

3.10 Calculate the transfer function of the system

$$
\dot{\mathbf{x}} = \begin{bmatrix} 0 & 1 \\ 1 & 0 \end{bmatrix} \mathbf{x} + \begin{bmatrix} 0 \\ 1 \end{bmatrix} u
$$

$$
y = \begin{bmatrix} 1 & 0 \end{bmatrix} \mathbf{x}.
$$

M   **3.11**   Repeat Problem 3.10 for the system

$$\dot{\mathbf{x}} = \begin{bmatrix} 0 & 1 \\ -3 & -4 \end{bmatrix} \mathbf{x} + \begin{bmatrix} 0 & 1 \\ -1 & 1 \end{bmatrix} \mathbf{u}$$

$$\mathbf{y} = \begin{bmatrix} 1 & 1 \\ 0 & 1 \end{bmatrix} \mathbf{x}.$$

    a. Calculate the transmission zeros, using the equation $\det H(s_0) = 0$.

    b. Repeat part (a), using MATLAB.

3.12   Suppose s_0 is not a pole of the matrix transfer function $H(s)$.

    a. Show that, if the input is $e^{s_0 t}\mathbf{w}u_{-1}(t)$, there is a component $H(s_0)\mathbf{w}e^{s_0 t}$ in the output.

Hint   Use partial-fraction ideas.

---

    b. Use the results of part (a) to write an interpretation of the concept of a transmission zero.

3.13   ***Servo, simplified model***   Calculate the transfer function θ/v for the dc servo of Problem 2.4 (Chapter 2).

M   **3.14**   ***Servo with flexible shaft***   Compute the transfer function θ_2/v and θ_1/v for the dc servo of Problem 2.5 (Chapter 2).

M   **3.15**   ***Drum speed control***   For the drum speed control model of Problem 2.6 (Chapter 2), compute the transfer functions ω_0/u_1, ω_0/u_2, and ω_0/T_0.

M   **3.16**   ***High-wire artist***   For the linearized system of Problem 2.18 (Chapter 2), compute the transfer functions θ/τ and ϕ/τ.

M   **3.17**   ***Blending tank***   For the linearized system of Problem 2.13 (Chapter 2), calculate the transfer functions relating the outputs $\Delta\ell$ and ΔC_A to inputs ΔF_A, ΔF_0, and ΔF_B.

M   **3.18**   ***Heat exchanger***   For the model of Problem 2.14 (Chapter 2), compute the transfer functions relating the output ΔT_{c3} to the inputs ΔF_H (control input), ΔF_c, ΔT_{H0}, and ΔT_{c0} (disturbances).

M   **3.19**   ***Chemical reactor***   For the linearized model of Problem 2.15 (Chapter 2), calculate the transfer functions between the inputs ΔQ (control), Δc_{A0}, ΔF (disturbances), and the outputs ΔT, Δc_A, Δc_B, and Δc_C. Compute both the pole-zero and ratio-of-polynomials forms.

M   **3.20**   ***Crane***   For the linearized crane system of Problem 2.17 (Chapter 2), compute the transfer functions x/F and θ/F.

M   **3.21**   ***Two-pendula problem***   For the model of Problem 2.19 (Chapter 2), with $\ell_1 = 1.5$ m and $\ell_2 = 1$ m, compute the transfer functions x/F, θ_1/F, and θ_2/F.

M   **3.22**   ***Maglev***   For the "Maglev" system of Problem 2.20 (Chapter 2), it is convenient to redefine the inputs as follows:

$$\text{Common-mode levitation:} \qquad \Delta u_{LC} = \frac{1}{2}(\Delta u_{L1} + \Delta u_{L2})$$

$$\text{Differential-mode levitation:} \qquad \Delta u_{LD} = \frac{1}{2}(\Delta u_{L1} - \Delta u_{L2})$$

$$\text{Common-mode guidance:} \qquad \Delta u_{GC} = \frac{1}{2}(\Delta u_{G1} + \Delta u_{G2})$$

$$\text{Differential-mode guidance:} \qquad \Delta u_{GD} = \frac{1}{2}(\Delta u_{G1} - \Delta u_{G2})$$

  a. Write the state equations in terms of the new inputs (only three will be required).

  b. Compute the transfer functions from each of the three inputs to Δz, Δy, and $\Delta \theta$.

3.23   An LTI system has the following A and C matrices:

$$A = \begin{bmatrix} 0 & 1 \\ 1 & 0 \end{bmatrix} \qquad C = \begin{bmatrix} 1 & -1 \end{bmatrix}.$$

  a. Calculate e^{At} and show that there exists a vector $\mathbf{x}^*$ such that $Ce^{At}\mathbf{x}^* = 0$, for all $t \geq 0$.

  b. Show that the observability matrix $\mathcal{O}$ is such that $\mathcal{O}\mathbf{x}^* = \mathbf{0}$.

  c. Show that $\mathbf{x}^*$ is an eigenvector of A and that $C\mathbf{x}^* = \mathbf{0}$.

3.24   A second-order system has complex conjugate eigenvalues with nonzero imaginary parts. Show that a system is unobservable if, and only if, the C matrix is identically zero.

3.25   An LTI system has the following A and B matrices:

$$A = \begin{bmatrix} 0 & 1 \\ 1 & 0 \end{bmatrix} \qquad B = \begin{bmatrix} 1 \\ -1 \end{bmatrix}.$$

  a. Calculate e^{At} and show that there exists a vector $\mathbf{x}^*$ such that $\mathbf{x}^{*^T} e^{At} B = \mathbf{0}$, $t \geq 0$.

  b. Show that the controllability matrix $\mathcal{C}$ is such that $\mathbf{x}^{*^T}\mathcal{C} = 0$.

  c. Show that $\mathbf{x}^*$ is an eigenvector of A^T and that $\mathbf{x}^{*^T} B = \mathbf{0}$.

3.26 The system of Figure 3.21 is composed of two identical LTI subsystems $\mathcal{S}$, both described by nth-order state equations

$$\dot{\mathbf{x}} = A\mathbf{x} + B\mathbf{u}$$

$$\mathbf{y} = C\mathbf{x} + D\mathbf{u}.$$

The state vector of the system has the state vector of the first subsystem as its first components and the state vector of the second subsystem as its last n components. Show that the composite system is not observable, and characterize the unobservable states, assuming that each subsystem is observable on its own.

*** H i n t*** For what initial states will the zero-input solution $\mathbf{y}(t)$ be identically zero?

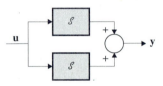

Figure 3.21 Parallel interconnection of two identical subsystems

3.27 Show that the system of Figure 3.21 is also uncontrollable. Characterize the uncontrollable states, assuming each subsystem to be controllable on its own.

*** H i n t*** What vectors of the composite state are reachable from the zero state?

3.28 *Servo, simplified model*

  a. Verify that the dc servo modeled in Problem 2.4 (Chapter 2) is controllable from the input v.

  b. Assess the observability from the outputs (i) θ and (ii) ω. Identify any unobservable modes.

(M) **3.29** *Servo with flexible shaft* For the servo modeled in Problem 2.5 (Chapter 2), assess controllability from the input v and observability from (i) θ_2, (ii) θ_1, (iii) θ_1 and θ_2. Identify any uncontrollable or unobservable modes. Relate your results to the transfer function calculated in Problem 3.14.

Ⓜ   **3.30   *Drum speed control***   For the drum speed control model of Problem 2.6 (Chapter 2):

   a. Assess controllability, with the following input combinations: (i) u_1, (ii) u_2, (iii) u_1, and u_2. Identify any uncontrollable mode.

   b. Assess observability with the following outputs: (i) ω_0, (ii) ω_1, (iii) Δ_1, (iv) ω_0, ω_1, and ω_2. Identify any unobservable modes.

Ⓜ   **3.31   *High-wire artist***   For the linearized system of Problem 2.18 (Chapter 2):

   a. Assess controllability. Identify any uncontrollable modes.

   b. Assess observability with the following outputs: (i) θ; (ii) ϕ; (iii) θ and ϕ. Identify any unobservable mode.

   Relate your results to the transfer function computed in Problem 3.16.

Ⓜ   **3.32   *Blending tank***   For the linearized system of Problem 2.13 (Chapter 2):

   a. Assess controllability with the following input combinations: ΔF_A alone, ΔF_0 alone, ΔF_A and ΔF_0. Find the uncontrollable mode(s), if any, and the controllable states.

   b. Assess observability with the following output combinations: $\Delta \ell$ alone, ΔC alone, $\Delta \ell$ and ΔC together. Find the unobservable mode(s), if any, and the unobservable states.

   c. Relate the results of parts (a) and (b) to the transfer functions calculated in Problem 3.17.

Ⓜ   **3.33   *Heat exchanger***   For the linearized system of Problem 2.14 (Chapter 2), assess controllability and observability with ΔF_H as the input and ΔT_{c3} as the output. Identify uncontrollable and unobservable modes and states, if any. Relate your results to the transfer function $\Delta T_{c3}/\Delta F_H$ in Problem 3.18.

Ⓜ   **3.34   *Chemical reactor***   A linearized model of the reactor of Problem 2.9 (Chapter 2) was developed in Problem 2.15.

   a. Investigate controllability, with ΔQ as input. Identify the uncontrollable modes, if any, and the uncontrollable states. Explain in terms of the model.

__* H i n t__   Compare the zero-state responses of Δc_A, Δc_B, and Δc_C.

---

   b. In the interest of minimizing sensor costs, it is of interest to assess observability for certain combinations of sensors. Try the following measurement schemes: (i) ΔT; (ii) ΔT and Δc_B; (iii) ΔT, Δc_B, and Δc_C. Identify the unobservable modes, if any, and the unobservable states.

Ⓜ **3.35** *Crane* For the linearized model of Problem 2.17 (Chapter 2):

    a. Assess controllability with F as input. Determine any uncontrollable modes and states.

    b. Assess observability with the following output combinations: Δx alone, θ alone, Δx and θ together. Determine any unobservable modes and states.

    c. Relate your results to the transfer functions of Problem 3.20.

Ⓜ **3.36** *Two-pendula problem* Using the linearized equations of Problem 2.19 (Chapter 2) for the two-pendula problem, with $M = m = 1$ kg:

    a. Show (algebraically) that the system is controllable from the input F if, and only if, $\ell_1 \neq \ell_2$.

    b. Using $\ell_1 = 1$ m and $\ell_2 = 0.8$ m, assess observability for the following sensor configurations: (i) x; (ii) θ_1 and θ_2; (iii) x, θ_1, and θ_2.

Ⓜ **3.37** *Maglev* For the linearized Maglev system of Problem 2.20 (Chapter 2):

    a. Compute the natural frequencies.

    b. Investigate controllability for the following input combinations: all control currents, all control currents except Δi_{L1}, all control currents except Δi_{G1}. (It is useful to know whether the vehicle can still be controlled if an actuator fails.) Identify uncontrollable modes and states, if any.

    c. Investigate observability for the following output combinations: all magnet gap measurements, i.e., all ΔS_{Li} and ΔS_{Gi}; all gap measurements except ΔS_{L1}; all gap measurements except ΔS_{G1}. Identify unobservable modes and states, if any.

Ⓜ **3.38** *Maglev* For the Maglev model as modified in Problem 3.22:

    a. Investigate controllability from the inputs Δu_{LC}, Δu_{LD}, and Δu_{GD}, taken separately. Identify uncontrollable modes and states, if any.

    b. Investigate observability for the outputs Δz, Δy, and $\Delta \theta$, taken separately. Identify unobservable modes and states, if any.

3.39 The diagonal form is sometimes used to compute e^{At}. Suppose there exists a T such that $T^{-1}AT = S$, where $S = \text{diag}[s_1, s_2, s_3, \ldots, s_n]$.

    a. Show that $e^{St} = \text{diag}[e^{s_1 t}, e^{s_2 t}, \ldots, e^{s_n t}]$.

    b. Show that $e^{At} = T e^{St} T - 1$.

    c. Verify by reducing the A matrix of Example 3.1 to diagonal form and using parts (a) and (b) to calculate e^{At}.

3.40 *Servo, simplified model* Express the dc servo model of Problem 2.4 (Chapter 2) in Jordan form.

Ⓜ **3.41** *Servo with flexible shaft* For the servo modeled in Problem 2.5:

    a. Transform the state equation to (complex) Jordan form;

    b. Transform the Jordan form to real block-diagonal form.

M   **3.42** *Chemical reactor*   A linearized model of the reactor of Problem 2.9 (Chapter 2) was developed in Problem 2.15. Transform the system to Jordan form, with the input and output combinations of Problem 3.34; investigate controllability and observability.

M   **3.43** Reduce to diagonal form the system

$$\dot{\mathbf{x}} = \begin{bmatrix} 0 & 1 & 0 & 0 & 0 \\ 0 & 0 & 1 & 0 & 0 \\ 0 & 0 & 0 & 1 & 0 \\ 0 & 0 & 0 & 0 & 1 \\ 0 & -4 & 0 & 5 & 0 \end{bmatrix} \mathbf{x} + \begin{bmatrix} 2.0625 \\ 0.1250 \\ 2.2500 \\ 0.5000 \\ 3.0000 \end{bmatrix} u$$

$$y = \begin{bmatrix} 0 & .6667 & -.6667 & -.6667 & 1.1667 \end{bmatrix} \mathbf{x}$$

and express it in terms of the canonical decomposition.

M   **3.44** Express in terms of the canonical decomposition the following Jordan form realizations:

$$\dot{\mathbf{x}} = \begin{bmatrix} 1 & 1 & 0 & 0 & 0 \\ 0 & 1 & 0 & 0 & 0 \\ 0 & 0 & 1 & 0 & 0 \\ 0 & 0 & 0 & -1 & 1 \\ 0 & 0 & 0 & 0 & -1 \end{bmatrix} \mathbf{x} + \begin{bmatrix} 1 \\ 0 \\ 1 \\ 0 \\ 1 \end{bmatrix} u$$

$$y = \begin{bmatrix} 0 & 1 & 1 & 1 & 0 \end{bmatrix} \mathbf{x}.$$

M   **3.45** Compute the transfer function y/u for the system of Problem 3.43, cancelling common factors. What can be concluded from pole-zero cancellations about the controllability and/or observability of the system modes?

M   **3.46** *Drum speed control*   It is sometimes possible to convert a multi-input, multi-output problem to a number of independent SISO problems. Consider the system of Problem 2.6 (Chapter 2). Define two new inputs: a common-mode input, $u_c = (u_1 + u_2)/2$, and a differential-mode input, $u_d = (u_1 - u_2)/2$. Clearly, $u_1 = u_c + u_d$ and $u_2 = u_c - u_d$.

   a. Redefine the B matrix using the new inputs.

   b. Define the two outputs $y_1 = \omega_0$ and $y_2 = \omega_1 - \omega_2$. Calculate the transfer function between the vector input $\begin{bmatrix} u_c \\ u_d \end{bmatrix}$ and the vector output $\begin{bmatrix} y_1 \\ y_2 \end{bmatrix}$ and show that it is diagonal (within numerical accuracy).

   c. Assess the controllability of the system from u_c, then from u_d, and relate the results to part (b).

3.47 *Blending tank*   For the linearized system of Problem 2.13 (Chapter 2):

   a. Calculate the transformation matrix T relating the state vectors

$$\begin{bmatrix} \Delta \ell \\ \Delta V_A \end{bmatrix} \quad \text{to} \quad \begin{bmatrix} \Delta \ell \\ \Delta C_S \end{bmatrix},$$

i.e., such that

$$\begin{bmatrix} \Delta \ell \\ \Delta Cs \end{bmatrix} = T^{-1} \begin{bmatrix} \Delta \ell \\ \Delta V_A \end{bmatrix}.$$

Use the symbolic form of the linearized equations, rather than the form with numerical values for the coefficients.

b. Show that the transformation matrix T diagonalizes the equations. (This is one case where the transformed variables have physical meaning.)

c. Use the transformed equations to interpret the results of Problem 3.32.

3.48 *Chemical reactor* For the linearized system of Problem 2.15 (Chapter 2), show that the differential equations for Δc_A and ΔT, taken together, are a minimal realization of the transfer function $\Delta T / \Delta Q$.

3.49 Show that the controllable canonical form is controllable.

3.50 Show that the observable canonical form is observable.

3.51 Give the controllable and observable canonical forms for a system with transfer function $H(s) = (s + 1)/(s^3 + s^2 + s + 1)$.

3.52 Repeat Problem 3.51 for $H(s) = (2s + 3)/(s^3 - 2s^2 + s + 2)$.

3.53 *Servo with flexible shaft* Give the controllable and observable canonical forms for the transfer function θ_2/v of Problem 3.14.

3.54 *Heat exchanger* Give the controllable and observable canonical forms for the transfer function $\Delta T_{c3}/\Delta F_H$ of Problem 3.18.

3.55 The system of Figure 3.22 consists of two cascaded subsystems, S_1 and S_2, with transfer functions $H_1(s)$ and $H_2(s)$, respectively. With

$$H_1(s) = \frac{1}{(s - 1)(s + 1)}, \qquad H_2(s) = \frac{2(s - 1)}{s^2 + s + 1}$$

a. Give realizations in controllable canonical form for S_1 and S_2.

b. If $x1$ and $x2$ are the state vectors for S_1 and S_2, respectively, give a realization for the composite system S_1 using $\mathbf{x} = \begin{bmatrix} x1 \\ x2 \end{bmatrix}$ as the state vector.

c. Is the realization of S derived in part (b) minimal? If not, find the uncontrollable and/or unobservable mode(s).

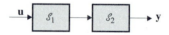

Figure 3.22 Series interconnection of two subsystems

3.56 Repeat Problem 3.55, but with $H_2(s)$ as the transfer function for S_1 and $H_1(s)$ as the transfer function for S_2.

3.57 Is the composite system $\mathcal{S}$ of Problem 3.55 internally stable? Is it input–output stable? Explain.

3.58 Is the composite system $\mathcal{S}$ of Problem 3.56 internally stable? Is it input–output stable? Explain.

3.59 Given the system

$$\dot{\mathbf{x}} = \begin{bmatrix} 0 & 1 & 0 \\ 0 & 0 & 1 \\ 1 & 0 & 0 \end{bmatrix} \mathbf{x} + \begin{bmatrix} 1 \\ 0 \\ -1 \end{bmatrix} u$$

$$y = \begin{bmatrix} 1 & -.5 & -1 \end{bmatrix} \mathbf{x}$$

a. Show that the system is not controllable, and find the uncontrollable mode. Is the system internally stable?

b. Calculate the transfer function y/u and show that the system is input–output stable.

c. Suppose a disturbance input w has been ignored in the modeling process, whose effect is taken into account by adding

$$\begin{bmatrix} 0 \\ 0 \\ 1 \end{bmatrix} w$$

to the RHS of the state equation. Show that the transfer function y/w is not input–output stable.

References

[1] Ward, R. C., "Numerical Computation of the Matrix Exponential with Accuracy Estimate," *SIAM J. Numerical Analysis*, vol. 14, pp. 971–981 (1977).

[2] Mohler, C. B., and C. Van Loan, "Nineteen Dubious Ways to Compute the Exponential for a Matrix," *SIAM Rev.*, vol. 20, pp. 801–836 (1978).

[3] Morris, J. L., *Computational Methods in Elementary Numerical Analysis*, Wiley (1983).

[4] Pugh, A. C., "Transmission and Systems Zeros," *Int. J. of Control*, vol. 26, pp. 315–324 (1977).

[5] Moler, C. B., and G. W. Stewart, "An Algorithm for Generalized Matrix Eigenvalue Problems," *SIAM J. Numerical Anal.*, vol. 10, pp. 241–256 (1973).

[6] Friedland, B., *Control System Design: An Introduction to State-Space Methods*, McGraw-Hill (1986).

[7] Kailath, T., *Linear Systems*, Prentice-Hall (1980).

[8] Ftrtmann, T. E., and K. L. Hitz, *An Introduction to Linear Control Systems*, Marcel Dekker (1977).

Chapter 4

Specifications, Structures, Limitations

4.1 INTRODUCTION

Figure 4.1 presents the system structure, using the terminology of Chapter 1. The plant $\mathcal{P}$ is driven by the control input $\mathbf{u}$ and the disturbance $\mathbf{w}$. The plant's output, $\mathbf{y}$, is measured by a sensor $\mathcal{S}$. $\mathcal{S}$ produces a signal $\mathbf{y}_m$, which is intended to be a reasonably faithful copy of the true output $\mathbf{y}$. Another sensor, $\mathcal{S}_1$, may be used to measure the plant disturbance $\mathbf{w}$. Finally, the controller, represented by $\mathcal{C}$, produces the plant input $\mathbf{u}$ as a function of three signals: the output set point or reference $\mathbf{y}_d$, the output measurement $\mathbf{y}_m$, and the disturbance measurement $\mathbf{w}_m$.

If the plant is linear and time-invariant (LTI), then zero-state linearity dictates that $\mathbf{y}$ is a linear combination of the two plant input vectors $\mathbf{u}$ and $\mathbf{w}$; i.e.,

$$\mathbf{y}(s) = P(s)\mathbf{u}(s) + P_w(s)\mathbf{w}(s). \tag{4.1}$$

Quite often, it is more convenient to work with $\mathbf{d}(s) = P_w(s)\mathbf{w}(s)$, the disturbance *referred to the output*, than with the physical disturbance input $\mathbf{w}$. An equivalent expression to Equation 4.1, then, is

$$\mathbf{y}(s) = P(s)\mathbf{u}(s) + \mathbf{d}(s). \tag{4.2}$$

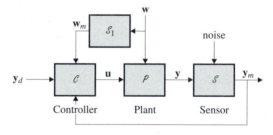

Figure 4.1 Operator blocks of a control system

Clearly, $\mathbf{d}(t)$ is the sum of the effects of all physical disturbances on the output $\mathbf{y}$. Figure 4.2 illustrates Equations 4.1 and 4.2.

The sensor $\mathcal{S}$ is assumed to have two inputs. The first is the plant output $\mathbf{y}$, and the second is a noise input. The sensor is represented by an LTI model, so

$$y_m(s) = P_s(s)\mathbf{y}(s) + \mathbf{v}(s) \tag{4.3}$$

where $\mathbf{v}$, like $\mathbf{d}$ in Equation 4.2, collects the effects on $\mathbf{y}_m$ of all noise signals. Ideally, $P_s = 1$ and $\mathbf{v} = 0$ so that $\mathbf{y}_m = \mathbf{y}$; i.e., the measurement is equal to the quantity measured. Equation 4.3 accounts for two types of errors. The transfer function P_s covers the sensor dynamics (i.e., the fact that sensors do not respond instantaneously), and $\mathbf{v}$ includes other discrepancies between $\mathbf{y}_m$ and $\mathbf{y}$.

If the controller is LTI, then, by linearity, its output $\mathbf{u}$ is a combination of its three inputs, so that $\mathbf{u}(s)$ satisfies an equation of the form

$$\mathbf{u}(s) = F_d(s)\mathbf{y}_d(s) + F_m(s)\mathbf{y}_m(s) + F_w(s)\mathbf{w}_m(s). \tag{4.4}$$

Not all three inputs need be used. Several control structures are defined according to whether $\mathbf{y}_d$, $\mathbf{y}_m$, or $\mathbf{w}_m$ is used to produce $\mathbf{u}$.

Figure 4.3 shows the controller structures to be studied in this chapter. *Open-loop control* uses only $\mathbf{y}_d$ (Fig. 4.3a); it corresponds to $F_m = F_w = 0$ in Equation 4.4. *Feedforward control* (Fig. 4.3b) is obtained by setting $F_d = F_m = 0$. *Feedback control* (Fig. 4.3c and d) uses both $\mathbf{y}_d$ and $\mathbf{y}_m$. In Figure 4.3c, $F_m(s) = -F_d(s)$, so that $\mathbf{u} = F_d(s)[\mathbf{y}_d(s) - \mathbf{y}_m(s)]$; that is known as *single-degree-of-freedom feedback* control. Figure 4.3d has the same feedback structure but allows the designer the independent choice of F_d or F_m; hence the name *two-degrees-of-freedom feedback* control. Combinations of the four structures are possible—for example, open loop with feedforward.

Figure 4.4 shows the system of Figure 4.1 in block-diagram form. For simplicity, it has been assumed that the measurement of $\mathbf{w}$ is noise-free. The system (within the dotted box) is seen to have three external inputs: $\mathbf{y}_d$, $\mathbf{w}$, and $\mathbf{v}$. By linearity, all signals in the system are generated by superposition, as

$$\mathbf{y}(s) = H_d(s)\mathbf{y}_d(s) + H_w(s)\mathbf{w}(s) + H_v(s)\mathbf{v}(s) \tag{4.5}$$

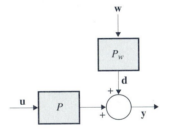

Figure 4.2   Collecting all disturbances into a disturbance referred to the output

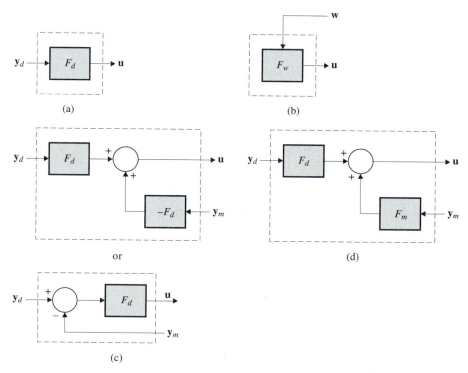

Figure 4.3   Control configurations: a) Open loop; b) Feedforward; c) Feedback, one degree of freedom; d) Feedback, two degrees of freedom

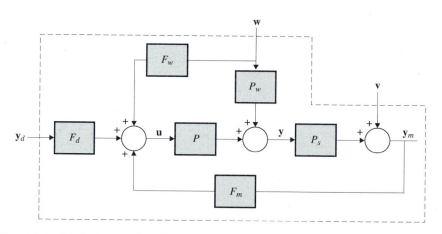

Figure 4.4   Block diagram for a linear control system

where

$$H_d = \text{set point to output closed-loop transfer function}$$

$$H_w = \text{disturbance to output closed-loop transfer function}$$

$$H_v = \text{sensor noise to output closed-loop transfer function.}$$

The error, defined as $\mathbf{e} = \mathbf{y}_d - \mathbf{y}$, satisfies

$$\mathbf{e}(s) = [I - H_d(s)]\mathbf{y}_d(s) - H_w(s)\mathbf{w}(s) - H_v(s)\mathbf{v}(s). \tag{4.6}$$

The transfer functions H_d, H_w, and H_v depend on the transfer functions of Equations 4.1, 4.3, and 4.4, associated respectively with the plant, the sensors, and the controller. The functional relationship depends on the controller structure. It is precisely this dependence that makes the study of different control schemes important and interesting. The design problem is to obtain H_d, H_w, and H_v with desirable properties via an appropriate controller design, i.e., by choosing the structure and dynamics of the controller.

In cases where it is convenient to work with $\mathbf{d}$, the disturbance referred to the output, rather than with $\mathbf{w}$, Equation 4.5 is replaced by

$$\mathbf{y}(s) = H_d(s)\mathbf{y}_d(s) + H_{wd}(s)\mathbf{d}(s) + H_v(s)\mathbf{v}(s) \tag{4.7}$$

with a corresponding change in Equation 4.6.

Ideally, the error would be zero; i.e., the output would track $\mathbf{y}_d$ exactly. To achieve this for all possible signals $\mathbf{y}_d$, $\mathbf{w}$, and $\mathbf{v}$, we would require $H_d(s) = I$ and $H_w(s)[H_{wd}(s)] = H_v(s) = 0$. That is not possible, unfortunately, so we are forced to choose between a number of nonideal solutions. In order to do that, we must be able to discriminate between acceptable and unacceptable departures from the ideal; i.e., we must have *performance specifications*.

The object of this chapter is the study of control systems specifications and factors that limit performance for a few control structures. The material is presented in terms of single-input, single-output (SISO) systems.

4.2 PERFORMANCE SPECIFICATIONS

4.2.1 Introduction

One absolute requirement must be satisfied before all other considerations: *internal stability*. The difference between internal stability and input–output stability was established in Chapter 3. As we saw there, it is possible for a system to be internally unstable and yet to have a stable transfer function, i.e., to be input–output stable. This happens when the system has unstable hidden modes. Therefore, internal stability must be ensured before the transfer functions that define the response to the system inputs are considered.

If the plant is expected to deviate from the design model, it is better represented by a *set* of models centered on the design model. The set of models could be generated, for example, by letting the model parameters vary over their tolerance intervals, with each parameter value defining a member of the set. For a control system to be acceptable, the design must be internally stable for every model in the set. This property is known as *robust stability*.

Once stability and robustness are assured, we can shift attention to response. It will not usually be possible to have good set-point tracking and disturbance and noise rejection for all functions $y_d(t)$, $w(t)$, and $v(t)$. That is why response specifications are normally given with reference to specific signals or classes of signals.

Sometimes we are concerned with the response to specific time functions—for example, the set of signals used for typical aircraft maneuvers. Specifications on acceptable errors are then defined in terms of features of the time response, in a manner that fits the application. We shall examine in some detail the response to the unit step, which is used as a test signal in many applications.

Even though the "proof of the pudding" is the actual time response, there are many instances where set-point, disturbance, or observation-noise signals are given not as specific time functions but only in terms of some average characteristics, most typically their frequency spectra. For example, ocean wave motion acting on a ship cannot be represented by a fixed, known function but is sufficiently regular to have a well-defined power spectrum. In such cases, specifications must also be based on average properties, such as the error spectrum or the root-mean-squared error.

Finally, a stable system with a satisfactory response may also be required to satisfy a *sensitivity* specification. For example, it is desirable that the response of an aircraft to the pilot's commands be the same at all speeds and altitudes in its flight envelope even though aircraft dynamics vary substantially over the envelope. In general, a change in plant transfer function from $P(s)$ to $P(s) + \Delta P(s)$ will cause a change in, say, $H_d(s)$ to $H_d(s) + \Delta H_d(s)$. A sensitivity measure will be developed, relating ΔP and ΔH_d.

4.2.2 Time-Domain Specifications

A time-domain specification describes the characteristics of the response (output or error) to a given set point or to disturbance time functions. Those functions may be specific to the particular application—for example, the set of signals used by an aircraft control system for typical maneuvers. In the absence of such specific signals, a unit step is often used as a test input. Figure 4.5 shows the output response $y(t)$ to a unit step in $y_d(t)$ (stability is assumed). Specifications are given in terms of the following features of the response:

1. The steady-state error, e_{ss}. This is the difference between the desired final value, 1.0, and the final value of $y(t)$.

2. The delay time T_d, defined as the time required for the response to reach 50% of its final value (the percentage may vary according to the application).

3. The rise time T_r, defined as the time required for the response to go from 10% to 90% of its final value.

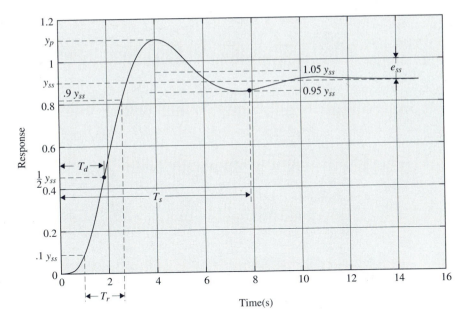

Figure 4.5 The step response and its main parameters

4. The percent peak overshoot, defined from the maximum value of the response as a percentage of the final value. More precisely, with reference to Figure 4.5, the percent overshoot is $100(y_p - y_{ss})/y_{ss}$.

5. The settling time T_s, defined as the time at which the response enters for the last time a $\pm 5\%$ band about the final value.

In the next chapter we shall address the question of time-domain specifications more fully and explore connections with the frequency-domain specifications that are used in the remainder of this chapter.

4.2.3 Frequency-Domain Specifications

Disturbance and observation-noise signals are usually characterized by their spectra; that is often the case for set-point signals as well. This characterization leads naturally to specifications in the frequency domain.

For the SISO case, Equations 4.5 and 4.6 are

$$y(s) = H_d(s)y_d(s) + H_w(s)w(s) + H_v(s)v(s)$$
$$e(s) = (1 - H_d(s))y_d(s) - H_w(s)w(s) - H_v(s)v(s).$$

For purposes of exposition, suppose $y_d = v = 0$, so that

$$e(s) = -H_w(s)w(s).$$

Assume that $w(t)$ is an "energy" signal, i.e., has a Fourier transform. If $H_w(s)$ is the transfer function of a causal, stable system, then the j-axis lies within the region of convergence of $H_w(s)$, and $H_w(j\omega)$ is its Fourier transform. By Parseval's theorem,

$$\int_0^\infty e^2(t)dt = \frac{1}{2\pi} \int_{-\infty}^\infty |H_w(j\omega)w(j\omega)|^2 dw$$

$$= \frac{1}{2\pi} \int_{-\infty}^\infty |H_w(j\omega)|^2 |w(j\omega)|^2 d\omega. \tag{4.8}$$

The left-hand side (LHS) of Equation 4.8 is the *integral-squared error* (ISE). It is the energy in the error signal and serves as a measure of signal magnitude. The quantity $|w(j\omega)|^2$ is the *energy density spectrum* of w.

In practice, disturbance signals are usually "power" signals, which do not have Fourier transforms. Many such signals exhibit stationarity, which, roughly speaking, means that their average properties do not change with time. For such signals it is possible to define a *power density spectrum*, which describes the distribution in frequency of the signal power in exactly the same way as the energy density spectrum describes the distribution of energy. For example, a sinusoid of frequency ω_0 is a power signal with a power density spectrum consisting of impulses at $\pm\omega_0$. Stationary random processes also have power density spectra. For power signals, the analog of Equation 4.8 is

$$(e_{rms})^2 = \frac{1}{2\pi} \int_{-\infty}^\infty |H_w(j\omega)|^2 \Phi_w(j\omega)d\omega \tag{4.9}$$

where e_{rms} = root-mean-squared value of e

Φ_w = power density spectrum of w (a nonnegative quantity).

The ISE (rms error) will be small if $|H_w|$ is small at frequencies where w has substantial energy (power). As for the other terms in Equation 4.6, if the sensor noise $v(t) \neq 0$, the transfer function H_v should also be small in magnitude at frequencies where $|v(j\omega)|$ is large. In the case of the reference to output transfer function H_d, Equation 4.6 shows that $|1 - H_d(j\omega)|$ should be small where $|y_d(j\omega)|$ is large; another way of saying this is that $H_d(j\omega) \sim 1$ where $|y_d(j\omega)|$ is relatively large.

The phase of $H_d(j\omega)$ is important. In Figure 4.6a, the circular arc is the locus of complex numbers $|H_d| = 1$; as the diagram shows, the complex number $1 - H_d$ is quite different from zero if $\angle H_d$ is not near $0°$. Therefore, a specification on $H_d(j\omega)$ must include magnitude and phase. The specification of phase is avoided if we work with $1 - H_d(j\omega)$; as Figure 4.6b shows, $H_d(j\omega)$ is near 1 if $|1 - H_d(j\omega)|$ is close to zero.

In practice, reference and disturbance signals are typically concentrated at low frequencies. Therefore, it is normally desirable to have $H_d(j\omega) \sim 1$ and $H_w(j\omega) \sim 0$ at low frequencies. It turns out to be undesirable to maintain these values at all frequencies. As we shall see, in a practical design, $H_d(j\omega)$ tends to zero and $H_w(j\omega)$ to one at high frequencies. The design objective is to make $H_d(j\omega) \sim 1$ and $H_w(j\omega) \sim 0$ over the frequency range 0 to ω_b, where y_d and w have significant

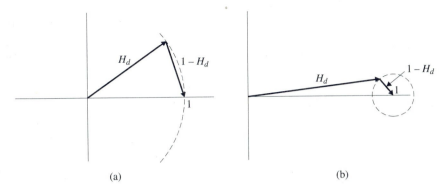

(a)   (b)

Figure 4.6   Showing the importance of phase differences

spectral constant. That frequency range is the system *passband*, and ω_b is the *bandwidth*, which can be defined in a number of ways, depending on the application.

Figure 4.7 illustrates frequency-domain specifications on $|H_d(j\omega)|$ and $|H_w(j\omega)|$ (or $|1 - H_d(j\omega)|$). The hatched areas are the permissible values. The dotted curve in Figure 4.7a dictates not only that $|H_d(j\omega)| \rightarrow 0$, but how rapidly it does so. It is often defined as a straight line on a Bode plot, and the magnitude of its slope is called the *rolloff rate*, expressed in decibels (db) per decade.

In addition to the limits of Figure 4.7, the value at $\omega = 0$ ($s = 0$) is often given. For a unit step in $y_d(t)$, $y_d(s) = 1/s$ and

$$\lim_{t \to \infty} y(t) = \lim_{s \to 0} s H_d(s) \cdot \frac{1}{s} = H_d(0).$$

If $H_d(0) = 1$, then $y(t) \rightarrow 1$ and the steady-state error e_{ss} is zero; it is usually desirable to specify the behavior of H_d at frequencies near zero.

It is often helpful to use a rational transfer function as a target shape for the frequency response $H_d(j\omega)$. The ideal response—unity in the passband and zero elsewhere—cannot be achieved by any causal system, let alone one described by a transfer function [1]. Various rational approximations have been derived by researchers in filter design [2]. One set is the Butterworth family, described by all-pole transfer functions of unit magnitude at $s = 0$. The poles of a kth-order Butterworth low-pass filter are located as shown in Figure 4.8, where $\phi_k = \frac{180°}{k}$. The first three Butterworth transfer functions are

$$B_1(s) = \frac{\omega_0}{s + \omega_0}$$

$$B_2(s) = \frac{\omega_0{}^2}{(s + \frac{\sqrt{2}}{2}\omega_2 + j\frac{\sqrt{2}}{2}\omega_0)(s + \frac{\sqrt{2}}{2}\omega_0 - j\frac{\sqrt{2}}{2}\omega_0)}$$

$$= \frac{\omega_0{}^2}{s^2 + \sqrt{2}\omega_0 s + \omega_0{}^2}$$

$$B_3(s) = \frac{\omega_0{}^3}{(s + \omega_0)(s + \frac{1}{2}\omega_0 + j\frac{\sqrt{3}}{2}\omega_0)(s + \frac{1}{2}\omega_0 - j\frac{\sqrt{3}}{2}\omega_0)}.$$

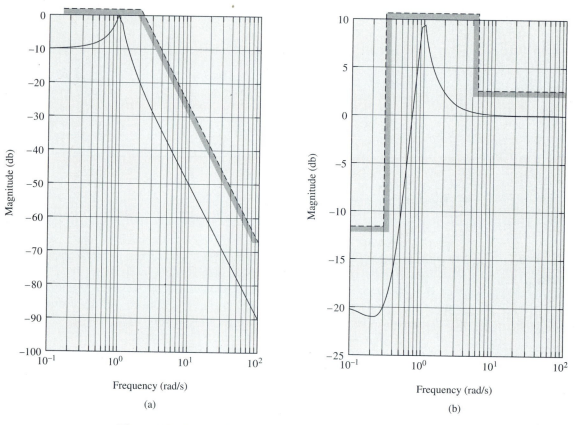

Figure 4.7 Typical frequency-domain specifications on: a) The reference output transmission and b) The disturbance-output transmission

Figure 4.9a shows the magnitudes $|B_k(j\omega)|$. It can be shown (see Problem 4.4) that

$$|B_k(j\omega)| = \frac{1}{[1 + (\frac{\omega}{\omega_0})^{2k}]^{1/2}} \tag{4.10}$$

so that

$$|B_k(j\omega_0)| = \frac{1}{\sqrt{2}}$$

for all k. Thus, ω_0 is the so-called 3-db bandwidth, because $20 \log(1/\sqrt{2}) = -3$. Note the rolloff rate, equal to 20 kdb/decade.

Figure 4.9b shows $|1 - B_k(j\omega)|$, the magnitude of the Butterworth high-pass filter. It is easy to show (see Problem 4.5) that the denominator of $1 - B_k(s)$ is the same as that of $B_k(s)$, and that the numerator is just the denominator without the constant term.

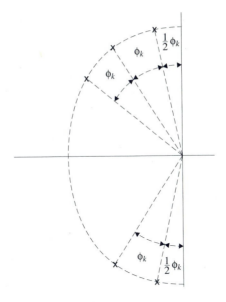

Figure 4.8   A Butterworth pole pattern

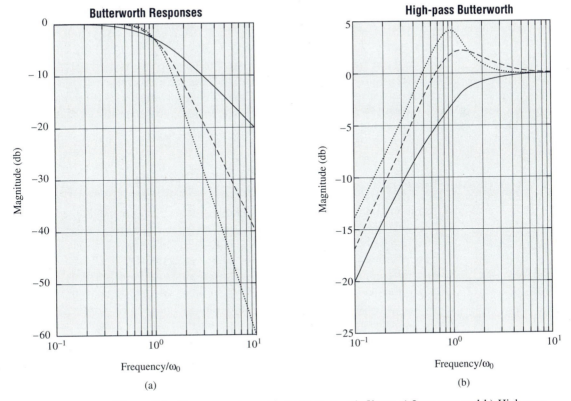

Figure 4.9   Frequency responses for Butterworth filters: a) Low-pass and b) High-pass

4.2.4 Sensitivity

The transfer functions associated with a control system are functions of the plant transfer function $P(s)$. If $P(s)$ changes from $P_0(s)$ to $P_0(s) + \Delta P(s)$, then $H_{d0}(s)$, for example, becomes $H_{d0}(s) + \Delta H_d(s)$. The *sensitivity* of H_d with respect to P is defined as

$$S_d{}^P(s) = \frac{\Delta H_d(s)/H_{do}(s)}{\Delta P(s)/P_0(s)} \tag{4.11}$$

or, basically, the fractional (or percent) change in $H_d(s)$ divided by the fractional (or percent) change in $P(s)$.

The formula is usually applied in the case of small—i.e., infinitesimal—changes, for which Equation 4.11 becomes

$$S_d{}^P(s) = \frac{\partial H_d(s)}{\partial P(s)} \frac{P_0(s)}{H_{d0}(s)}. \tag{4.12}$$

(Students of economics may recognize Equation 4.12 as the *elasticity* of H_d with respect to P.)

A sensitivity "yardstick" is obtained by applying the signal y_d directly to the plant input. In that case, $y(s) = P(s)y_d(s)$, so $H_d = P$, $\Delta H_d = \Delta P$, and, from Equation 4.11, $S_d{}^P = 1$. If a control system results in $|S_d{}^P(j\omega)| < 1$, then sensitivity at ω has been reduced by the control system; the opposite is true if $|S_d{}^P(j\omega)| > 1$. In general, sensitivity is reduced at some frequencies, increased at others. Specifications on sensitivity are expressed in the frequency domain, and generally attempt to limit $|S_d{}^P(j\omega)|$ at frequencies where $\Delta P/P$ is relatively large and where the input has significant spectral content.

4.3 OPEN-LOOP CONTROL

Open-loop control is the simplest control structure. It is rather limited in performance and is usually reserved for special applications where feedback control is either impossible or unnecessary. Because of its simplicity, open-loop control is a good starting point for the study of control structures. If feedback control is to be appreciated, then the limitations of open-loop control must be fully understood. Furthermore, concepts such as stability conditions and performance limitations due to actuation limits and system dynamics appear in relatively simple form. Those same concepts apply to other control structures but in more complex surroundings.

4.3.1 Basic Ideas and Expressions

In *open-loop control*, the manipulated input is derived from the set-point function y_d only, as shown in Figure 4.10; this corresponds to the case $F_m(s) = F_w(s)$ in

Equation 4.4. More often than not, $y_d(t)$ is supplied as a function in a computer, so that open-loop control requires no measurements. From Figure 4.10 it follows directly that

$$y = FPy_d + d \qquad (4.13)$$

and

$$e = (1 - FP)y_d - d. \qquad (4.14)$$

With reference to Equation 4.7, for the open-loop structure,

$$H_d(s) = F(s)P(s) \qquad (4.15)$$

$$H_{wd}(s) = 1. \qquad (4.16)$$

Perfect tracking of y_d occurs if $H_d(s) = F(s)P(s) = 1$, i.e., if $F(s) = P^{-1}(s)$. The practical objective is to make $F(j\omega)P(j\omega) \approx 1$ in the system passband, i.e., the frequency range over which y_d has significant spectral content. Since $H_{wd}(s) = 1$, open-loop control does nothing to attenuate the effects of disturbance inputs; the best that can be said is that it does not amplify them, either.

The sensitivity of $H_d(s)$ with respect to $P(s)$ is calculated from Equation 4.11. Since $\Delta H_d = F(P_0 + \Delta P) - FP_0 = F \Delta P$,

$$S_d{}^P(s) = \frac{F \Delta P / FP_0}{\Delta P / P_0} = 1. \qquad (4.17)$$

A sensitivity of 1 implies that a given percent change in $P(j\omega)$ translates into an equal percent change in the transmission $H_d(j\omega)$. Open-loop control neither decreases nor increases sensitivity.

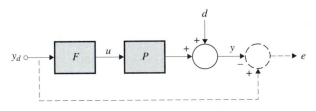

Figure 4.10 An open-loop control system

4.3.2 Stability Conditions

It was established in Chapter 3 that internal stability depends solely on the A matrix of the realization—more specifically, on the eigenvalues of A. The B and C matrices enter into the transfer function of the system; they may play a role in input–output stability, but not in internal stability. Internal stability is unaffected by the addition of new external inputs or the deletion of existing ones, because such changes modify only the B matrix, through the addition or deletion of columns. Similarly, outputs can be added or deleted with no effect on internal stability, because that affects only the C matrix. We shall add inputs and outputs, and view this as injecting test inputs into the system and taking extra measurements, neither of which would be expected to change the stability properties of the system.

The "test" inputs and outputs are chosen so that the resulting modified system is controllable and observable. As we saw in Chapter 3, such a system has no hidden modes, and internal stability is then guaranteed by input–output (i.e., transfer function) stability. The system of Figure 4.11 is the same as that of Figure 4.10 but with one additional input, v, and one additional output, z.

■ **Theorem 4.1**   Assume that $F(s)$ and $P(s)$ are realized by controllable and observable state representations. Then the system of Figure 4.11 is controllable and observable.

Proof: The proof relies on the basic definitions of controllability and observability rather than on the tests of Chapter 3. Let $\mathbf{x}_F$ and $\mathbf{x}_P$ be the states of the realizations of $F(s)$ and $P(s)$, respectively. The state of the composite system is $\mathbf{x} = \begin{bmatrix} \mathbf{x}_F \\ \mathbf{x}_P \end{bmatrix}$.

Controllability: The system is controllable if, starting from the zero state, there exist input functions $y_d^*(t)$ and $v^*(t)$ such that $\mathbf{x}_F(T) = \mathbf{x}_F^*$ and $\mathbf{x}_P(T) = \mathbf{x}_P^*$, for all $\mathbf{x}_F^*$, $\mathbf{x}_P^*$, and $T > 0$. Since the realization for F is controllable, some $y_d^*(t)$ will take $\mathbf{x}_F$ to the desired value at T. In the process, an output $z^*(t)$ will be generated.

By the same token, since the realization for P is controllable, there exists a $u^*(t)$ that takes $\mathbf{x}_P$ to $\mathbf{x}_P^*$ at time T. Now, u is not an input of the system of Figure 4.11 but an internal variable. The output of the first block, $z^*(t)$, contributes a part of $u(t)$. We need only add $v^*(t) = u^*(t) - z^*(t)$ to ensure that $u(t) = u^*(t)$ and $\mathbf{x}_P(T) = \mathbf{x}_P^*$. Since there exist inputs $y_d^*(t)$ and $v^*(t)$ capable of taking the system to any desired state, the system is controllable.

Observability: It is necessary to show that, given $z(t)$ and $y(t)$, $0 < t \leq T$, for a zero-input solution with arbitrary $\mathbf{x}_F(0)$ and $\mathbf{x}_P(0)$, it is possible uniquely to

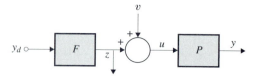

Figure 4.11   Open-loop system with additional input and output

recover $\mathbf{x}_F(0)$ and $\mathbf{x}_P(0)$. Let $y_d(t) = v(t) = 0$ (zero inputs). Since the realization for F is observable, and since the input to F is zero, $\mathbf{x}_F(0)$ is obtainable from observation of $z(t)$. Since $v(t) = 0$, $u(t) = z(t)$ and both the input and output of P are available, guaranteeing unique recovery of $\mathbf{x}_P(0)$ since the realization of P is observable. The composite system is thus observable. ∎

The next step is to derive the stability conditions.

■ **Theorem 4.2**   The system of Figure 4.11 is internally stable if, and only if, the transfer functions $F(s)$ and $P(s)$ are stable.

Proof: For the two-input, two-output system of Figure 4.11,

$$y(s) = FPy_d(s) + Pv(s)$$

$$z(s) = Fy_d(s)$$

or, in vector-matrix form,

$$\begin{bmatrix} y(s) \\ z(s) \end{bmatrix} = \begin{bmatrix} FP & P \\ F & 0 \end{bmatrix} \begin{bmatrix} y_d(s) \\ v(s) \end{bmatrix}. \tag{4.18}$$

Because the realization is controllable and observable, it is internally stable if, and only if, it is input–output stable, i.e., if all elements of the matrix transfer function of Equation 4.18 are stable. Thus, $P(s)$, $F(s)$, and $F(s)P(s)$ must have only LHP poles.

If $P(s)$ and $F(s)$ are stable transfer functions, so is $F(s)P(s)$, because all poles of FP are necessarily poles of either F or P. Therefore, if $P(s)$ and $F(s)$ are stable, the condition that FP be stable is redundant, and the theorem is established. ∎

Since the systems of Figures 4.10 and 4.11 differ only in the matter of inputs and outputs, they have identical internal stability properties. Therefore, Theorem 4.2 applies as well to the system of Figure 4.10 as to that of Figure 4.11.

4.3.3 Performance Limitations

As we saw in Section 4.3.1, open-loop control neither attenuates disturbances nor reduces sensitivity. The stability theorem just derived shows that open-loop control cannot stabilize an unstable plant, since $P(s)$ is required to be stable at the outset. It appears, then, that open-loop control is suitable only in cases where the plant is stable, well modeled, and not very perturbed, and where the objective is to adjust the set-point response.

Even then, there are limitations on the achievable transmission $H_d(j\omega)$. The first is due to the actuators. From Figure 4.10, with $d = 0$,

$$u(s) = \frac{y(s)}{P(s)} = \frac{H_d(s)}{P(s)} y_d(s)$$

and

$$u(j\omega) = \frac{H_d(j\omega)}{P(j\omega)} y_d(j\omega). \tag{4.19}$$

Typically, P is low-pass, so that, for ω greater than some ω_1, $|P(j\omega)|$ is less than 1 and becomes smaller as ω increases. Loosely speaking, ω_1 is the natural bandwidth of the plant. If the set-point signal $y_d(t)$ has significant spectral content up to $\omega_b > \omega_1$, it is desirable to make $H_d(j\omega) \approx 1$ for frequencies up to ω_b, in order to transmit y_d faithfully. But this makes the ratio $|H_d(j\omega)/P(j\omega)|$ large for frequencies between ω_1 and ω_b, and hence $|u(j\omega)|$ is relatively large at those frequencies. The actuator is then called upon to compensate for the low high-frequency gain of the plant and deliver considerable power at high frequencies; in most cases, this leads to higher costs. To summarize: pushing bandwidth beyond its natural plant level requires high-performance (and expensive) actuators.

The other limitation is more subtle and is a consequence of the stability result. It appears when $P(s)$ has one or more RHP zeros. It is possible to make $H_d(s)$ equal 1 by choosing $F(s) = P^{-1}(s)$. If $P(s)$ has no zeros in the closed RHP, such an $F(s)$ is stable and therefore satisfies the condition of internal stability. That condition is violated if $P(s)$ has an RHP zero, because that zero becomes an RHP pole of $F(s) = P^{-1}(s)$. Therefore, it is not possible to have both $H_d(s)$ equal to 1 *and* internal stability if $P(s)$ has one or more zeros in the closed RHP.

In fact, *closed RHP zeros of P must be retained as zeros of $H_d = FP$*, because F is not allowed to contain the RHP poles that would be required to cancel them out. In other words, H_d must have the zeros of P in the closed RHP to be admissible.

The presence of RHP zeros turns out to be the fundamental limitation imposed by the system dynamics on the performance of *all* types of control systems. The term *minimum-phase* is used to describe a stable rational transfer function with a stable inverse, i.e., with LHP zeros only; other transfer functions are said to be *non-minimum-phase*. The term comes from the fact that a transfer function with a given frequency magnitude characteristic has the least phase lag, at any frequency, if it is minimum-phase.

Example 4.1   For $P(s) = (2s + 1)/(s^2 + 3s + 2)$, design an open-loop compensator $F(s)$ such that

$$H_d(s) = \frac{1}{s^2 + 1.4s + 1}.$$

This is a second-order Butterworth function.

Solution

$$F(s)P(s) = H_d(s)$$

or

$$F(s) = \frac{H_d(s)}{P(s)} = \frac{1}{s^2 + 1.4s + 1} \cdot \frac{s^2 + 3s + 2}{2s + 1}$$

$$F(s) = \frac{s^2 + 3s + 2}{(2s + 1)(s^2 + 1.4s + 1)}.$$

Note that P is stable and minimum-phase, so $H_d(s)$ could have been any stable transfer function.

Example 4.2   Repeat Example 4.1 for $P(s) = (-s + 1)/(s^2 + 3s + 2)$, and $H_d(s) = (-s + 1)/(s^2 + 1.4s + 1)$.

Solution

$$F(s)P(s) = H_d(s)$$

$$F(s) = \frac{H_d(s)}{P(s)} = \frac{-s + 1}{s^2 + 1.4s + 1} \cdot \frac{s^2 + 3s + 2}{-s + 1}$$

$$F(s) = \frac{s^2 + 3s + 2}{s^2 + 1.4s + 1}.$$

Note that $H_d(s)$ needs to have the zero at $s = 1$; otherwise, $F(s)$ would have an unstable pole and, from Theorem 4.2, the overall system would be internally unstable.

4.3.4 Effect of RHP Zeros on Frequency Response

The presence of RHP zeros in the plant transfer function cuts down on our freedom to assign a transfer function and limits our choices to those that have the same RHP zeros. The impact of this on performance—particularly on frequency response—is now made explicit.

Rather than work with H_d, it proves easier to work with $W(s) = 1 - H_d$, the set-point to error-transfer function. With

$$e = y_d - y$$

$$= y_d - H_d y_d \tag{4.20}$$

we have

$$\frac{e}{y_d} = 1 - H_d(s) = W(s). \tag{4.21}$$

Clearly, making $|W(j\omega)|$ small is the same as requiring $H_d(j\omega) \approx 1$. The technical advantage of working with W rather than with H_d is that we need not worry about its phase, as long as its magnitude is small. Note that the poles of $W(s)$ are the same as those of $H_d(s)$, so that $W(s)$ is a stable transfer function. We shall also assume that $W(s)$ is proper and has no zeros in the open RHP. The following is derived from the theory of complex variables [3].

Let $W(s)$ be proper, analytic, and nonzero in the open RHP. Then, at each RHP point $z_0 = x_0 + jy_0, x_0 > 0$,

$$\frac{1}{\pi} \int_{-\infty}^{\infty} \log |W(j\omega)| \frac{x_0}{x_0{}^2 + (y_0 - \omega)^2} d\omega = \log |W(z_0)|. \tag{4.22}$$

Now, suppose $H_d(s)$ has an RHP zero at $s = z_0$. Then, by Equation 4.20,

$$W(z_0) = 1 - H_d(z_0) = 1. \tag{4.23}$$

From Equation 4.22,

$$\frac{1}{\pi} \int_{-\infty}^{\infty} \log |W(j\omega)| f(z_0, \omega) d\omega = \log 1 = 0 \tag{4.24}$$

where $f(z_0, \omega) = x_0 / [x_0{}^2 + (y_0 - \omega)^2]$.

Figure 4.12 illustrates the weighting function $f(z_0, \omega)$. It is positive for all ω and has its maximum at $\omega = y_0$. The peak is high and narrow if x_0 is small (the RHP zero is close to the j-axis), and low and broad if x_0 is large (an RHP zero is farther from the j-axis).

Recall that $\log |W(j\omega)|$ is positive if $|W(j\omega)| > 1$ and negative if $|W(j\omega)| < 1$. From Equation 4.24, the net area under the curve $\log |W(j\omega)| f(z_0, \omega)$ must be zero; i.e., negative and positive areas must cancel out. In particular, if $|W(j\omega)|$ is much less than 1 ($\log |W(j\omega)|$ is large and negative) over the frequency region where $f(z_0, \omega)$ is large (say, where $\omega = y_0 \pm x_0$), a large negative area is accumulated and it is necessary to have large values of $|W(j\omega)|$ at other frequencies to generate an equal positive weighted area. If the weight $f(z_0, \omega)$ is small at those other frequencies, then $\log |W(j\omega)|$ must be relatively large over a broad frequency range to accumulate enough positive area. In short, if the frequency band where $|W(j\omega)|$ is small covers most of the range where $f(z_0, \omega)$ is large, $|W(j\omega)|$ must be large somewhere else. This is illustrated in Example 4.3.

Example 4.3

The plant (and hence H_d) has a real RHP zero at $s = x_0$. The function $W(j\omega)$ is to have the form shown in Figure 4.13.[1] Derive, and discuss, the relationship enforced by Equation 4.24 between A, B, and ω_b.

---

[1]This magnitude function is not that of a rational $W(s)$, but could be approximated to an arbitrary degree by a rational function.

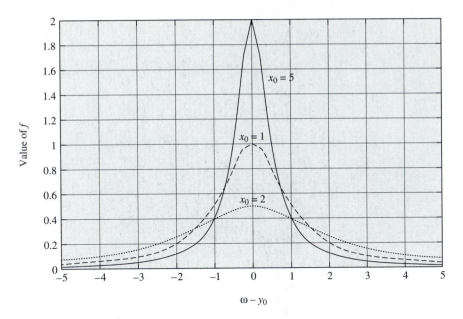

Figure 4.12   Frequency weighting function

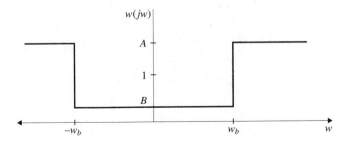

Figure 4.13   Frequency response function used to illustrate frequency-domain tradeoffs

Solution   First, note that

$$f(z_0, \ \omega) = \frac{x_0}{x_0{}^2 + (y_0 - \omega)^2} = \frac{-d}{d\omega} \tan^{-1} \frac{y_0 - \omega}{x_0}$$

so that Equation 4.24 becomes

$$\frac{2}{\pi} \left[-\int_0^{\omega_b} \log B \frac{-d}{d\omega} \tan^{-1} \left(\frac{y_0 - \omega}{x_0} \right) d\omega \right.$$

$$\left. -\int_{\omega_b}^{\infty} \log A \frac{d}{d\omega} \tan^{-1} \left(\frac{y_0 - \omega}{x_0} \right) d\omega \right] = 0$$

where the fact that both $\log |W(j\omega)|$ and $f(z_0, \omega)$ are even has been used. Using also the fact that $y_0 = 0$ yields

$$\log B \tan^{-1} \frac{\omega_b}{x_0} + \log A \left(\frac{\pi}{2} - \tan^{-1} \frac{\omega_b}{x_0} \right) = 0$$

or

$$\log A = \frac{\tan^{-1} \omega_b/x_0}{\pi/2 - \tan^{-1} \omega_b/x_0} (- \log B)$$

which is the required expression. Note that $B < 1$, so that $- \log B$ is positive. Making B smaller ($- \log B$ larger) forces $\log A$ to increase; a small $|W(j\omega)|$ in the passband must be compensated by a large value outside the passband.

At constant B, A varies with ω_b. As ω_b increases, $\tan^{-1} \omega_b/x_0$ increases toward $\pi/2$. The numerator increases, the denominator decreases, and therefore $\log A$ increases with bandwidth; thus, to keep B low over a bandwidth appreciably greater than x_0, it is necessary to tolerate a large value outside the passband.

As a rule of thumb, an RHP zero at $s = x_0$ restricts the system bandwidth to about x_0 rad/s. This rule, as we shall see, also holds for closed-loop control. Non-minimum-phase plants are simply difficult to control—with any control scheme.

4.3.5 Feedforward Control

Feedforward control is a variation of open-loop control. It is applicable when the disturbance input is measured. From Figure 4.14a it is easy to see that the effect of the disturbance on the output y will be nullified if $y' = -d$. We set up an open-loop problem in Figure 4.14b, where the output is y' and $y_d = -d$. The open-loop controller F is chosen, as in the case of open-loop design, to make y' as close as possible to y_d. The final step is to generate $y_d = -d$, as in Figure 4.14c; note that $P_w(s)$, the transfer function relating the measured disturbance w to the output, is assumed to be known.

For feedforward control, with reference to Equation 4.5,

$$H_w(s) = [1 - F(s)P(s)]P_w(s). \tag{4.25}$$

Example 4.4   **(Level Control)**

In the level-control system of Example 2.8, the transfer function from the valve stroke u to the level is $\Delta \ell / \Delta u = (-2.0)/(s + .005)$, and the transfer function relating the disturbance input to the level is $\Delta \ell / \Delta F_{\text{in}} = 1/(s + .005)$. Design a feedforward compensator to cancel out the effect of the disturbance.

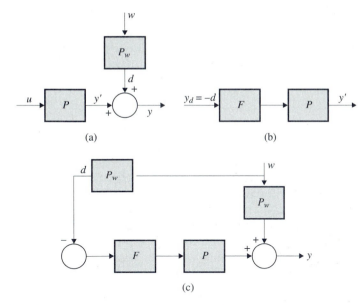

Figure 4.14   Development of the feedforward system

Solution   The plant is stable and minimum-phase, so we pick $H_d(s) = 1$, and therefore

$$F(s) = \frac{s + .005}{-2}.$$

The final step is to generate $y_d = -d$, or

$$y_d = \frac{-1}{s + .005} \Delta F_{in}$$

and therefore

$$\Delta u = F(s)y_d = \frac{1}{2} \Delta F_{in}.$$

In this simple case, feedforward control requires the valve to open immediately upon sensing a change of flow into the tank. If the models are exact, the compensation is perfect.

4.4   CLOSED-LOOP CONTROL, ONE DEGREE OF FREEDOM

This section introduces the most important feedback control structure, the single-degree-of-freedom (1-DOF) structure. Its name means that design freedom can be exercised to synthesize a compensator, or controller, in the feedforward path preceding the plant. A two-degrees-of-freedom (2-DOF) system includes the design freedom of synthesizing a compensator in the feedback path as well. This section unfolds in a manner parallel to that of the previous section. Basic expressions are derived, followed by stability conditions. Here again, actuators limit performance, and so do sensors—which, of course, did not appear in the open-loop case. Through the stability conditions, the plant dynamics limit performance in a manner quite analogous to the open-loop case.

4.4.1 Basic Ideas and Expressions

Figure 4.15 shows the basic configuration for 1-DOF control. Since feedback is used, the control input u is generated from e_m, the *measured error*, the difference between the desired output y_d and the measured output y_m. The transfer function $P_s(s)$ represents the sensor dynamics; the noise v represents additive sensor errors, such as bias and random noise. The dotted line is not part of the actual system but serves to define the error e between y_d and y.

For the time being, the sensor is assumed to measure y exactly, so $P_s(s) = 1$ and $v = 0$. This leads to Figure 4.16. From that figure,

$$y(s) = FPe(s) + d(s)$$
$$= FP[y_d(s) - y(s)] + d(s)$$

so that

$$y(s) = \frac{FP}{1 + FP}y_d(s) + \frac{1}{1 + FP}d(s) \qquad (4.26)$$

and

$$e(s) = y_d(s) - y(s)$$
$$= \frac{1}{1 + FP}y_d(s) - \frac{1}{1 + FP}d(s) \qquad (4.27)$$

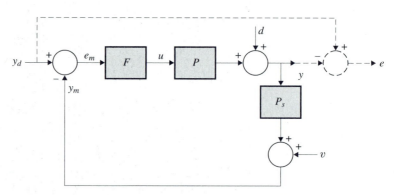

Figure 4.15   Block diagram for a 1-DOF feedback system

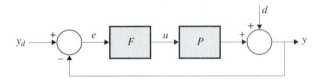

Figure 4.16   Simplified block diagram for a 1-DOF feedback system

In terms of Equation 4.5, for the 1-DOF system,

$$H_d(s) = \frac{FP}{1 + FP} \tag{4.28}$$

$$H_{wd}(s) = \frac{1}{1 + FP}. \tag{4.29}$$

The transfer function $F(s)P(s)$ is called the *loop gain*. If $|F(j\omega)P(j\omega)| \gg 1$, then

$$\frac{FP}{1 + FP} \approx 1$$

$$\frac{1}{1 + FP} \approx \frac{1}{FP}. \tag{4.30}$$

A high loop gain is therefore desirable at frequencies in the passband, since it makes $H_d = y/y_d$ approximately equal to 1 and $H_{wd} = y/d$ small.

If $|F(j\omega)P(j\omega)| \ll 1$, then

$$\frac{FP}{1 + FP} \approx FP$$

$$\frac{1}{1 + FP} \approx 1 \tag{4.31}$$

so set-point signals are attenuated and disturbance signals go through almost undisturbed.

In almost all cases, the loop gain has a magnitude greater than 1 for all $\omega < \omega_c$, and less than 1 for $\omega > \omega_c$. The frequency ω_c at which the loop gain has unit magnitude is called the *crossover frequency*. In view of Equations 4.29 and 4.30, ω_c is roughly the same as the system bandwidth. Figures 4.17 a) and b) illustrate the crossover frequency and the approximate expressions of Equations 4.30 and 4.31. The crossover frequency is approximately 2 rad/s, where $|FP(j\omega)| = 0$ db. For large loop gain (roughly, more than 20 db, for $\omega < 0.4$ rad/s), Figure 4.17 shows that $FP/(1 + FP)$ is close to 1 (0 db, 0° of phase); from Figure 4.18, $1/(1 + FP)$ is nearly $1/FP$ (log magnitude and phase are the negatives of those of FP). For small loop gain (roughly, less than -20 db, for $\omega > 6$ rad/s), we see that $FP/(1 + FP)$ approximates FP in magnitude and phase, while $1/(1 + FP)$ is near 1.

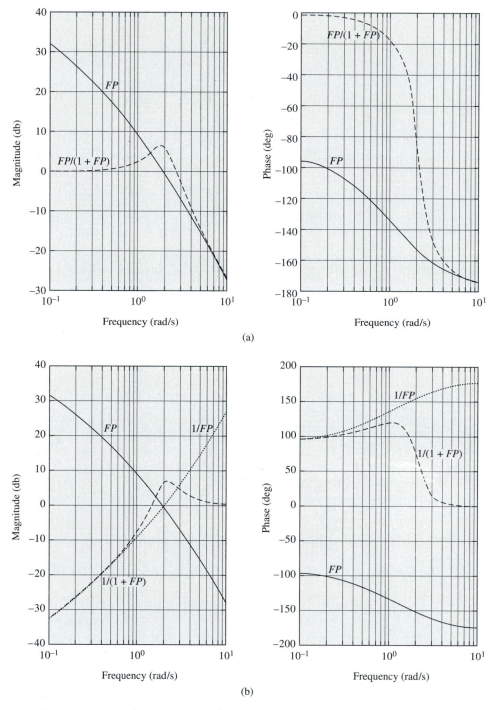

Figure 4.17 Illustration of closed-loop frequency response as a function of open-loop response reference output: a) Reference output; b) Disturbance output

The recipe appears simple enough: make the loop gain large in the passband. Unfortunately, as we shall see, stability complicates the picture and restricts our ability to follow the simple recipe. Naive attempts to apply it almost invariably result in unstable designs.

We also need sensitivity expressions to complete our survey of basics. Let

$$T = H_d = \frac{FP}{1 + FP}.$$ (4.32)

Then, from Equation 4.12,

$$\text{sensitivity} = S = \frac{\partial T}{\partial P} \frac{P}{T}$$

$$= \frac{(1 + FP)F - F^2 P}{(1 + FP)^2} \cdot P \cdot \frac{1 + FP}{FP}$$

$$= \frac{1}{1 + FP}.$$ (4.33)

The sensitivity, $S(s)$, is seen to be small if the loop gain is large in magnitude. For large loop gain FP, this means that changes in the plant P do not significantly affect the closed-loop transfer function T. This is not surprising, in hindsight. For example, let $F = 1$ and $P = 100$. Then

$$T = \frac{100}{101} = 0.990099.$$

Now try

$$F = 1, P = 200:$$

$$T = \frac{200}{201} = \frac{100}{100.5} = 0.99502.$$

Despite a doubling of P, T changes by only about half a percent. As long as $|FP| \gg 1$, the denominator of T will be close to FP and T will be near 1.

By comparison of Equations 4.29 and 4.33, the sensitivity S is equal to H_{wd}, the response to disturbance. Thus, making S small simultaneously achieves two positive results: it desensitizes the system *and* produces disturbance attenuation.

From Equations 4.32 and 4.33,

$$T(s) + S(s) = \frac{FP}{1 + FP} + \frac{1}{1 + FP} = 1.$$ (4.34)

Equation 4.34 is the reason $T(s)$ is called the *complementary sensitivity*.

Finally, note that Equations 4.26 and 4.27 can be written

$$y(s) = T(s)y_d(s) + S(s)d(s)$$ (4.35)

$$e(s) = S(s)y_d(s) - S(s)d(s).$$ (4.36)

Since $H_d = T$ and $H_{wd} = S$, the design objective is to make $T \approx 1$ and S small over the desired passband. Fortunately, since $S = 1 - T$, the two desiderata are compatible.

It is useful to think of closed-loop control in terms of T and S. Given S and $T = 1 - S$, the compensator F is easily derived as follows. From Equations 4.32 and 4.33,

$$1 + FP = \frac{1}{S}$$

$$F = \frac{1}{P}\left(\frac{1}{S} - 1\right) = \frac{1}{PS}(1 - S)$$

or

$$F(s) = \frac{T(s)}{S(s)P(s)}. \tag{4.37}$$

This result implies that a design may be carried out by specifying S (or T). Equation 4.38 then finds, uniquely, a controller that yields the desired S (or T).

Since we can only realize controllers that have proper or strictly proper transfer functions, we must impose a condition to ensure that this will be so. We shall normally require $\lim_{s \to \infty} T(s) = 0$, so that $\lim_{s \to \infty} S(s) = 1$. Suppose that, for large s,

$$T(s) \to \frac{k_T}{s^{N_T}}, \qquad P(s) \to \frac{k_P}{s^{N_P}}$$

where $N_T (N_P)$ is the number of poles of $T(P)$ minus the number of zeros of $T(P)$, i.e., the excess of poles over zeros of $T(P)$. Then

$$F(s) \to k s^{N_P - N_T}.$$

For $F(s)$ to be at least proper, we must have $N_P - N_T \leq 0$. To summarize: *Let $T(s)$ be strictly proper. Then $F(s)$ is proper (strictly proper) if the excess of poles over zeros of $T(s)$ is equal to (greater than) that of $P(s)$.*

It is possible to express this as a requirement on $S(s)$, should it be the design parameter. Let

$$T(s) = \frac{b_m s^m + b_{m-1} s^{m-1} + \cdots + b_0}{s^n + a_{n-1} s^{n-1} + \cdots + a_0}$$

where $m = n - N_T, \ 0 \leq N_T \leq n$. Then

$$S(s) = 1 - T(s)$$
$$= \frac{s^n + a_{n-1} s^{n-1} + \cdots + a_{m+1} s^{m+1} + (a_m - b_m)s^n + \cdots + (a_0 - b_0)}{s^n + a_{n-1} s^{n-1} + \cdots + a_0}.$$

$$\tag{4.38}$$

Equation 4.38 reveals that the leading N_T coefficients of the numerator and denominator of $S(s)$ are identical; a transfer function $S(s)$ constructed in this way will lead to a $T(s)$ with a pole excess of N_T.

4.4.2 Stability Conditions

Internal stability is proved using the same methodology as in the case of open-loop control. Figure 4.19 shows a modified system structure with somewhat different inputs and outputs, chosen to ensure controllability and observability. Since the choice of inputs and output variables has no effect on internal stability, stability of the system of Figure 4.19 implies stability of the original system.

■ **Theorem 4.3**   Assume that F and P in Figure 4.19 are represented by controllable and observable realizations. Then the composite system is controllable and observable.

Proof:
Controllability: We must show that, starting from the zero state at $t = 0$, it is possible to take the composite state vector $\begin{bmatrix} \mathbf{x}_F \\ \mathbf{x}_P \end{bmatrix}$ to $\begin{bmatrix} \mathbf{x}_F^* \\ \mathbf{x}_P^* \end{bmatrix}$ at $t = T$, where $T > 0$ and $\mathbf{x}_F^*$ and $\mathbf{x}_P^*$ are arbitrary. Since F and P are controllable, there exist functions $e^*(t)$ and $u^*(t)$ that, respectively, take $\mathbf{x}_F(t)$ to $\mathbf{x}_F^*$ and $\mathbf{x}_P(t)$ to $\mathbf{x}_P^*$, at time T. The zero-state outputs resulting from the application of $e^*(t)$ and $u^*(t)$ are $z^*(t)$ and $y^*(t)$, respectively.

Choose the system inputs $y_d^*(t)$ and $w^*(t)$ as follows:

$$y_d^*(t) = y^*(t) + e^*(t)$$
$$w^*(t) = u^*(t) - z^*(t)$$

Then it is easy to see that all relationships between the signals are satisfied with $e(t) = e^*(t)$ and $u(t) = u^*(t)$, so that the state at time T is indeed the desired one. Thus, the system is controllable.
Observability: Let the inputs $y_d(t) = w(t) = 0$; let $z(t)$ and $y(t)$ be observed. Since $e(t) = -y(t)$, the input and the output of F are accessible; since F is observable, the initial state $\mathbf{x}_F(0)$ can be uniquely deduced. Since $u(t) = z(t)$, the input and output of P are also given, so, by the same reasoning, $\mathbf{x}_P(0)$ can be obtained and the composite system is observable.   ■

Since the system is controllable and observable, internal stability is guaranteed by input–output stability, leading to the following theorem.

■ **Theorem 4.4**   The system of Figure 4.16 is internally stable if, and only if, $T(s)$, $P^{-1}(s)T(s)$, and $P(s)S(s)$ are stable.

Proof: From Figure 4.18,

$$z(s) = F(y_d - y)$$
$$= Fy_d - FP(z + w)$$

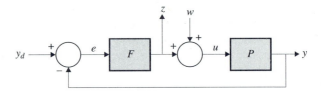

Figure 4.18   The 1-DOF system with additional input and output

and

$$z(s) = \frac{F}{1 + FP} y_d - \frac{FP}{1 + FP} w. \tag{4.39}$$

Furthermore,

$$y(s) = P(z + w)$$

$$= P\left(\frac{F}{1 + FP} y_d - \frac{FP}{1 + FP} w + w \right)$$

$$= \frac{FP}{1 + FP} y_d + \frac{P}{1 + FP} w. \tag{4.40}$$

Using the definitions of T and S, Equations 4.39 and 4.40 are rewritten as

$$\begin{bmatrix} z \\ y \end{bmatrix} = \begin{bmatrix} P^{-1}T & -T \\ T & PS \end{bmatrix} \begin{bmatrix} y_d \\ w \end{bmatrix}. \tag{4.41}$$

Since stability of all elements of the matrix transfer function is necessary and sufficient for input–output stability, it follows that the system is input–output stable if, and only if, T, $P^{-1}T$, and PS are stable transfer functions. Since input–output and internal stability are equivalent for controllable and observable realizations, the system of Figure 4.18 is internally stable. Finally, since the systems of Figures 4.16 and 4.18 differ only in the matter of inputs and outputs, the system of Figure 4.16 is also stable.   ∎

◆ ◆ ◆ **REMARK**
Since $S(s) = 1 - T(s)$, S and T have identical poles, so the stability of one implies the stability of the other. Thus, the condition that T be stable may be replaced by the condition that S be stable.   ◆

If $P(s)S(s)$ is to be stable, S must cancel out the RHP poles of $P(s)$. If P has a pole of multiplicity m at $s = p_0$, S must have a zero of the same multiplicity at the same point. It can be shown that this implies that

$$S(p_0) = \frac{dS}{ds}(p_0) = \cdots = \frac{d^{m-1}}{ds^{m-1}} S(p_0) = 0. \tag{4.42}$$

This condition may also be expressed in terms of $T(s) = 1 - S(s)$, as follows:

$$T(p_0) = 1 - S(p_0) = 1$$

$$\frac{d^i T}{ds^i}(p_0) = 0, \qquad i = 1, 2, \ldots, m - 1. \tag{4.43}$$

Similarly, if $P^{-1}(s)T(s)$ is to be stable, T must cancel the RHP zeros of P (i.e., the RHP poles of P^{-1}). If z_0 is a zero of P with multiplicity m, then

$$T(z_0) = \frac{dT}{ds}(z_0) = \cdots = \frac{d^{m-1}T}{ds^{m-1}}(z_0) = 0 \tag{4.44}$$

and

$$S(z_0) = 1$$

$$\frac{d^i S}{ds^i}(z_0) = 0, \qquad i = 1, 2, \ldots, m - 1. \tag{4.45}$$

These so-called *interpolation* conditions are important, because they show how the system dynamics constrain the choices of S and T. That is, for simple RHP poles and zeros, S is 0 at RHP poles and 1 at RHP zeros; T is 1 at RHP poles and 0 at RHP zeros.

We now have in place the theoretical elements required to define a synthesis procedure.

a. If $P(s)$ has neither poles nor zeros in the closed RHP (i.e., including the j-axis), arbitrarily choose either T or S, and calculate the other through the relation $T + S = 1$. The choice must be such as to ensure that F is at least proper (because our stability result was proved for the case where F has a realization). Then use Equation 4.37 to calculate the compensator F from P, T, and S.

b. If $P(s)$ has closed RHP poles or zeros, the choice of T or S is not free but must satisfy the interpolation conditions. If T is the design parameter, it must have all RHP zeros of P as its zeros and must satisfy Equation 4.42 at the RHP poles of P. The problem is to satisfy those conditions, while at the same time making $T(j\omega) \approx 1$ over the desired passband. If S is the design parameter, it must have all RHP poles of P as its zeros and satisfy Equation 4.45 at the RHP poles of P. The problem is to choose an $S(s)$ that satisfies these conditions *and* has $|S(j\omega)|$ small in the passband. Once T or S has been chosen, the rest of the design proceeds as in **a**.

Example 4.5   **(Level Control)**

For the level-control problem of Example 2.8 (Chapter 2), $P(s) = (-2.0)/(s + .005)$ (see Eq. 3.25 in Chapter 3). Design a compensator $F(s)$ so that $T(s) = \omega_0^2/(s^2 + \sqrt{2}\omega_0 s + \omega_0^2)$; i.e., T is a second-order Butterworth response.

Solution   We have

$$S(s) = 1 - T(s) = \frac{s^2 + \sqrt{2}\omega_0 s}{s^2 + \sqrt{2}\omega_0 s + \omega_0^2}.$$

Using Equation 4.37,

$$F(s) = \frac{T}{PS} = \frac{\omega_0^2(s + .005)}{s(s + \sqrt{2}\omega_0)(-2)}.$$

In this example, P has neither RHP poles nor zeros, so the choice of T (or S) is unconstrained, save for the requirement that F be proper.

Example 4.6   **(dc Servo)**

For the dc servo of Example 2.1 (Chapter 2),

$$P(s) = \frac{88.76}{s(s + 21.53)(s + 2.474)}.$$

Choose an $S(s)$ such that (i) the interpolation conditions are satisfied; (ii) $S(0) = 0$; (iii) $S(s)$ has all its poles as multiple poles at $s = -2$; and (iv) the order of S is the least required to ensure a proper $F(s)$. Calculate $F(s)$.

Solution   Since $P(s)$ has a pole excess of 3, the leading three coefficients in the denominator and numerator of $S(s)$ must be identical if $F(s)$ is to be proper. On the other hand, P has an unstable pole at $s = 0$, so $S(0) = 0$ [which automatically takes care of (ii)]. This means that the numerator of S has no term s^0. We write

$$S(s) = \frac{a_3 s^3 + a_2 s^2 + a_1 s}{(s + 2)^3}$$

$$= \frac{a_3 s^3 + a_2 s^2 + a_1 s}{s^3 + 6s^2 + 12s + 8}.$$

We must have $a_3 = 1$, $a_2 = 6$, and $a_1 = 12$, so

$$S(s) = \frac{s^3 + 6s^2 + 12s}{(s + 2)^3}$$

$$T(s) = 1 - S = \frac{8}{(s + 2)^3}$$

and

$$F = \frac{T}{PS} = \frac{8}{s(s^2 + 6s + 12)} \frac{s(s + 21.53)(s + 2.474)}{88.76}$$

$$= .0901 \frac{(s + 21.53)(s + 2.474)}{s^2 + 6s + 12}.$$

Example 4.7 For $P(s) = (-s + 1)/[(-s + 2)(s + 3)]$, design a $T(s)$ such that: (i) the interpolation conditions are satisfied; (ii) $T(0) = 1$; (iii) $T(s)$ has all its poles as multiple poles at $s = -a$, $a > 0$; and (iv) the order of $T(s)$ is the least required to ensure a proper $F(s)$. Calculate $F(s)$.

Solution The pole excess of $P(s)$ is 1, so we might try

$$T(s) = \frac{k}{(s + a)}.$$

Clearly, this will not do: we can satisfy (ii) by setting $k = a$, but we have no freedom left to satisfy $T(2) = 1$ and $T(1) = 0$. Try

$$T(s) = \frac{k(s + b)(s - 1)}{(s + a)^3}$$

which incorporates the condition $T(1) = 0$.

We must have

$$T(0) = \frac{-kb}{a^3} = -1$$

$$T(2) = \frac{k(2 + b)(+1)}{(2 + a)^3} = 1$$

so that

$$+2k + kb = +2k - a^3 = (2 + a)^3$$

or

$$k = a^3 + 3a^2 + 6a + 4$$

$$b = \frac{-a^3}{a^3 + 3a^2 + 6a + 4}.$$

Then

$$\begin{aligned} S &= 1 - T \\ &= \frac{s^3 + (3a - k)s^2 + (3a^2 + k - kb)s + a^3 + kb}{(s + a)^3} \\ &= \frac{s^3 + (3a - k)s^2 + (3a^2 + k + a^3)s + a^3 - a^3}{(s + a)^3} \\ &= \frac{s^3 - (a^3 + 3a^2 + 3a + 4)s^2 + (2a^3 + 6a^2 + 6a + 4)s}{(s + a)^3}. \end{aligned}$$

$S(s)$ has a zero at $s = 2$ and is factored as

$$S(s) = \frac{s(s-2)(s-c)}{(s+a)^3}$$

where $c = a^3 + 3a^2 + 3a + 2$.

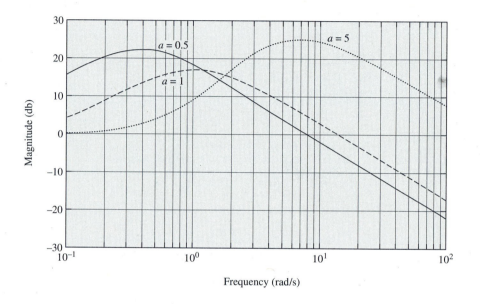

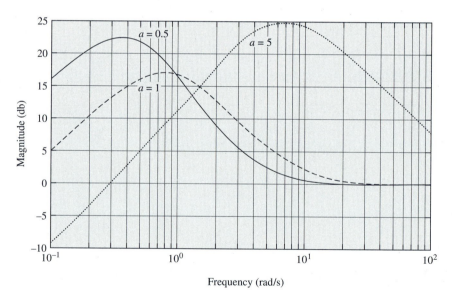

Figure 4.19 Sensitivity and complementary sensitivity

The controller is

$$F(s) = \frac{T}{PS} = \frac{k(s+b)(s-1)}{s(s-c)(s-2)} \frac{(s-2)(s+3)}{(s-1)}$$

$$= \frac{k(s+b)(s+3)}{s(s-c)}$$

where

$$k = a^3 + 3a^2 + 6a + 4$$

$$b = \frac{-a^3}{k}$$

$$c = a^3 + 3a^2 + 3a + 2.$$

Figure 4.19 shows the magnitude of $S(j\omega)$ and $T(j\omega)$ for $a = 0.5$, 1, and 5. Obviously, the performance is quite poor. We shall soon see why this plant is so difficult to control.

4.4.3 Performance Limitations: Sensors and Actuators

Limitations on performance originate from actuators, sensors, stability conditions, and robustness requirements. The first two are addressed in this section.

Figure 4.20 restores the measurement noise input, $v(t)$. Equations 4.26 and 4.27 are still valid, but the effect of v is to be added. From Figure 4.20, with $y_d = d = 0$,

$$y(s) = -FP(y+v)$$

$$y(s) = -\frac{FP}{1+FP}v(s) = -T(s)v(s)$$

and

$$e(s) = y_d - y(s) = T(s)v(s).$$

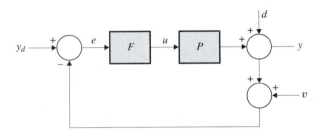

Figure 4.20   The 1-DOF system with observation noise

Therefore, with all inputs, y_d, d, and v, we use superposition to modify Equations 4.35 and 4.36:

$$y(s) = T(s)y_d(s) + S(s)d(s) - T(s)v(s) \tag{4.46}$$

$$e(s) = S(s)y_d(s) - S(s)d(s) + T(s)v(s). \tag{4.47}$$

With reference to Equation 4.7, the transfer functions describing the 1-DOF system are

$$H_d = T, \qquad H_{wd} = S, \qquad H_v = -T.$$

In the absence of sensor noise, it is possible to simultaneously pursue two objectives: making the set-point transmission close to 1 and the disturbance transmission close to 0. That is achieved by making $|S|$ small; hence, $T = 1 - S \approx 1$.

In the presence of sensor noise, making $T(s)$ close to 1 also has the effect of transmitting the sensor noise directly to the output and the error. In hindsight, this is not surprising. After all, the control system strives to reduce the difference between the desired output y_d and the *measured* output y_m. If $y_m = y + \Delta$, then $y_d - y_m = y_d - y - \Delta$; even if the control system succeeds in making $y_d - y_m = 0$, that still means y differs from y_d by the measurement error Δ. In short, the control system is only as good as the information it receives from sensors.

If the sensor noise has high spectral content in the system passband, its contribution to the error may be dominant. In some cases, it may even be necessary to reduce the design bandwidth to cut down the effect of the sensor noise. One conclusion is that, for good control, sensor noise should be low within the design bandwidth. Another is that the bandwidth of T should be no greater than necessary to meet specifications on set-point response and disturbance attenuation; it is desirable that $|T(j\omega)|$ be attenuated, or *rolled off*, outside the passband. Often, a *rolloff rate*, in decibels per decade, specifies the (negative) slope at the log magnitude of T outside the passband.

To address the actuator questions, we write, from Figure 4.20,

$$y(s) = Pu(s) + d(s)$$

$$u(s) = P^{-1}(y - d).$$

Using Equation 4.46,

$$u(s) = P^{-1}[Ty_d + (S-1)d - Tv]$$
$$= P^{-1}T(y_d - d - v). \tag{4.48}$$

The reasoning developed in the study of open-loop control also holds here. If $T(j\omega) \approx 1$ at frequencies beyond ω_1, the natural bandwidth of P, then $|P(j\omega)|$ will decrease well below 1 while $|P^{-1}(j\omega)T(j\omega)|$ will be large for $\omega > \omega_1$, and the actuator may not have enough bandwidth to provide the required power. In some cases, the actuator may be damaged.

Example 4.8   (Level Control)

For the system of Example 4.5, compute the plant input generated by a step change of 0.1 m in the desired level.

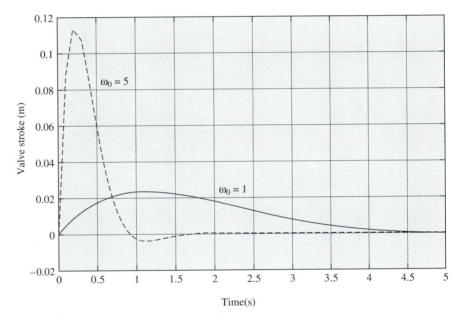

Figure 4.21   Valve stroke for different bandwidths

Solution   From Example 4.5,

$$\frac{T}{P} = \frac{\omega_0^2}{s^2 + \sqrt{2}\omega_0 s + \omega_0^2} \frac{s + .005}{-2}.$$

Now,

$$P(j\omega) = \frac{-2}{.005\left(\frac{j\omega}{.005} + 1\right)} = \frac{-400}{\left(\frac{j\omega}{.005} + 1\right)}.$$

Now, $400 = 52$ db, so $|P(j\omega)| > 1$ $(= 0$ db$)$ up to between 2 and 3 decades above the break frequency .005 rad/s, i.e., between .5 and 5 rad/s. Since $|T(j\omega)| \approx 1$ up to about ω_0, we can expect the control effort (i.e., the valve stroke) to increase substantially as the bandwidth ω_0 is increased beyond a few radians per second. This is shown in Figure 4.21. The figure shows clearly that fast response requires a valve capable of large excursions. At equal precision, such a valve is more costly than a valve designed for lower flow rates.

4.4.4 Performance Limitations: Stability

In the absence of RHP poles and zeros of $P(s)$, there are no restrictions on S or T other than stability. As illustrated in Example 4.5, any stable S or T is allowable.

Recall that $S = 0$ at RHP poles of P, and $S = 1$ at RHP zeros of P. RHP poles by themselves do not limit our ability to make $|S(j\omega)|$ small; it is not difficult to make S small everywhere in the complex plane and 0 at the RHP poles. On the other hand, RHP zeros do limit our ability to make $|S(j\omega)|$ small, because they force $S(s)$ to be 1 at some points in the plane. The consequence is that the overall sensitivity is increased everywhere. To illustrate, we may think of a tight rubber membrane stretched over the complex plane: if we hold the membrane up at some points, it is lifted everywhere else as well.

The problem is the same as in the case of open-loop control except for the fact that, unlike the function $W(s)$, $S(s)$ may have RHP zeros. Since Equation 4.22 does not apply to functions with zeros in the open RHP, a preliminary step is required.

Let $S(s)$ have RHP zeros at $p_1, p_2, \ldots, p_k$, the open RHP poles of P. We assume that $S(s)$ has no other RHP zeros. We define the *Blaschke product*, $B_p(s)$, as

$$B_p(s) = \frac{-p_1 + s}{p_1 + s} \cdot \frac{-p_2 + s}{p_2 + s} \cdots \frac{-p_k + s}{p_k + s}.$$

Note that B_p is stable because its poles are at $-p_1, -p_2, \ldots, -p_k$. The zeros of B_p are all in the RHP, at $p_1, p_2, \ldots, p_k$.

We claim that $S(s)$ may be written as

$$S(s) = \tilde{S}(s)B_p(s) \tag{4.49}$$

where $\tilde{S}(s)$ has no zeros in the open RHP. To see this, consider the example

$$S(s) = \frac{s(-s + 1)}{(s + 2)^2}.$$

Then let

$$B_p(s) = \frac{-s + 1}{s + 1}$$

and

$$S(s) = \frac{s(s+1)}{(s+2)^2}\frac{-s+1}{s+1}$$

$$= \widetilde{S}(s) \cdot B_p(s). \tag{4.50}$$

Clearly, $\widetilde{S}(s) = [s(s+1)]/(s+2)^2$ has no zeros in the open RHP.

It is not difficult to see that, in general, $\widetilde{S}(s)$ is generated by "pulling out" the open RHP zeros of $S(s)$ and replacing them with zeros in the open LHP. Of course, if P has no RHP poles, S has no RHP zeros and this procedure is not necessary: Equation 4.49 can still be used, with $B_p(s) = 1$.

One of the important properties of a Blaschke product is that $|B_p(j\omega)| = 1$ for all ω; i.e., it is an *all-pass* transfer function.

To show this, let p_1 be real. Then

$$\left|\frac{-p_1 + j\omega}{p_1 + j\omega}\right|^2 = \frac{p_1{}^2 + \omega^2}{p_1{}^2 + \omega^2} = 1.$$

If p_1 is complex, then

$$\left|\frac{-p_1 + j\omega}{p_1 + j\omega}\right|^2 \left|\frac{-p_1{}^* + j\omega}{p_1{}^* + j\omega}\right|^2 = \frac{(Rep_1)^2 + (\omega - I_m p_1)^2}{(Rep_1)^2 + (\omega + I_m p_1)^2} \cdot \frac{(Rep_1)^2 + (\omega + I_m p_1)^2}{(Rep_1)^2 + (\omega - I_m p_1)^2}$$

$$= 1.$$

Since $B_p(s)$ is a product of such terms, it follows that $B_p(j\omega) = 1$.

If $S(s)$ is proper, then so is $\widetilde{S}(s)$, since $B_p(s)$ has equal numbers of poles and zeros. Thus, with $S(s)$ assumed proper, $\widetilde{S}(s)$ is a proper, stable transfer function with no zeros in the open RHP, and Equation 4.22 applies:

$$\frac{1}{\pi}\int_{-\infty}^{\infty} \log|\widetilde{S}(j\omega)| f(s_0, \omega)d\omega = \log|\widetilde{S}(s_0)| \tag{4.51}$$

where s_0 is any RHP point and $f(s_0, \omega)$ is as in Equation 4.24.

From Equation 4.50,

$$\widetilde{S}(s) = \frac{S(s)}{B_p(s)}.$$

It follows that

$$\widetilde{S}(s_0) = \frac{S(s_0)}{B_p(s_0)}$$

and also that

$$|\widetilde{S}(j\omega)| = \frac{|S(j\omega)|}{|B_p(j\omega)|} = |S(j\omega)|$$

since $|B_p(j\omega)| = 1$.

Therefore, Equation 4.51 becomes

$$\frac{1}{\pi} \int_{-\infty}^{\infty} \log |S(j\omega)| f(s_0, \ \omega) d\omega = \log \frac{|S(s_0)|}{|B_p(s_0)|}. \tag{4.52}$$

Equation 4.52 is informative if there exists an RHP point (or points) where S has a known value. That is the case when P has RHP zeros: by the interpolation condition, $S(z_0) = 1$ if z_0 is a zero of P in the open RHP. Thus,

$$\frac{1}{\pi} \int_{-\infty}^{\infty} \log |S(j\omega)| f(z_0, \omega) d\omega = \log \frac{1}{|B_p(z_0)|}. \tag{4.53}$$

In the absence of RHP poles, $B_p(z_0) = 1$ and the right-hand side (RHS) of Equation 4.53 is zero. This equation is the same as Equation 4.23 in the open-loop case.

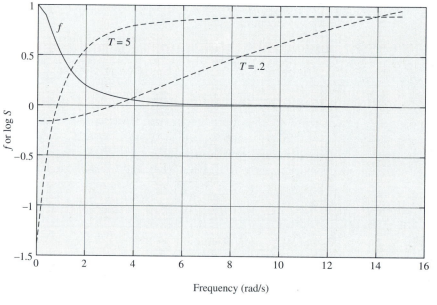

Figure 4.22 Admissible sensitivity frequency responses

Figure 4.22 shows, for $z_0 = 1$ and no RHP poles, the weighting function $f(z_0, \ \omega)$ and two admissible sensitivity log magnitude curves. For purposes of demonstration, the form $S(s) = k[(T_s + 1)/(.1T_s + 1)]$ is chosen, with $T_s = .2$ and 5, and k calculated to meet the condition $S(z_0) = 1$. The net weighted area under each sensitivity log magnitude curve must be zero, because we accumulate negative area when $|S(j\omega)| < 1$ (log $|S(j\omega)| < 0$) and compensate with positive area where $|S(j\omega)| > 1$. We accumulate net negative area at low frequencies, and the positive contribution comes at higher frequencies. The larger the bandwidth, the greater the negative area and, furthermore, the smaller the weight given to the positive contribution. It follows that a large bandwidth is obtained at the cost of large $|S(j\omega)|$ at higher frequencies. As a rule of thumb, for real zeros, the achievable bandwidth in radians per second is roughly equal to the real RHP zero closest to the origin.

If there are RHP poles as well as RHP zeros, the situation is worse. Suppose there is one RHP pole, p_0, and an RHP zero, z_0. Then

$$B_p(z_0) = \frac{-z_0 + p_0}{z_0 + p_0}.$$

Since z_0 and p_0 are positive, $|B_p(z_0)| < 1$ and $\log 1/|B_p(z_0)| > 0$. The net area represented by the LHS of Equation 4.53 must be positive, which means that the curves $\log|S(j\omega)|$ must be generally more positive than in the absence of RHP poles. The situation is particularly bad if p_0 and z_0 are close, because $|B_p(z_0)|$ is small and $1/|B_p(z_0)|$ is large.

By a process entirely analogous to that which led to Equation 4.53, an equation is written for the complementary sensitivity:

$$\frac{1}{\pi} \int_{-\infty}^{\infty} \log |T(j\omega)| f(p_0, \omega) d\omega = \log \frac{1}{|B_z(p_0)|} \tag{4.54}$$

where B_z is the Blaschke product associated with the *zeros* of P in the open RHP, and p_0 is a pole of P in the open RHP. In the absence of RHP zeros, $B_z(s) = 1$.

Figure 4.23 shows, for $p_0 = 1$ and no RHP zeros, two complementary sensitivity magnitude curves that satisfy Equation 4.54. For demonstration purposes, the form $T(s) = (T_1 s + 1)/(T_2 s + 1)^2$ has been selected, with $T_2 = 2$ and 5, and T_1 calculated to satisfy the constraint $T(p_0) = 1$. In contrast with the case of $S(j\omega)$, we would ideally like to have $|T(j\omega)| = 1$ [$\log |T(j\omega)| = 0$] at low frequencies, with rolloff at frequencies beyond the bandwidth supported by the sensors and actuators. Clearly, it is necessary to accumulate some positive area at low frequencies, to compensate for the negative area generated at higher frequencies. If $\log |T(j\omega)|$ starts rolling off at some frequency that is appreciably less than p_0 (= 1 here), the negative area is given appreciable weight, and $\log |T(j\omega)|$ must be significantly greater than 0 over part of the passband. Conversely, if $\log |T(j\omega)|$ is to be near zero in the passband (thus accumulating a small positive area), then the rolloff must start at a frequency sufficiently high for the weight placed on the negative contribution to be small. The conclusion is that the control of an unstable system requires a bandwidth at least equal to the magnitude of the largest unstable pole.

Thus, whereas RHP zeros impose an upper limit on the bandwidth that *can* be achieved, RHP poles impose a lower limit on the bandwidth that *must* be achieved. The implication for actuators and sensors is that it is useless for the bandwidths of these components to extend much beyond the limit imposed by the RHP zeros, but that they must have enough bandwidth to control the unstable poles.

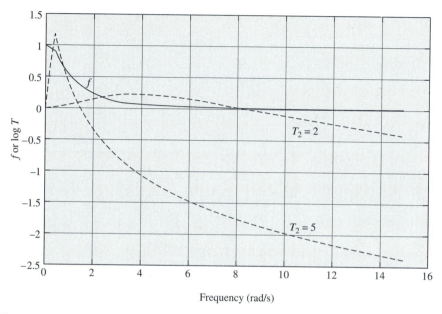

Figure 4.23   Admissible complementary sensitivity functions

4.5   AN INTRODUCTION TO H^∞ THEORY

As a followup to the preceding section, we present an introduction to H^∞ theory. This theory originated in the work of Zames only a few years ago, and is having a significant impact on the development of control methodology. We shall use the theory to derive for a given plant the best (in some sense) $S(s)$ that satisfies the interpolation conditions. The result, which does not take into account the need to roll off $T(s)$ or to have a proper or strictly proper $F(s)$, is not so much a design as a yardstick against which to evaluate a design. If a control system sensitivity were close to optimal in this sense at frequencies within the desired bandwidth, the design would have little room for improvement.

At its most basic level, H^∞ theory addresses the following problem: given a stable, proper, rational weighting function $W(s)$ with all zeros in the open LHP, find the proper, stable sensitivity function $S(s)$ that (i) satisfies given interpolation constraints at some RHP points and (ii) minimizes

$$\mu = \sup_{\omega} |S(j\omega)W(j\omega)|. \tag{4.55}$$

The interpolation constraints are precisely the ones discussed earlier. The optimization problem is of the "mini-max" or worst-case variety: we seek to minimize the supremum, over the whole frequency range, of the weighted sensitivity. The weighting function reflects the fact that low sensitivity is more important at some frequencies (normally low frequencies) than at other frequencies. Since it is the magnitude of the *product* that is minimized, the sensitivity tends to be small at frequencies where the weight is relatively great.

The solution, derived from functional analysis, is of the form

$$S(s)W(s) = kB(s) \tag{4.56}$$

where k is a constant and $B(s)$ is a Blaschke product.

The Blaschke product $B(s)$ in Equation 4.56 will be chosen to satisfy interpolation constraints, if any. We saw that there are no interpolation constraints on $S(s)$ if the plant $P(s)$ is stable and minimum-phase; in that case, $B(s) = 1$ and k may be any real number, as small as desired. If $P(s)$ has poles $p_1, p_2, \ldots, p_N$ in the closed RHP, then $S(p_i) = 0, i = 1, 2, \ldots, N$. If $P(s)$ has zeros $z_1, z_2, \ldots, z_M$ in the closed RHP, then $S(z_i) = 1, i = 1, 2, \ldots, M$.

We shall assume for the moment that $P(s)$ may have poles in the open RHP but not on the j-axis (those will be handled separately). Therefore, let $p_i, i = 1, 2, \ldots, N$, be poles of $p(s)$ in the open RHP. To satisfy the interpolation constraints,

$$S(p_i) = k\frac{B(p_i)}{W(p_i)} = 0, \qquad i = 1, 2, \ldots, N. \tag{4.57}$$

Since $W(s)$ has no RHP poles, $W(p_i) \neq \infty$ and therefore we must have $B(p_i) = 0, i = 1, 2, \ldots, N$. This forces $B(s)$ to have the form

$$B(s) = B_p(s)B'(s) \tag{4.58}$$

where $B_p(s) = [(-s + p_1)/(s + p_1)] \cdots \cdot [(-s + p_N)/(s + p_N)]$. In Equation 4.58, B' is a Blaschke product, to be determined presently.

Note that, in the absence of RHP zeros, the term B' is not required, since all interpolation constraints are satisfied with $B_p(s)$. In that case, k may be any real nonzero constant, as small as desired.

Given RHP zeros, B' is found by using the interpolation conditions at the RHP zeros of p. If these zeros are $z_1, z_2, \ldots, z_M$, stability forces $S(z_1) = S(z_2) = \cdots = S(z_M) = 1$. From Equations 4.57 and 4.58,

$$S(z_i)W(z_i) = W(z_i) = kB'(z_i)B_p(z_i), \qquad i = 1, 2, \ldots, M$$

or

$$kB'(z_i) = \frac{W(z_i)}{B_p(z_i)}, \qquad i = 1, 2, \ldots, M. \tag{4.59}$$

Equation 4.59 represents M equations; it is to be expected that M free parameters will be needed to satisfy them. The gain k is one parameter; B' must have the remaining $M - 1$. Indeed, it can be shown that B' has the form

$$
B'(s) = \begin{cases} \left(\dfrac{-s + a_1}{s + a_1}\right) \cdot \left(\dfrac{-s + a_2}{s + a_2}\right) \cdots \left(\dfrac{-s + a_{M-1}}{s + a_{M-1}}\right) & \text{if } M > 1 \\[2mm] 1 & \text{if} \qquad M = 1 \end{cases}
\tag{4.60}
$$

where the a_i are complex numbers with the positive real part chosen, together with k, to satisfy Equation 4.59.

Once k and the a_i have been determined, S is calculated as

$$
S(s) = \frac{k B'(s) B_p(s)}{W(s)}
\tag{4.61}
$$

which is stable, since B' and B_p are stable and W has no RHP zeros. The compensator design then proceeds as before.

Note that, since $|B(j\omega)| = 1$,

$$
|S(j\omega)| = \frac{|k|}{|W(j\omega)|}.
\tag{4.62}
$$

This equation reveals that the shape of the sensitivity magnitude curve is just the inverse of that of W. If $|W(j\omega)|$ is relatively high at low frequencies, $|S(j\omega)|$ will be relatively low. Thus, the shape of the optimal $|S(j\omega)|$ is independent of P; the influence of the plant dynamics is exerted entirely through the gain k.

Example 4.9   For $P(s) = (-s + 1)/[(-s + 2)(s + 3)]$ (see Example 4.7), calculate the optimal $S(s)$, given the weight function $W(s) = (0.1\tau s + 1)/(\tau s + 1)$. The magnitude $|W(j\omega)|$ is shown in Figure 4.24, where it is seen that low frequencies, up to the bandwidth $1/\tau$, are given more weight than high frequencies.

Solution   The RHP pole gives

$$
B_p(s) = \frac{-s + 2}{s + 2}.
$$

Since there is one RHP zero, $M = 1$ and $B'(s) = 1$. From Equation 4.59,

$$
\begin{aligned}
k &= \frac{W(1)}{B_p(1)} \\[2mm]
&= \frac{.1\tau + 1}{\tau + 1} \frac{3}{1} \\[2mm]
&= 3 \frac{.1\tau + 1}{\tau + 1}.
\end{aligned}
$$

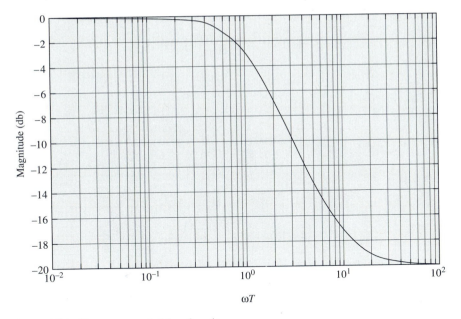

Figure 4.24 Frequency weighting function

Using Equation 4.61,

$$S = 3\frac{.1\tau + 1}{\tau + 1}\frac{-s + 2}{s + 2}\frac{\tau s + 1}{.1\tau s + 1}.$$

Note that, using Equation 4.62,

$$|S(j\omega)| = 3\frac{.1\tau + 1}{\tau + 1}\sqrt{\frac{\omega^2\tau^2 + 1}{.01\omega^2\tau^2 + 1}}.$$

For $\tau \gg 1$ (high bandwidth $= 1/\tau$), $k \approx 3$. For $\tau \gg 1$ (low bandwidth), $k \approx .3$. Clearly, the results are better at low frequencies if we do not demand high bandwidth. Figure 4.25 shows magnitude curves for a few values of τ.

The sensitivity function of Example 4.9 does not satisfy the condition that F be proper. This is not surprising, since that requirement was not imposed on the solution. One way to modify the solution is to define

$$T'(s) = \frac{T(s)}{(\tau s + 1)^n} \tag{4.63}$$

where n is such as to ensure that the pole excess of T' is at least as great as that of P, and τ is a small time constant. In effect, $T'(j\omega) \approx T(j\omega)$ for $\omega \ll 1\tau$, so

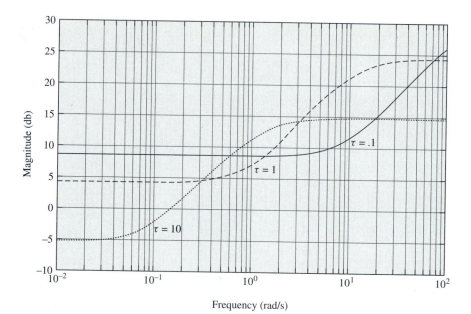

Figure 4.25   Optimal sensitivity magnitudes for different bandwidths

that $T' \approx T$ at low frequencies. Then

$$S'(s) = 1 - \frac{T(s)}{(\tau s + 1)^n}. \tag{4.64}$$

The rest of the design is as before.

We now wish to enlarge the allowable class of plants to include those with j-axis poles and/or zeros. Let $p(s)$ be a polynomial whose roots are the j-axis poles of $P(s)$; i.e.,

$$P(s) = \frac{P'(s)}{p(s)}$$

where $P'(s)$ has no j-axis poles. We restrict $W(s)$ to the form

$$W(s) = \frac{W'(s)}{p(s)}$$

where W is a proper or strictly proper transfer function, and W' has all its poles and zeros in the open LHP. Since S must be zero at the unstable poles of P, including the roots of $p(s)$, we write

$$S = pS'.$$

Clearly, $WS = W'S'$, so that minimization of the ∞ norms of WS and $W'S'$ is the same problem. The solution, as before, is

$$W'S' = kB_p(s)B'(s).$$

The interpolation conditions come from the fact that, at all z_i such that $P(z_i) = 0$,

$$S(z_i) = p(z_i)S'(z_i) = 1$$

so that

$$W'(z_i)S'(z_i) = \frac{W'(z_i)}{p(z_i)} = W(z_i) = kB_p(z_i)B'(z_i)$$

which is precisely Equation 4.59.
We illustrate with an example.

Example 4.10   The cart-and-inverted-pendulum system of Example 3.18 (Chapter 3) has the linearized transfer function

$$\frac{x}{F} = \frac{(s + 3.13)(s - 3.13)}{s^2(s - 4.43)(s + 4.43)}.$$

Find the compensator that solves the H^∞ problem, with $W(s) = (s + 1)^2/s^2$. (Note that W, like P, has a double pole at $s = 0$.)

Solution   To satisfy the interpolation conditions at $s = 4.43$, we use

$$B_p(s) = \frac{-s + 4.43}{s + 4.43}.$$

Because $M = 1$ (one RHP zero), $B'(s) = 1$ and

$$k = \frac{W(3.13)}{B_p(3.13)}$$

$$= \frac{(4.13)^2}{(3.13)^2} \cdot \frac{(7.56)}{(1.30)} = 10.12.$$

Thus,

$$S(s) = 10.12 \cdot \frac{-s + 4.43}{s + 4.43} \cdot \frac{s^2}{(s + 1)^2}.$$

To summarize the H^∞ procedure:

1. Choose a minimum-phase weighting function $W(s)$ (i.e., no zeros in the closed RHP). $W(s)$ *must* have poles at the j-axis poles of $P(s)$, if any, and *may not* have j-axis poles at zeros of $P(s)$. It may have other j-axis poles.

2. From the poles of P in the *open* RHP, form the Blaschke product $B_p(s)$. If P has no open RHP poles, $B_p(s) = 1$.

3. Form the Blaschke product $B'(s)$, having $M - 1$ poles (and zeros), where M is the number of zeros of P in the *closed* RHP. If $M = 1$, then $B'(s) = 1$.

4. Use the interpolation condition of Equation 4.59 to write M equations for k and the poles of B'. Solve to obtain k and $B'(s)$.

5. Write $S(s)$ as in Equation 4.61.

4.6   CLOSED-LOOP CONTROL, TWO DEGREES OF FREEDOM

In the 1-DOF configuration, $H_d = T$ and $H_{wd} = S = 1 - T$. If the design is focused on shaping H_{wd} to provide attenuation for a given disturbance spectrum, then $H_d = T = 1 - H_{wd}$ falls out as a by-product. There are cases in which it is desirable to choose H_d and H_{wd} independently; for example, the disturbance and set-point signals may have very different spectra. The two-degrees-of-freedom (2-DOF) configuration provides an extra degree of flexibility to make that possible. Figure 4.26 shows two independently adjustable controllers, one in the forward path and one in the feedback path. There are two transfer functions, $F(s)$ and $R(s)$, to be used as design parameters. The configuration of Figure 4.26 is only one of several 2-DOF possibilities. (See Problems 4.35 and 4.36.)

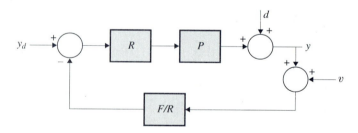

Figure 4.26   A 2-DOF feedback system

From the figure,

$$y = d + PR\left[y_d - \frac{F}{R}(y + v)\right]$$

$$y = \frac{PR}{1 + FP}y_d + \frac{1}{1 + FP}d - \frac{FP}{1 + FP}v$$

or

$$y = RPSy_d + Sd - Tv \tag{4.65}$$

and

$$e = y_d - y$$
$$e = (1 - RPS)y_d - Sd + Tv. \tag{4.66}$$

From Equations 4.65 and 4.66, for the 2-DOF configuration,k

$$H_d = RPS, \qquad H_{wd} = S, \qquad H_v = -T.$$

The transmissions H_{wd} and H_v, from d and v, respectively, to y, are exactly the same as in the case of the 1-DOF design. The transmissions from y_d to y and e are different; $H_d = RPS$ is independently adjustable by the additional parameter R. It is easy to show that the sensitivity of the transmission H_d to changes in P is S, as in the 1-DOF case.

This suggests the following procedure. First, do a 1-DOF design for purposes of disturbance reduction and sensitivity. Then choose R to adjust the response to set point.

Stability is studied from Figure 4.27, with test inputs and outputs. The proof of controllability and observability is left to the reader under the assumption that P, R, and F/R are realized minimally.

From the figure,

$$y = P\left\{v_1 + R\left[y_d - \frac{F}{R}(y + v_2)\right]\right\}$$
$$= Pv_1 + PRy_d - PF(y + v_2) \tag{4.67}$$
$$y = \frac{PR}{1 + FP}y_d + \frac{P}{1 + FP}v_1 - \frac{FP}{1 + FP}v_2$$

or

$$y = RPSy_d + PSv_1 - Tv_2. \tag{4.68}$$

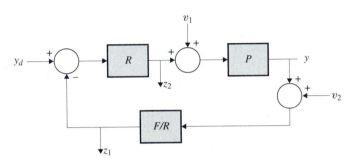

Figure 4.27 A 2-DOF feedback system with additional inputs and outputs

Also,

$$z_1 = \frac{F}{R}(y + v_2) \tag{4.69}$$

$$= \frac{FP}{1 + FP}y_d + \frac{FP}{R(1 + FP)}v_1 + \frac{F}{R(1 + FP)}v_2$$

or

$$z_1 = T y_d + R^{-1}T v_1 + R^{-1}P^{-1}T v_2. \tag{4.70}$$

Finally,

$$z_2 = P^{-1}y - v_1$$
$$z_2 = RS y_d - T v_1 - P^{-1}T v_2. \tag{4.71}$$

■ **Theorem 4.5**   The 2-DOF configuration of Figure 4.27 is internally stable if, and only if, the transfer functions T, PS, $P^{-1}T$, RS, $R^{-1}T$, RPS, and $R^{-1}P^{-1}T$ are all stable.

Proof: Those are the transfer functions of the three-input, three-output system of Figure 4.27. Since the system is controllable and observable, the stability of these functions is both necessary and sufficient. Of course, the conditions of Theorem 4.5 are also applicable to Figure 4.26, because the choices of inputs and outputs do not affect internal stability.   ■

Since H_{wd} and H_v are the same as in the 1-DOF case, the following is a convenient design procedure:

a. Create a 1-DOF design by selecting F, concentrating on the responses to disturbances and observation noise. This part is independent of R. It fixes S and T, which satisfy the first three conditions of Theorem 4.5; those conditions are precisely the stability conditions for the 1-DOF case.

b. With S in hand, choose R to satisfy some design objective for $H_d = RPS$.

From Theorem 4.5, we know that $H_d = RPS$ must be stable, but we must also see what restrictions, if any, are placed on H_d by the requirements that RS, $R^{-1}T$, and $R^{-1}P^{-1}T$ also be stable.

It is desirable to cancel out RHP zeros of PS by poles of R, so as to make $H_d = RPS$ a minimum-phase transfer function. The sensitivity S has zeros at the RHP poles of P; those do not appear in PS, due to the pole-zero cancellation. S may have other RHP zeros, which may be cancelled by poles of R, as this would satisfy the condition that both RS and RPS be stable. The RHP zeros of P may *not* be cancelled out. Such zeros cannot also be zeros of S, because $S = 1$ at RHP zeros of P. If we try to cancel an RHP zero of P with a pole of R, RPS will be stable, but not RS. It follows that $H_d = RPS$, the set-point to output-transfer functions, must retain the RHP zeros of P. These are the only obligatory RHP zeros of H_d, since the RHP zeros of S that are left in the product PS may be cancelled by R.

If one starts from the stability of $R^{-1}T$ and $R^{-1}P^{-1}T$, a similar argument can be constructed about the RHP *poles* of R^{-1}, i.e., the RHP zeros of R. The conclusion is that R may have RHP zeros only at those RHP zeros of T that are not also zeros of P. That result is somewhat academic, because it would rarely be desirable to give R any RHP zeros, as these would appear in $H_d = RPS$.

Given T (and S), the second phase of a 2-DOF design is carried out by choosing a stable $H_d(s)$ with zeros at the closed RHP zeros of P, and such that $H_d(j\omega) \approx 1$ in the desired passband for the set-point signal. Once $H_d = RPS$ is known, R can be calculated since P and S are given. Note that the problem of selecting a stable H_d with given RHP zeros is basically the open-loop design problem. Also note that the unstable poles of P do not matter. Although such poles (in the presence of RHP zeros) have a deleterious effect on the sensitivity, they do not restrict our freedom to choose the set-point transmission.

To complete the discussion of 2-DOF design, we note that

$$u = \frac{1}{P}(y - d)$$

$$= \frac{H_d}{P}y_d - \frac{T}{P}(d + v) \qquad (4.72)$$

so that the control effort tends to be large if $|T(j\omega)|$ or $|H_d(j\omega)|$ is large over a bandwidth greatly exceeding that of $P(j\omega)$. Thus, T should be rolled off outside the disturbance-stop bandwidth, and H_d should attenuate outside the set-point passband.

Example 4.11 (Level Control)

Consider the level-control system of Example 4.5. Suppose that the disturbance d referred to the output is relatively small and has a spectrum extending to about 2 rad/s. Suppose also that the system is called upon to handle relatively large set-point changes, and to do so without undue valve motion. To handle the disturbances, let $T(s) = \omega_0^2/(s^2 + \sqrt{2}\omega_0 s + \omega_0^2)$, with $\omega_0 = 2$ rad/s. To avoid large inputs in response to set-point changes, let y/y_d be similar in form to $T(s)$, but with $\omega_0 = 0.5$ rad/s. The response will be slow, but the input will be less likely to "kick." Design the required 2-DOF system.

Solution   First, design the 1-DOF loop to handle disturbances. We have

$$S(s) = 1 - T(s) = \frac{s^2 + 2\sqrt{2}s}{s^2 + 2\sqrt{2}s + 4}$$

and

$$F(s) = \frac{T}{SP} = \frac{4}{s(s + 2\sqrt{2})}\frac{s + .005}{-2}$$

$$= \frac{-2(s + .005)}{s(s + 2.82)}.$$

Next, design $R(s)$. We have

$$RPS = \frac{.25}{s^2 + .5\sqrt{2}s + .25}.$$

Note that *any* stable transfer function RPS is allowed, because P has no RHP zeros. Going on,

$$R = \frac{0.25}{s^2 + .707s + .25} \cdot \frac{s^2 + 2.82s + 4}{s(s + 2.82)} \cdot \frac{s + .005}{-2}$$

$$= \frac{-0.125(s^2 + 2.82s + 4)(s + .005)}{s(s + 2.82)(s^2 + .707s + .25)}.$$

The compensator in the feedback path is

$$\frac{F}{R} = \frac{16(s^2 + .707s + .25)}{s^2 + 2.82s + 4}.$$

Problems

M **4.1** Compute the unit step response of $y/y_d = H_d(s) = (.2s + 1)/(s + 1)^3$. From the graph of the response, obtain the steady-state error, e_{ss}; the delay time, T_d; the rise time, T_r; the peak overshoot (if any); and the settling time, T_s.

M **4.2** Repeat Problem 4.1 for $H_d(s) = [.9(.2s + 1)]/(s^2 + .7s + 1)$.

M **4.3** Repeat Problem 4.2 for $H_d(s) = [9(.2s + 1)]/[(s^2 + .7s + 1)(s^2 + s + 9)]$.

4.4 Show that the magnitude of the kth-order Butterworth low-pass filter is

$$|B_k(j\omega)| = \frac{1}{\left[1 + \left(\frac{\omega}{\omega_0}\right)^{2k}\right]^{1/2}}.$$

4.5 Show that the Butterworth high-pass filter $1 - B_k(s)$ has the same denominator as $B_k(s)$, and that its numerator is that of $B_k(s)$, minus the constant term.

4.6 You are given the following plants:

a. $P(s) = \frac{s+3}{(s+1)(s+2)}$

b. $P(s) = \frac{-s+3}{(s+2)(s+2)}$

c. $P(s) = \frac{1}{s(s+1)}$

d. $P(s) = \frac{2}{(-s+1)(s+2)}$

For each plant, design (if possible) an open-loop compensator $F(s)$ such that $H_d(s) = 4/(s^2 + 2s + 4)$ and the system is internally stable. If that is impossible, explain why.

M   **4.7**   To further explore the significance of RHP zeros, let $H_d(s) = (-Ts + 1)/(Ts + 1)$. Calculate $W(s) = 1 - H_d(s)$, and sketch the magnitude Bode plot of $W(j\omega)$, showing relevant features as functions of T. How does the location of the RHP zero affect the low-frequency behavior of $W(j\omega)$?

M   **4.8**   *Active suspension*   For the suspension system of Example 2.2 (Chapter 2), the transfer functions $P(s)$ (plant) and $P_w(s)$ (disturbance) are given by Equations 3.23 and 3.24 (Chapter 3), respectively. Suppose the roadway deviation y_R can be measured and is to be used as a feedforward input, as in Figure 4.28.

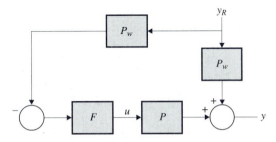

Figure 4.28   Feedforward design for the active suspension

a. Show that $y/y_R = P_w(s)[1 - F(s)P(s)]$.

b. Design the compensator $F(s)$ so that

$$FP = \frac{k(s^2 + 85.861)}{s^2 + 1.4\omega_0 s + \omega_0^2}.$$

Choose k such that $F(0)P(0) = 1$.

c. Obtain the magnitude Bode plots of y/y_R for $\omega_0 = 4, 10$, and 50 rad/s.

d. Calculate u/y_R for the values of ω_0 in part (c), and obtain magnitude Bode plots. What can you conclude about the behavior of the control u as a function of bandwidth?

M   **4.9**   Given $FP = k/[s(s + 1)(.1s + 1)]$:

a. Compute the Bode plots (magnitude and phase) of $F(s)P(s)$ for $k = 1$.

b. Compute the Bode plots of $T = \frac{FP}{1+FP}$ and $S = \frac{1}{1+FP}$.

c. Over what range of frequencies do the high-gain approximations $T(s) \approx 1$ and $S(s) \approx 1/FP$ yield magnitudes that are within 3 db of the true values? Over what range are the approximate phases within $10°$ of the true values?

d. Repeat part (c) for the low-gain approximations $T \approx FP$ and $S \approx 1$.

e. Over what range of frequencies is neither approximation satisfactory?

4.10 Given $P(s) = [4(s+2)]/(s^2 + .1s + 1)$, design a 1-DOF feedback system such that $T(s) = 1/(s^2 + 1.4s + 1)$.

4.11 Given $P(s) = 4/(s^2 + 4)$:

a. Show that it is not possible to design a 1-DOF system that is stable and has a proper controller F and S of the form $(s^2 + b_1 s + b_0)/(s^2 + a_1 s + a_0)$.

b. Show that it is possible to satisfy the requirements of part (a) with $S = (s^3 + a_2 s^2 + a_1 s + a_0)/(s+2)^3$.

Calculate a_1, a_2, a_3, and $F(s)$.

4.12 A 1-DOF system is to be designed for a stable plant with a single RHP zero, at $s = a > 0$. The complementary sensitivity $T(s) = [(-\tau_1 s + 1)/(\tau_1 s + 1)] \cdot [1/(\tau_2 s + 1)]$ is close to 1 at low frequencies, and its magnitude frequency response is that of $1/(\tau_2 s + 1)$.

a. Calculate τ_1 so as to ensure stability.

b. Calculate $S(s)$.

c. Plot the Bode magnitude plot of $S(s)$ for $\tau_2 = 1$ and $a = .1$, 1, and 10. Discuss the effect of a on the sensitivity.

M **4.13** *Active suspension*   With reference to Problem 4.8, the suspension system is to be controlled by a 1-DOF feedback system.

a. Show that $y/y_R = P_w(s)S(s) = P_w(s)[1 - T(s)]$.

b. Choose $T(s) = [k(s^2 + 85.861)]/(s^2 + 1.4\omega_0 s + \omega_0^2)$. Choose k such that $T(0) = 1$. Calculate the required compensator $F(s)$.

c. Obtain the magnitude plots of y/y_R for $\omega_0 = 4$, 10, and 50 rad/s.

d. Calculate u/y_R for the values of ω_0 in part (c), and obtain magnitude Bode plots.

M **4.14** For the plant $P(s) = [2(s+1)]/[(s+3)(-s+6)]$, design a stable 1-DOF system with $T(s) = (b_1 s + b_0)/(s+a)^2$, for $a = 0.1$, 1, and 10. As an additional requirement, $T(0) = 1$. For each value of a, plot $|T(j\omega)|$ and $|S(j\omega)|$.

M **4.15** Repeat Problem 4.14 for $P(s) = [2(-s+1)]/[(s+3)(s+6)]$, with $S = (s^2 + b_1 s + b_0)/(s+a)^2$.

4.16 For $P(s) = [2(-s+1)]/[(s+3)(-s+6)]$ and $T(s) = [k(-s+b)]/(s^2 + a_1 s + a_0)$, calculate b and k (possibly as functions of a_0 and a_1) to ensure internal stability. What is the lower bound on the magnitude $|S(0)|$?

4.17 We wish to modify the control of Problem 4.16 to ensure that $T(0) = 1$, in addition to ensuring stability. A T of order 3 is proposed.

  a. Show that this cannot be done if T has either no zeros or one zero.

  b. With $T(s) = [k(s + b_1)(s + b_2)]/(s + a)^3$, determine k_1, b_1, and b_2 as functions of a. Plot $|T(j\omega)|$ for $a = .1$, 1, and 10.

4.18 For stable plants, there are many similarities between a 1-DOF design and an open-loop/feedforward design, as Problems 4.8 and 4.13 readily show. Figure 4.29 shows a design with both open-loop and feedforward control, and with a noise signal v added to account for errors in the measurement of the disturbance.

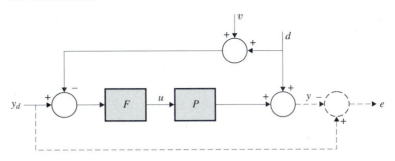

Figure 4.29 Open-loop and feedforward control

  a. Calculate $y(s)$ and $e(s)$ in terms of the three inputs y_d, d, and v. Use $H_d(s) = F(s)P(s)$.

  b. Compare with Equations 4.46 and 4.47 for a 1-DOF feedback system. Show that the two sets of equations are identical if $H_d(s) = T(s)$.

  c. Write $u(s)$ in terms of the three inputs, and show that with $H_d(s) = T(s)$, the expression is identical to Equation 4.48 for the 1-DOF case.

  d. If $P(s)$ has an RHP zero at $s = z_0$, what conditions must be satisfied by $H_d(s)$ and $T(s)$?

  e. Compare the two schemes from the following points of view: response to set-point inputs, response to disturbances, and sensitivity to plant variations.

4.19 There are cases in which sensor dynamics may not be neglected.

  a. In Figure 4.30, calculate the transfer functions y/y_d, y/d, and y/v. Discuss the effect of a low-pass $P_s(s)$ on performance.

  b. It is proposed that the dynamics of $P_s(s)$ be "undone" by cascading the measurement with $P_s^{-1}(s)$, inserted at point x in Figure 4.30. Suppose a stable 1-DOF design has been achieved under the assumption that $P_s(s) = 1$. If $P_s(s)$ is stable, what condition(s) must $P_s(s)$ satisfy for the closed-loop system to remain stable? Calculate the same transfer functions as in part (a), and discuss the effect on setpoint tracking performance.

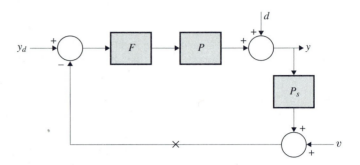

Figure 4.30   System with sensor dynamics

4.20   The open-loop problem is also amenable to treatment by H^∞ theory. Let

$$S'(s) = 1 - F(s)P(s).$$

a. Show that, for internal stability, it is necessary that $S'(z_0)$ equal 1 at the zeros of P in the closed RHP and that S' be stable.

b. For the $P(s)$ of Problem 4.6, parts (a) and (b), calculate that S' minimizes

$$\sup_\omega |S'(j\omega)||W(j\omega)|$$

where $W(s) = (0.1s + 1)/(s + 1)$.

(M)   **4.21**   *Active suspension*   Let us use H^∞ theory to redesign the feedforward system of Problem 4.8. Let the weight function be $W(s) = 0.1(s + 10\omega_0)^2/(s^2 + 1.4\omega_0 s + \omega_0{}^2)$.

a. Choose $F(s)$ to minimize

$$\sup_\omega |W(j\omega)||H_{wd}(j\omega)||1 - F(j\omega)P(j\omega)|$$

for $\omega_0 = 4$ rad/s.

b. For $\omega_0 = 4,\ 10,$ and 50 rad/s, compute the magnitude Bode plot of y/y_R.

4.22   Assume $P(s)$ has no j-axis poles or zeros, and let $B_p(s)$ and $B_z(s)$ be the Blaschke products corresponding, respectively, to the open RHP poles and zeros of P. Show that the optimum sensitivity, in the H^∞ sense, for a weighting function $W(s)$ is the same for $P(s)$ and for the plant $P'(s) = B_z(s)/B_p(s)$.

(M)   **4.23**   As a consequence of Problem 4.22, the effect of RHP poles and zeros on the optimum sensitivity can be studied for the transfer function $B_z(s)/B_p(s)$, rather than for $P(s)$.

Given $W(s) = (.1Ts + 1)/(Ts + 1)$, obtain the minimum sensitivity for a stable plant with a single RHP zero, at $s = 1$. Plot $|S(j\omega)|$ for $Ts = 5, 1,$ and 0.1. What can you conclude about the effect of bandwidth on performance?

(M) **4.24** Repeat Problem 4.23 for $T = 1$ and a stable plant with RHP zeros at $s = 1$ and $s = 1 + a$. Plot $|S(j\omega)|$ for $a = 0.1$, 1, and 2. What can you conclude about the effect on performance of the separation between zeros?

(M) **4.25** Repeat Problem 4.23 for $T = 1$ and a stable plant with RHP zeros at $s = \zeta \pm j\sqrt{1 - \zeta^2}$. Plot $|S(j\omega)|$ for $\zeta = .1$, .5, .9. What effect does the damping factor of the zeros have on performance?

(M) **4.26** Repeat Problem 4.23 for $T = 1$ and a plant with an RHP pole at $s = 1$ and an RHP zero at $s = 1 + a$, for $a = -.5$, .1, .5, and 2. What effect does the separation between RHP poles and zeros have on performance?

4.27 For the plant of Problem 4.14, show that the H^∞ norm of the weighted sensitivity may be as small as we like, for any proper minimum-phase weighting function.

4.28 Calculate the optimal sensitivity for the plant of Problem 4.15, with $W(s) = (0.1Ts + 1)/(Ts + 1)$ and $T = 0.1$, 1, and 10.

4.29 Repeat Problem 4.28 for the plant of Problem 4.16.

4.30 Calculate the optimum sensitivity, in the H^∞ sense, for the plant $P(s) = (s^2 + 1)/[s(s + 1)^2]$ and $W(s) = (s + 1)/s$.

4.31 *High-wire artist* Calculate the optimum sensitivity, in the H^∞ sense, for the plant whose transfer function is ϕ/τ in Problem 3.16 (Chapter 3), and for a weight $W(s) = (s + 12)^2/s^2$.

(M) **4.32** *Two-pendula problem* Calculate the optimum sensitivity, in the H^∞ sense, for the transfer function X/F of Problem 3.21 (Chapter 3) and the weighting function $W(s) = (s + 1)^2/[s^2(s + 5)^2]$. Plot $|S(j\omega)|$. (The pole at $s = -5$ is chosen so as to ensure that $|W(j\omega)|$ is relatively large up to about $\omega = 5$, because the plant has its largest RHP pole at approximately $s = 5$.) (See Section 4.4.4.)

(M) **4.33** For a plant $P(s) = (.5s + 1)/(.1s + 1)^3$, the sensitivity is to be small up to about 20 rad/s because of the expected disturbance spectrum. On the other hand, fairly large step inputs to y_d are expected, and we are willing to give up some speed of response (i.e., bandwidth) in order to keep the control effort at reasonable levels.

We choose

$$S(s) = \frac{.05s}{.05s + 1} \quad \text{and} \quad \frac{y}{y_d} = \frac{1}{Ts + 1}.$$

a. Design a 2-DOF system that yields the required $S(s)$ and y/y_d.

b. Calculate the transfer function u/y_d.

c. Obtain $u(t)$ in response to $y_d = u_{-1}(t)$, for $T = .05$, .5, and 1.

4.34   Given $P(s) = (.5s - 1)/[s(s + 1)]$, design a stable 2-DOF control system such that $S(s)$ is of the form $k[(s + a)/(s + 1)]$, and y/y_d is of the form $(Ts + 1)/(s^2 + 1.4s + 1)$.

4.35   Figure 4.31 shows a 2-DOF structure different from that in Figure 4.21.

    a. Show that the matrix transfer function between the inputs y_d, d, and v and the outputs y and u are the same as for the configuration of Figure 4.27.

    b. The configuration of Figure 4.27 allows cancellation of the RHP zeros of S by R. Show that that is not possible for this configuration if the system is to be internally stable.

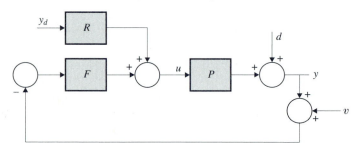

Figure 4.31   An alternate 2-DOF structure

4.36   Figure 4.32 shows yet another 2-DOF structure.

    a. Show that, if R, F, and P are minimally realized, the system is controllable from the inputs y_d, v_1, and v_2, and observable from the outputs y, z_1, and z_2.

    b. Calculate as functions of R, P, P^{-1}, T, and S the elements of the 3×3 matrix transfer function with inputs and outputs as in part (b).

    c. Show that the system is internally stable if, and only if, T, S, PS, $P^{-1}T$, and R are stable.

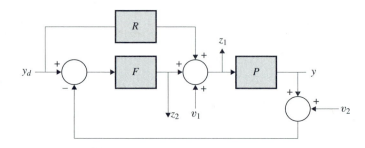

Figure 4.32   An alternate 2-DOF structure

References

[1] Zadeh, L. A., and C. Desoer, *Linear Systems—A State-Space Approach*, McGraw-Hill (1963).

[2] Guillemin, E. A., *Synthesis of Passive Networks*, Wiley (1957).

[3] Freudenberg, J. S., and D. P. Looze, "Right Half Plane Poles and Zeros and Design Trade-offs in Feedback Systems," *IEEE Trans. on Automatic Control*, vol. AC-30, pp. 555–565 (1985).

[4] Francis, B. A., J. W. Helton, and G. Zames, "H^∞-Optimal Feedback Controllers for Linear Multivariable Systems," *IEEE Trans. on Automatic Control*, vol. AC-29, pp. 888–900 (1984).

Feedback System Stability in Terms of the Loop Gain

5.1 INTRODUCTION

In Chapter 4, we explored the factors that limit performance. In so doing, we saw that it is possible to choose the sensitivity S (or the complementary sensitivity T), subject to certain constraints, and derive the compensator F that realizes the chosen S (or T). The compensator generally turns out to be at least as complex as the plant model unless a rather special S or T is selected.

The implementation by computer algorithm of a compensator of relatively high order is not too difficult. There was a time, however, when all controllers were electric or pneumatic analog devices. Those controllers were limited to a few off-the-shelf mathematical structures with at most three adjustable parameters. In the vast majority of cases, an analog controller with a special-purpose mathematical structure was too expensive to contemplate.

As a result of technological limitations, the controller structure was fixed, and there evolved a body of design methods based on the use of the controller $F(s)$ as the design parameter, as opposed to working directly with T or S. This body of knowledge is known as *classical control design*. It is based on a complex plane and frequency-domain analysis, largely because these lent themselves to the use of graphical methods.

In Chapter 4, stability was studied from the viewpoint of the design parameter S or T. Since stability is paramount in any control system, the first order of business in this chapter is to rephrase the stability conditions in terms of the loop gain, which is now the design parameter. That will take the form of conditions on pole-zero cancellations, followed by the Routh, Root Locus, and Nyquist criteria. The latter will be used to study robustness with respect to loop gain variations. The Kharitonov polynomials will be introduced to study stability when system parameters are allowed to vary.

5.2 STABILITY CONDITIONS CONCERNING POLE-ZERO CANCELLATIONS

The first conditions are necessary conditions that forbid certain pole-zero cancellations between F and P. Let $F(s)$ and $P(s)$ be written as ratios of polynomials, i.e., as

$$F(s) = \frac{N_F(s)}{D_F(s)}$$

$$P(s) = \frac{N_P(s)}{D_P(s)}. \tag{5.1}$$

It is assumed that the pairs (N_F, D_F) and (N_P, D_P) are *coprime*, i.e., have no common factors. This is assured, in particular, if $F(s)$ and $P(s)$ are transfer functions of minimal realizations.

The loop gain is

$$L(s) = F(s)\ P(s) = \frac{N_L(s)}{D_L(s)} = \frac{N_F(s)}{D_F(s)}\frac{N_P(s)}{D_P(s)}. \tag{5.2}$$

In the writing of Equation 5.2, it is assumed that common factors between N_F and D_P (or N_P and D_F) are cancelled, so that N_L and D_L are coprime. Thus, it is *not* true, in general, that $N_L = N_F N_P$ and $D_L = D_F D_P$; that holds only in the absence of cancellations.

For example, if F and P are given by

$$F(s) = \frac{-s - 1}{(s + 2)(s + 3)} \qquad P(s) = \frac{(s + 3)}{(s + 1)(s - 1)}$$

then

$$L(s) = F(s)P(s) = \frac{1}{(s + 1)(s + 2)}.$$

Note that $N_L(s) = 1$ and $D_L(s) = (s + 1)(s + 2)$ are indeed coprime, and that $N_L \neq N_F N_P$ and $D_L \neq D_F D_P$.

By Theorem 4.4 (Chapter 4), the necessary and sufficient conditions for internal stability, given minimal realizations of F and P, are that the transfer functions S, PS, and $P^{-1}T$ be stable. We write these as ratios of polynomials.

$$S(s) = \frac{1}{1 + L} = \frac{D_L}{N_L + D_L} \tag{5.3}$$

$$P(s)S(s) = \frac{N_P}{D_P}\frac{D_L}{N_L + D_L} \tag{5.4}$$

and

$$T(s) = \frac{L}{1+L} = \frac{N_L}{N_L + D_L} \tag{5.5}$$

$$P^{-1}(s)T(s) = \frac{D_P}{N_P} \frac{N_L}{N_L + D_L}. \tag{5.6}$$

There can be no pole-zero cancellation between D_L and $N_L + D_L$. To show this, assume that there is, i.e., that s_0 is a root of both D_L and $D_L + N_L$. Then,

$$D_L(s_0) = 0$$

and

$$0 = N_L(s_0) + D_L(s_0) = N_L(s_0).$$

which implies that $N_L(s_0) = 0$. Now, because D_L and N_L are coprime, they may not have a common root s_0; therefore, the existence of a pole-zero cancellation violates the assumption of coprimeness.

Since unstable roots of $N_L + D_L$ cannot be cancelled out by the numerator D_L, it follows that $S(s)$ will be stable if, and only if, the polynomial $D_L(s) + N_L(s)$ has all its roots in the open LHP. Note that this polynomial is the characteristic polynomial of the closed-loop system.

Next, consider PS. Because S is stable, the denominator factor $N_L(s) + D_L(s)$ has only LHP roots. For stability of PS, it follows that the unstable roots of D_P, if any, must be cancelled out by the numerator; i.e., the unstable roots of D_P must be roots of either N_P or D_L. Since N_P and D_P are coprime, they have no common roots. Therefore, the unstable roots of D_P must also be roots of D_L. As Equation 5.2 shows, the roots of D_P *will* be roots of D_L *unless* they are cancelled by N_F. It follows that, if $S(s)$ is stable, a necessary and sufficient condition for stability of PS is that $N_F(s)$ and $D_P(s)$ have no common root in the closed RHP.

The same reasoning applied to $P^{-1}T$ leads to the conclusion that, given $S(s)$ stable, the transfer function $P^{-1}(s)T(s)$ is stable if, and only if, $N_P(s)$ and $D_F(s)$ have no common roots in the closed RHP.

To summarize, the following conditions are necessary and sufficient for internal stability, given minimality of the realizations for F and P:

1. There are no cancellations by $F(s)$ of closed RHP poles or zeros of $P(s)$.

2. The polynomial $D_L(s) + N_L(s)$ has all its roots in the open LHP.

The special case $F(s) = k$, called *pure-gain control*, will be used presently as a vehicle for the study of stability. It satisfies the first condition automatically, of course, since it has neither poles nor zeros. Stability is established by checking the second condition.

5.3 THE ROUTH CRITERION

It was stated in the preceding section that the polynomial $D_L(s) + N_L(s)$ must be stable. This can always be verified by computing the roots and checking that the real parts are all negative. In a design context, where D_L and N_L are also functions of some design parameters, this could become tedious, as roots would have to be computed for several values of the design parameters.

Routh's contribution [1] was a test to ascertain, *without computing roots*, whether or not all roots of a polynomial had negative real parts. Given the difficulty of calculating roots in the precomputer era, Routh's work was an important step forward. The result is presented algorithmically, without proof, as it is difficult to establish without a good deal of background material.

We begin with a polynomial,

$$Q(s) = a_n s^n + a_{n-1} s^{n-1} + \cdots + a_1 s + a_0$$

with $a_n \neq 0$.

We form the *Routh array* as follows. The first two rows, labeled n and $n - 1$, are:

$$
\begin{array}{c|cccc}
n & a_n & a_{n-2} & a_{n-4} & \cdots \\
n - 1 & a_{n-1} & a_{n-3} & a_{n-5} & \cdots
\end{array}
$$

The process continues until we run out of coefficients. If the last entry, a_0, is in row n, a zero is placed below a_0 in row $n - 1$.

Row $n - 2$ is formed as follows:

$$
\begin{array}{c|ccc}
n - 2 & \dfrac{a_{n-1} a_{n-2} - a_n a_{n-3}}{a_{n-1}} & \dfrac{a_{n-1} a_{n-4} - a_n a_{n-5}}{a_{n-1}} & \cdots
\end{array}
$$

The numerators are determinant-like quantities; they are formed in a manner similar to the determinants of 2×2 matrices, but with the order reversed. The process of forming row $n - 2$ continues until we run out of elements.

Subsequent rows are formed in exactly the same way, each time from the previous two rows. This process is carried out until the row label is 0.

The Routh–Hurwitz test is as follows. All roots of the polynomial $Q(s)$ have negative real parts if, and only if, the elements of the leftmost column of the array are nonzero and all have the same sign. Furthermore, the number of sign reversals encountered while scanning the column is the number of RHP roots of $Q(s)$.

Example 5.1   Set up the Routh array for the polynomial $Q(s) = s^4 - s^3 + 3s^2 + 0s + 2$ and calculate the number of RHP roots.

Solution   The first three rows are

$$
\begin{array}{c|ccc}
4 & 1 & 3 & 2 \quad 0 \\
3 & -1 & 0 & 0 \quad 0 \\
2 & \frac{(-1)(3)-(1)(0)}{-1} & \frac{(-1)(2)-(1)(0)}{-1} & 0
\end{array}
$$

The complete array is

$$
\begin{array}{c|cccc}
4 & 1 & 3 & 2 & 0 \\
3 & -1 & 0 & 0 & 0 \\
2 & 3 & 2 & 0 & 0 \\
1 & \frac{2}{3} & 0 & & \\
0 & 2 & & &
\end{array}
$$

Two sign changes occur in the first column, between rows 4 and 3 and between rows 3 and 2. The polynomial has two RHP roots.

There are two special cases of interest.

1. If a row has a zero in the first column and has at least one nonzero element, replace the zero in the first column with ϵ, an infinitesimally small positive quantity, and continue building the array.

2. An entire row of zeros may be encountered, always in an odd-numbered row— say, $k-1$. To take care of this situation, use the preceding row (row k) to form a polynomial in even powers of s, as follows:

$$p(s) = A_1 s^k + A_2 s^{k-2} + \cdots.$$

Here, $A_1, A_2, \ldots$ are the entries of row k. The entries of the new row $(k-1)$ are just the coefficients of the derivative of $p(s)$.

There is a bonus in this last case: the roots of $p(s)$ are also roots of the original polynomial.

Example 5.2   Repeat Example 5.1 for $Q(s) = s^4 - s^2 + 2s + 2$.

Solution   The Routh array begins with the two lines

$$
\begin{array}{c|ccc}
4 & 1 & -1 & 2 \\
3 & 0 & 2 & 0
\end{array}
$$

We replace the zero in the first column with ϵ, and go on:

$$
\begin{array}{c|ccc}
4 & 1 & -1 & 2 \\
3 & \epsilon & 2 & 0 \\
2 & \frac{-\epsilon-2}{\epsilon} & 2 & 0 \\
1 & \left(\frac{-\epsilon}{2+\epsilon}\right)\left(-2\frac{(\epsilon+2)}{\epsilon}-2\epsilon\right) & 0 & \\
0 & 2 & &
\end{array}
$$

For small, positive ϵ, the leading term in row 2 is approximately $-2/\epsilon$ and the leading term in row 1 is approximately 2. There are two sign changes and hence two roots in the RHP.

In control systems design, the Routh–Hurwitz criterion is most often used to ascertain the ranges of design parameters that lead to stability. An example illustrates this.

Example 5.3   **(dc Servo)**

A pure-gain controller $F(s) = k$ is proposed for the dc servo of Example 2.1 (Chapter 2). For what values of k is the closed-loop system stable?

Solution   The plant transfer function is

$$
P(s) = \frac{88.76}{s(s+21.526)(s+2.474)} = \frac{N_P(s)}{D_P(s)}.
$$

Since $F(s) = k$, $L(s) = kP(s)$ and the characteristic polynomial is

$$
D_L(s) + N_L(s) = s(s+21.526)(s+2.474) + 88.76k
$$
$$
= s^3 + 24.00s^2 + 53.255s + 88.76k.
$$

The Routh array is

$$
\begin{array}{c|ccc}
3 & 1 & 53.255 & 0 \\
2 & 24.00 & 88.76k & 0 \\
1 & \frac{1278.12-88.76k}{24.00} & 0 & \\
0 & 88.76k & &
\end{array}
$$

There are no sign changes in the first column if

$$
1278.12 - 88.76k > 0 \quad \text{or} \quad k < 14.400
$$

and

$$
88.76k > 0 \quad \text{or} \quad k > 0.
$$

Therefore, stability is attained if

$$0 < k < 14.400.$$

Note that if $k = 14.400$, the elements of row 1 are all zero. From row 2, with $k = 14.400$, we form the polynomial

$$p(s) = 24.00s^2 + 88.76k$$
$$= 24.00s^2 + 1278.12$$

whose roots are $s = \pm j7.298$. The case $k = 14.400$ is the borderline case, and two roots are imaginary at $\pm j7.298$. For that value of k, row 1 is constructed from dp/ds, and the array becomes

$$
\begin{array}{c|ccc}
3 & 1 & 53.255 & 0 \\
2 & 24.00 & 1278.12 & 0 \\
1 & 48.00 & 0 & 0 \\
0 & 1278.12 & & \\
\end{array}
$$

The positivity of the elements of the first column shows that the other roots are stable.

5.4 THE ROOT LOCUS METHOD

Example 5.3 is typical of a large class of problems in which one seeks to ascertain the effect of variations in a single parameter. The Routh criterion yields only a yes–no answer with respect to stability. It tells us what values of the parameter result in closed-loop stability, but has nothing to say about the behavior or the closed-loop poles in response to changes in the parameter. The Root Locus, originally developed by Evans [2], fills this gap.

As its name suggests, the Root Locus is the geometric locus traced out in the complex plane by the roots of the closed-loop characteristic equation as a specific parameter k varies from 0 to ∞. The Root Locus methods described here are applicable when the characteristic polynomial is of the form $D(s) + kN(s)$, where $D(s)$ and $N(s)$ are coprime polynomials. This is satisfied notably, in the case of the pure-gain controller $F(s) = k$, as in Example 5.3, since the closed-loop characteristic polynomial is then $D_P(s) + kN_P(s)$.

The characteristic equation is

$$D(s) + kN(s) = 0. \tag{5.7}$$

With $G(s) = N(s)/D(s)$, this is also written as

$$1 + kG(s) = 0. \tag{5.8}$$

Note that, in the case of pure-gain control, $N(s) = N_P(s)$, $D(s) = D_P(s)$, and $G(s) = P(s)$. As we shall see, there are cases in which the Root Locus parameter is not the control gain k, and therefore $G(s) \neq P(s)$.

Let $G(s)$ be a proper transfer function of degree n. Then the left-hand side (LHS) of Equation 5.7 is a polynomial of order n which, for a given value of k, has n roots. Those roots are plotted as points in the complex plane. If that is done for each value of k as k varies continuously from 0 to infinity, the result is a locus with n branches, corresponding to the n roots. This is clarified in Example 5.4.

Example 5.4   Let $P(s) = (s + 1)/(s^2 + s + 1)$. Plot the locus of roots of $1 + kP(s)$ as k varies from 0 to ∞.

Solution   The roots satisfy

$$1 + k\frac{s + 1}{s^2 + s + 1} = 0$$

or

$$s^2 + s + 1 + ks + k = 0.$$

This is solved analytically; the roots $s^*(k)$ are

$$s^*(k) = -\frac{1}{2}(k + 1) \pm \frac{1}{2}\sqrt{k^2 - 2k - 3}.$$

The roots are complex when $k^2 - 2k - 3 < 0$, which occurs if $-1 < k < 3$; they are real for other values of k.

For $k = 0$,

$$s^*(0) = -\frac{1}{2} \pm j\frac{\sqrt{3}}{2}$$

which happen (not by coincidence!) to be the roots of the denominator of $P(s)$. For $k = 3$, there is a double root at

$$s^*(3) = -\frac{1}{2}(3 + 1) = -2.$$

To study the behavior for large k, we write

$$s^*(k) = -\frac{1}{2}(k + 1) \pm \frac{1}{2}\sqrt{k^2 - 2k + 1 - 4}$$

$$= -\frac{1}{2}(k + 1) \pm \frac{1}{2}\sqrt{(k - 1)^2 - 4}$$

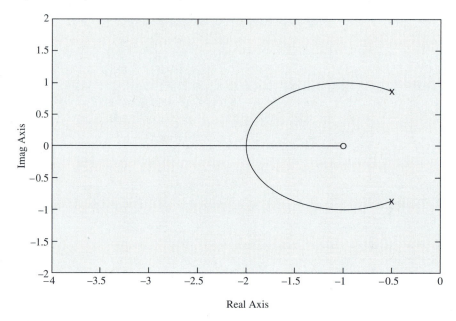

Figure 5.1 Root locus plot

which, for large k, becomes

$$s^*(k) \approx -\frac{1}{2}(k+1) \pm \frac{1}{2}(k-1) = -1, -k$$

so that one root tends to -1, the zero of $P(s)$, while the other tends to $-\infty$. The locus is plotted in Figure 5.1.

To gain some insight into the nature of the Root Locus, write Equation 5.7 as

$$G(s) = -\frac{1}{k}. \tag{5.9}$$

As $G(s)$ is a complex quantity, this leads to

$$\measuredangle\, G(s) = \measuredangle -\frac{1}{k} = 180° \tag{5.10}$$

$$|G(s)| = \frac{1}{k} \tag{5.11}$$

for $k > 0$. Equation 5.10 is the *angle condition*: any point s on the Root Locus is such that $\measuredangle\, G(s) = 180°$. If a point s satisfies the angle condition, then it is a closed-loop pole for $k = 1/|G(s)|$ according to the *magnitude condition*, Equation 5.11.

It is sometimes useful to plot the Root Locus for $k < 0$; in that case, the angle condition is

$$\not\angle G(s) = 0°. \tag{5.12}$$

Good computer packages are now available for the Root Locus. Nevertheless, to develop insight about a specific problem, a control engineer must be able to determine the main features of the Root Locus without recourse to a computer. A number of rules have been developed to help sketch the locus. We shall now describe some of them.

Rule 1   The Root Locus branches start at the poles of $G(s)$ and end at its zeros.

From Equation 5.7, the equation

$$D(s) + kN(s) = 0$$

collapses into $D(s) = 0$ if $k = 0$. Since the roots of $D(s)$ are the poles of $G(s)$, those are the closed-loop poles for $k = 0$. Writing

$$\frac{1}{k}D(s) + \frac{N(s)}{D(s)} = 0$$

we see that, for large k, s is such that $\frac{N(s)}{D(s)} \to 0$. Thus, for large k, the closed-loop poles tend to the roots of $N(s)$—i.e., to the open-loop zeros—and also to infinity if $\frac{N}{D}$ is strictly proper.

Rule 2   For $k > 0$, a point on the real axis belongs to the Root Locus if the total number of real-axis poles and zeros to its right is odd.

To show this, write the angle condition as

$$\not\angle G(s^*) = 180° \pm k360° \tag{5.13}$$

where k is an integer. Let s^* be a point on the real axis.

From Figure 5.2 we observe that: (i) the net contribution of the complex conjugate poles and zeros to $\not\angle G(s^*)$ is zero; (ii) real poles and zeros to the left of s^* contribute $0°$; (iii) each zero to the right of s^* contributes $180°$, and each pole $-180°$. Since $+180°$ and $-180°$ are the same,

$$\not\angle G(s^*) = \text{contributions of complex poles and zeros}$$

$$+ \text{ contributions of real poles and zeros left of } s^*$$

$$+ \text{ contributions of real poles and zeros right of } s^*$$

$$= 180° \text{ [number (N.) of real poles right of } s^*$$

$$+ \text{ N. of real zeros right of } s^*].$$

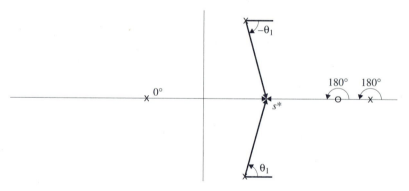

Figure 5.2 Pertaining to the angle of a transfer function at a point on the real axis

To satisfy Equation 5.13, the total of the number of real poles plus the number of real zeros to the right of s^* must be odd.

For $k < 0$, it can be shown that a real-axis point belongs to the Root Locus if the total number of poles and zeros to its right is even (including zero).

Rule 3   As $k \to \infty$, the branches tend to straight-line asymptotes radiating from a common point, or centroid, on the real axis. The location of the centroid is given by

$$\sigma = \frac{\Sigma \text{ poles of } G - \Sigma \text{ zeros of } G}{\text{N. of poles} - \text{N. of zeros}} \tag{5.14}$$

and the asymptote angles are

$$\angle s = \frac{180° \pm k360°}{\text{N. of zeros} - \text{N. of poles}} \tag{5.15}$$

where k is an integer. If the number of poles equals the number of zeros, there are no asymptotes.

For the proof of the result on the centroid, we refer the reader to Shinners [3]. The result on angles is established from Equation 5.7. From Figure 5.3, if s^* is a Root Locus point far from the origin, the angles of the vectors from the poles and zeros to s^* are all approximately equal to $\angle s^*$. Using Equation 5.13,

$$\angle G(s^*) = (\text{N. of zeros} - \text{N. of poles}) \angle s^* = 180° \pm k360°$$

and the result follows.

For $k < 0$, the centroid result also holds, and $180°$ in the numerator of Equation 5.15 is removed.

Other rules exist concerning the points where the locus leaves the real axis (breakaway points) or enters the real axis (breaking points) and the angles at which the branches leave poles or arrive at zeros. With the development of computer packages, those rules have lost some of their former importance. (The reader interested

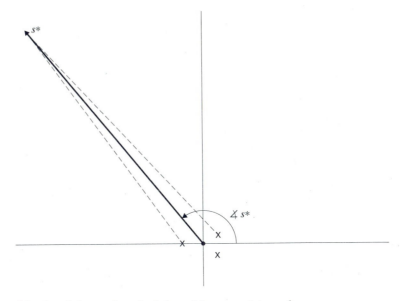

Figure 5.3 Pertaining to the calculation of the asymptote angles

in the details is referred to any of the classical textbooks on control, such as Dorf [4] and Shinners.)

Example 5.5   Use the rules to obtain the Root Locus for $P(s)$ as in Example 5.4.

Solution   Figure 5.1 shows the Root Locus. We begin with a pole-zero plot of $P(s)$. Next, we apply Rule 2 to fill in the real-axis portions. In this case, the real axis part to the left of -1 is on the locus, and the total number of real-axis poles plus zeros to the right of any point on that portion is 1, an odd number.

Rule 3 is used to calculate the asymptotes. The centroid is given by Equation 5.14:

$$\sigma = \frac{(-.5 + j.866 - .5 - j.866) - (-1)}{2 - 1}$$

$$= 0.$$

The asymptotic directions are

$$\measuredangle s = \frac{180° \pm k360°}{1 - 2}$$

$$= -180° \mp k360°$$

$$= -180°.$$

For $k = \pm 1, \pm 2, \ldots$, the angles are equivalent to $180°$, e.g., $-540°$, $180°$. There is therefore but one asymptotic direction, along the negative real axis. In general, the number of asymptotic directions is equal to the number of poles minus the number of zeros.

To draw the Root Locus, we apply Rule 1. There are two branches, starting from the two poles. The branches meet on the real axis; one branch heads for the zero at -1, the other along the asymptote to the "zero" at infinity.

Example 5.6 **(dc Servo)**

Draw the Root Locus for the dc servo of Example 2.1 (Chapter 2). Select the gain so that the complex closed-loop poles are those of a second-order Butterworth filter.

Solution From Equation 3.25, the open-loop transfer function is

$$\frac{\theta}{v} = \frac{88.76}{s(s + 21.576)(s + 2.424)}.$$

Figure 5.4 illustrates the construction of the Root Locus, starting with the pole-zero plot. The rules are applied as follows.
Rule 2: Real-axis portions are -2.424 to 0, and left of -21.576.

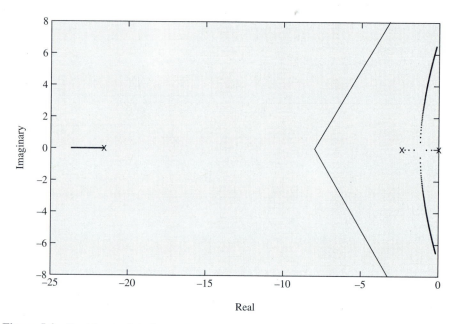

Figure 5.4 Root locus plot, dc servo

Rule 3:

$$\sigma = \frac{(0 - 21.576 - 2.424) - 0}{3 - 0} = -8.000$$

$$\angle s = +\frac{180° \pm k360°}{0 - 3}$$

$$= -60° \mp k120°$$

$$= -60°, +60°, -180°.$$

The asymptotes are drawn in Figure 5.4.

Rule 1: The three branches emanate from the poles and go to the three "zeros" at infinity along the asymptotes. This requires that the branches starting at 0 and -2.424 meet and break away from the real axis, in order to head along the asymptotes at $\pm60°$.

The j-axis crossing points of the Root Locus are obtained from the Routh criterion. The locus crosses the j-axis when the gain takes the marginal stability value. As was shown in Example 5.3, that gain is 14.400 and results in a pair of roots at $\pm j7.298$. Clearly, the points $\pm j7.298$ are closed-loop poles for $k = 14.400$, so they belong to the Root Locus.

Figure 5.4 shows the Root Locus (MATLAB rlocus). Figure 5.5 shows the portion of the Root Locus near the origin. The poles of a Butterworth transfer function of order 2 lie on radial lines at $\pm45°$ from the negative real axis, so the poles lie where these lines intersect the Root Locus, at $s^* = -1.15 \pm j1.15$. The

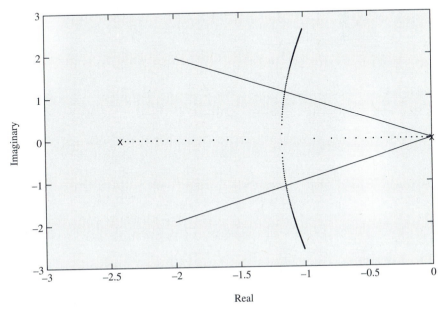

Figure 5.5 Root locus near the origin, dc servo

gain required to place the poles there is calculated from the magnitude condition, Equation 5.11:

$$k = \frac{1}{|P(s)|}$$

$$= \left| \frac{s^*(s^* + 21.576)(s^* + 2.424)}{88.76} \right|$$

$$= 0.657.$$

Most computer packages will also calculate this. (MATLAB rlocfind)

Example 5.7   (Pendulum on a Cart)

Obtain the Root Locus for the cart-and-pendulum system of Example 2.9 (Chapter 2), with x as the output and pure-gain control. Can the system be stabilized with that pure-gain control system?

Solution   The transfer function is given by Equation 3.27 (Chapter 3). It is

$$\frac{x}{F} = \frac{(s + 3.1305)(s - 3.1305)}{s^2(s + 4.4272)(s - 4.4272)}.$$

The Root Locus is displayed in Figure 5.6. Since there is always one unstable pole and a pair of poles on the j-axis for any positive k, the system cannot be stabilized by pure-gain feedback with $k > 0$. (The reader may verify that $k < 0$ also fails to stabilize.)

Example 5.8   (Active Suspension)

To demonstrate the application of the Root Locus method to study the effect of a parameter other than the control gain, consider the suspension system of Example 2.2 (Chapter 2), without the active input force u. Calculate the spring constant K_1 such that a 30-kg increase in the mass M results in a 3-cm change in x_1. For that value of K_1 and for the values of M, m, and K_2 given in the example, plot the Root Locus of the system poles as the damping factor D varies. Find the value of D for which the least damped complex poles have the minimum (most negative) real part.

Solution   The steady-state calculation is performed in Example 2.7, where it is shown that

$$x_1{}^* = x_{10} + x_{20} = \frac{Mg}{K_1} - \frac{m + M}{K_2} g.$$

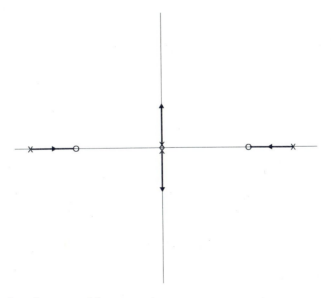

Figure 5.6   Root locus, pendulum on cart

The specifications require that

$$0.03m = \frac{(30 \text{ kg})(9.8 \text{ m/s}^2)}{K_1}$$

or $K_1 = 9800 \text{ N/m}.$

The study of the system natural frequencies requires only the homogeneous system, which is

$$\frac{d}{dt}\begin{bmatrix} x_1 \\ x_2 \\ v_1 \\ v_2 \end{bmatrix} = \begin{bmatrix} 0 & 0 & 1 & 0 \\ 0 & 0 & 0 & 1 \\ \frac{-K_1}{M} & \frac{K_1}{M} & \frac{-D}{M} & \frac{D}{M} \\ \frac{K_1}{m} & \frac{-(K_1+K_2)}{m} & \frac{D}{m} & \frac{-D}{m} \end{bmatrix}\begin{bmatrix} x_1 \\ x_2 \\ v_1 \\ v_2 \end{bmatrix}$$

Substitution of the known values yields

$$\frac{d}{dt}\begin{bmatrix} x_1 \\ x_2 \\ v_1 \\ v_2 \end{bmatrix} = \begin{bmatrix} 0 & 0 & 1 & 0 \\ 0 & 0 & 0 & 1 \\ -32.667 & 32.667 & \frac{-D}{300} & \frac{D}{300} \\ 196 & -796 & \frac{D}{50} & \frac{-D}{50} \end{bmatrix}\begin{bmatrix} x_1 \\ x_2 \\ v_1 \\ v_2 \end{bmatrix}$$

At this point, it is possible to calculate the characteristic polynomial $\det(sI - A)$ in the form $Q(s) + DN(s)$. This is not too onerous in this 4×4 case, but a larger matrix would probably need a symbolic manipulation program. We choose to follow a different route and write the state equations in such a way as to make

D the control gain in a single-input, single-output (SISO) feedback system. note that D appears linearly in the A matrix, which is written as

$$
\begin{bmatrix}
0 & 0 & 1 & 0 \\
0 & 0 & 0 & 1 \\
-32.667 & 32.667 & 0 & 0 \\
196 & -796 & 0 & 0
\end{bmatrix}
+ D
\begin{bmatrix}
0 & 0 & 0 & 0 \\
0 & 0 & 0 & 0 \\
0 & 0 & \frac{-1}{300} & \frac{1}{300} \\
0 & 0 & \frac{1}{50} & \frac{-1}{50}
\end{bmatrix}.
$$

The second matrix, the one multiplying D, is of rank 1. It can therefore be expressed as a column vector times a row vector, i.e., a dyad:

$$
\begin{bmatrix}
0 & 0 & 0 & 0 \\
0 & 0 & 0 & 0 \\
0 & 0 & \frac{-1}{300} & \frac{1}{300} \\
0 & 0 & \frac{1}{50} & \frac{-1}{50}
\end{bmatrix}
=
\begin{bmatrix}
0 \\
0 \\
\frac{1}{300} \\
\frac{-1}{50}
\end{bmatrix}
\begin{bmatrix} 0 & 0 & -1 & +1 \end{bmatrix}.
$$

This is used to write the state equations as

$$
\frac{d}{dt}
\begin{bmatrix}
x_1 \\ x_2 \\ v_1 \\ v_2
\end{bmatrix}
=
\begin{bmatrix}
0 & 0 & 1 & 0 \\
0 & 0 & 0 & 1 \\
-32.667 & 32.667 & 0 & 0 \\
196 & -796 & 0 & 0
\end{bmatrix}
\begin{bmatrix}
x_1 \\ x_2 \\ v_1 \\ v_2
\end{bmatrix}
+
\begin{bmatrix}
0 \\ 0 \\ \frac{1}{300} \\ \frac{-1}{50}
\end{bmatrix}
(-Dy)
$$

$$
y = \begin{bmatrix} 0 & 0 & 1 & -1 \end{bmatrix}
\begin{bmatrix}
x_1 \\ x_2 \\ v_1 \\ v_2
\end{bmatrix}.
$$

Figure 5.7 illustrates this system. The transfer function $G(s)$ is that of a linear system, with the following matrices:

$$
A =
\begin{bmatrix}
0 & 0 & 1 & 0 \\
0 & 0 & 0 & 1 \\
-32.667 & 32.667 & 0 & 0 \\
196 & -796 & 0 & 0
\end{bmatrix};
\quad
B =
\begin{bmatrix}
0 \\ 0 \\ \frac{1}{300} \\ \frac{-1}{50}
\end{bmatrix};
\quad
C = \begin{bmatrix} 0 & 0 & 1 & -1 \end{bmatrix}.
$$

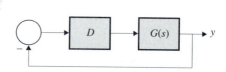

Figure 5.7 Block diagram where the parameter D enters as a gain, active suspension

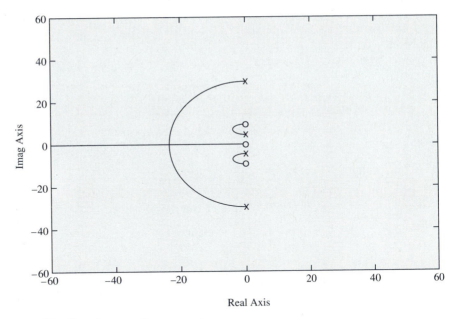

Figure 5.8   Root locus, active suspension

The transfer function is

$$G(s) = \frac{0.02333s(s + j9.2582)(s - j9.2582)}{(s + j4.9365)(s - j4.9365)(s + j28.36)(s - j28.36)}.$$

Figure 5.8 shows the Root Locus (MATLAB rlocus). The two least damped poles have their minimum real parts for $D = 2950$ n/m/s. Those poles are then at $-3.449 \pm j6.7075$.

◆ ◆ ◆ **R E M A R K**

In Example 5.8, we exploited two specific properties of the system A matrix. First, the matrix is linear in the Root Locus parameter, D. Second, the matrix formed by retaining only those terms of A that depend on D is of rank 1. It turns out, in general, that those conditions are necessary and sufficient for the characteristic polynomial to be expressible as $Q(s) + kN(s)$. The Root Locus rules cannot be used if this is not the case. Of course, we can always compute closed-loop poles for different values of k, and plot the result. ◆

5.5 THE NYQUIST CRITERION

5.5.1 Introduction

Early designers of feedback amplifiers, including Bode and Nyquist, found that they could conveniently measure the frequency responses of their devices before closing the loop. They needed to predict the stability properties of their amplifiers after closing the loop. Using the Theory of Complex Variables, Nyquist [5] developed the criterion that bears his name.

The Nyquist criterion, like the Routh criterion, establishes conditions under which the polynomial $N_L(s) + D_L(s)$ has all its roots in the open LHP. The Nyquist criterion does not directly use the polynomials N_L and D_L; rather, it uses the open-loop frequency function $L(j\omega)$. Furthermore, unlike the Routh criterion, the Nyquist criterion yields more than a yes–no answer to the question of stability; it actually leads the way to the very rich lode of frequency-domain techniques.

5.5.2 The Principle of the Argument

The roots of the closed-loop characteristic polynomial satisfy the characteristic equation

$$D_L(s) + N_L(s) = 0.$$

Suppose s_0 is such a root. Then

$$D_L(s_0) + N_L(s_0) = 0. \tag{5.16}$$

We claim that, if N_L and D_L are coprime, s_0 also satisfies

$$1 + \frac{N_L(s_0)}{D_L(s_0)} = 1 + L(s_0) = 0. \tag{5.17}$$

The progression from Equation 5.16 to Equation 5.17 is simple if $D_L(s_0) \neq 0$; we need only divide by $D_L(s_0)$. To understand that $D_L(s_0)$ must be nonzero, note that, if it were zero, $N_L(s_0)$ would also have to be zero to satisfy Equation 5.16, and s_0 would then be a common root of D_L and N_L. That is excluded because D_L and N_L are coprime. Therefore, division by $D_L(s_0)$ is allowed, and Equation 5.17 is justified.

Henceforth in this section, we shall seek roots of $1 + L(s)$ rather than $N_L(s) + D_L(s)$.

Figure 5.9 illustrates the concept of a mapping from one complex plane to another. A closed contour C_s is defined in the s-plane. For every point (e.g., s_1, s_2, s_3) on C_s, a complex number $L(s)$ is calculated [e.g., $L(s_1)$, $L(s_2)$, $L(s_3)$] and plotted in a second complex plane, the L-plane. The mapping of the points on C_s defines a contour C_L in the L-plane. If $L(s)$ is a single-valued function of s, C_L must begin and end at $L(s_1)$; i.e., C_L is also a closed contour.

The contour C_s is traversed in the clockwise direction. As drawn in Figure 5.9, its image C_L is also traversed in the clockwise direction, but that need not be the

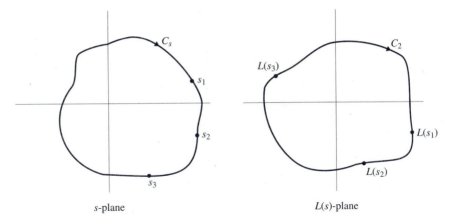

Figure 5.9 Mapping a contour from the s-plane to the $L(s)$-plane

case; it is quite possible for C_L to be traversed in the counterclockwise direction as C_s is traversed clockwise.

Example 5.9 The rectangular contour of Figure 5.10 is to be mapped through the function $L(s) = 1/(s + 1)$. Calculate $L(s)$ for the points s_1, s_2, ..., s_6, and sketch the mapping in the L-plane of the rectangular s-plane contour.

Solution It is not difficult to calculate $L(s)$ by straightforward numerical calculations, but more insight is generated if we use the vector representation of complex numbers. Recall (see, for instance, Bélanger [6]) that $s + a$ is a complex number represented by the vector drawn from point a to point s. The vectors $s_1 + 1$ and $s_2 + 1$ are shown in Figure 5.10.

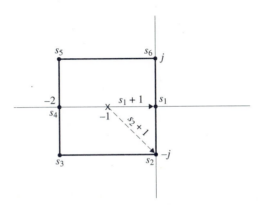

Figure 5.10 Contour in the s-plane

Now, we know that

$$\left| \frac{1}{s+1} \right| = \text{(length of vector } s+1)^{-1}$$

$$\measuredangle \frac{1}{s+1} = -\text{(angle of vector } s+1).$$

By inspection, we have

$$L(s_1) = 1; \qquad L(s_2) = \frac{1}{\sqrt{2}} \measuredangle 45°; \qquad L(s_3) = \frac{1}{\sqrt{2}} \measuredangle 135°$$

$$L(s_4) = 1 \measuredangle 180°; \qquad L(s_5) = \frac{1}{\sqrt{2}} \measuredangle -135°; \qquad L(s_6) = \frac{1}{\sqrt{2}} \measuredangle -45°.$$

Figure 5.11 shows the mapping C_L. Note that the right-angle turns of C_s also appear in C_L; that angle-preservation property is general to a conformal mapping. [A mapping through $L(s)$ is conformal at all points when L is analytic and has a nonzero derivative.]

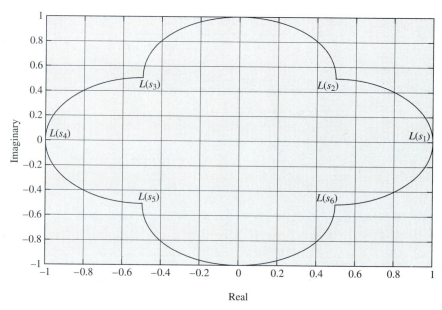

Figure 5.11 Map of s-plane contour in the $L(s)$-plane

Example 5.10   Repeat Example 5.9 for $L(s) = 1/(s + 3)$.

Solution   From Figure 5.12,

$$L(s_1) = \frac{1}{3}; \qquad L(s_2) = \frac{1}{\sqrt{10}} \angle \tan^{-1}\frac{1}{3}; \qquad L(s_3) = \frac{1}{\sqrt{2}} \angle -45°$$

$$L(s_4) = 1; \qquad L(s_5) = \frac{1}{\sqrt{2}} \angle -45°; \qquad L(s_6) = \frac{1}{\sqrt{10}} \angle -\tan^{-1}\frac{1}{3}.$$

Figure 5.13 shows the contour C_L.

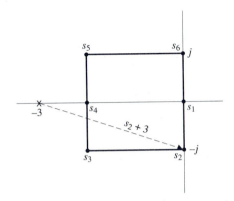

Figure 5.12   Contour in the s-plane

The contour C_L in Figure 5.9 encircles the origin once, clockwise; that is defined as one *positive encirclement*. In contrast, the contour C_L in Figure 5.13 encircles the origin once, counterclockwise; that is tallied as one *negative encirclement*.

In general, the *principle of the argument* relates the (algebraic) number of encirclements of the origin by C_L to the number of poles and zeros of $L(s)$ within the contour C_s. It is stated as follows:

Let C_s be a closed contour in the s-plane, and let $L(s)$ be a rational function of s, analytic and nonzero on the contour C_s and with finite numbers P of poles and Z of zeros within C_s. Also, let N be the number of positive encirclements of the origin of the $L(s)$-plane by the contour C_L. Then

$$N = Z - P. \tag{5.18}$$

In Example 5.9, $L(s)$ has one pole and no zeros within C_s, so that $Z = 0$ and $P = 1$. Therefore, $N = -1$; the origin is encircled once in the negative (counterclockwise) direction in the L-plane. In Example 5.10, $Z = P = 0$, so $N = 0$; the origin is not encircled.

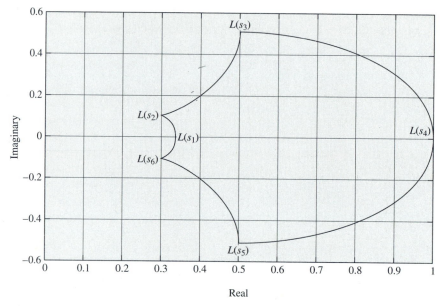

Figure 5.13 Map of s-plane contour in the $L(s)$-plane

A heuristic explanation of the principle of the argument is gleaned from Examples 5.9 and 5.10. As s goes around C_s, the angle of the complex number $1/(s+1)$ in Example 5.9 goes through all positive values from 0 to 360°, so C_L describes a counterclockwise path around the origin. On the other hand, as s goes around C_s, the angle of $1/(s+3)$ in Example 5.10 increases from zero, reaches a 45° maximum, goes back through zero to −45°, and finally ends at zero. The angle never "flips around," because the pole at −3 is outside the contour.

If, in Example 5.9, $L(s)$ is changed to $s + 1$, the diagram of Figure 5.10 does not change. In this case, however,

$$\not{\angle} (s+1) = \text{ angle of vector } (s+1)$$

so that, for example, $\not{\angle} L(s_2) = -45°$ (instead of +45°). The result is that C_L is traversed clockwise instead of counterclockwise; i.e., $N = 1$.

We see that every pole of $L(s)$ inside C_s generates one negative encirclement, while every zero inside C_s produces one positive encirclement. The poles and zeros of $L(s)$ that lie outside C_s produce no encirclements. The formula $N = Z - P$ follows.

5.5.3 The Nyquist Contour

The Nyquist criterion is an application of the principle of the argument to the following problem: find the number of RHP zeros of $1 + L(s)$.

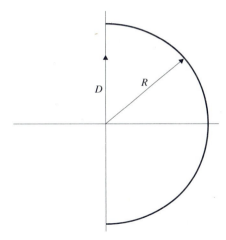

Figure 5.14   The Nyquist contour

Figure 5.14 shows the Nyquist contour D. The semicircle has a radius R tending to infinity, so D encompasses the whole RHP. If Z is the (unknown) number of RHP zeros of $1 + L(s)$, then

$$Z = N + P \tag{5.19}$$

according to Equation 5.18.

Now, P is the number of RHP poles of $1 + L(s)$. The poles of $1 + L(s)$ are also the poles of $L(s)$; the values of s that make $L(s)$ "blow up" are also the ones that make $1 + L(s)$ "blow up." Therefore,

$$P = \text{ number of poles of } L(s) \text{ in the open RHP.}$$

It is assumed that $L(s)$ has no poles or zeros on the j-axis, as the principle of the argument does not allow poles or zeros on D.

To apply Equation 5.19, we need to know N. The Nyquist contour D is mapped through $1 + L(s)$ in a second complex plane. Figure 5.15 shows the mapping C_L through $L(s)$ and $C_{L'}$ through $1 + L(s)$: the second is just the first shifted to the right by 1. The number N is the number of encirclements of the origin by $C_{L'}$. From Figure 5.15, it is easy to see that N is also the number of encirclements of the point $(-1, 0)$ by C_L.

To apply the Nyquist criterion, we map the Nyquist contour through $L(s)$ and count encirclements of the point $(-1, 0)$ by C_L to obtain N. It is assumed that P, the number of RHP poles of $L(s)$, is known, so Equation 5.19 can be applied. For stability, $Z = N + P = 0$, so N, the net (algebraic) number of encirclements, must be $-P$. In particular, if $L(s)$ is stable ($P = 0$), N must be zero for closed-loop stability.

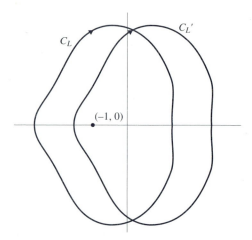

Figure 5.15   Maps through the mappings $L(s)$ and $1 + L(s)$

5.5.4 Construction of the Nyquist Locus

The map of a point $s = j\omega$ is simply $L(j\omega)$. Therefore, the map of the imaginary axis is simply the frequency response plotted in the complex plane with ω as a parameter. A point $L(j\omega)$ is plotted as a point with x-component Re $L(j\omega)$ and y-component Im $L(j\omega)$.

It is also possible, and more helpful, to view the locus as a *polar* plot, with $|L(j\omega)|$ as the magnitude and $\not< L(j\omega)$ as the angle. If the phase angle of $L(j\omega)$ increases as ω increases, the Nyquist locus is moving counterclockwise (in the positive-angle direction); if the phase angle is decreasing, the Nyquist locus is moving clockwise. As ω increases, the locus moves away from the origin if the magnitude is increasing with frequency, and toward the origin if it is decreasing.

Since $L(-j\omega) = L^*(j\omega)$, a negative-frequency point has the same real part (x-coordinate) as $L(j\omega)$, but an imaginary part (y-coordinate) of opposite sign. The map of the negative part of the j-axis is obtained by simply reflecting the map for positive ω about the real axis.

Before proceeding to an example, we introduce one helpful modification. It is often the case that $L(s) = kL'(s)$, where k is a gain parameter to be determined in the design process. In such cases, the equations

$$1 + kL'(s) = 0$$

and

$$\frac{1}{k} + L'(s) = 0 \tag{5.20}$$

are equivalent.

Equation 5.20 suggests that encirclements of the point $(-\frac{1}{k}, 0)$ by the locus $L'(j\omega)$ be considered, rather than encirclements of $(-1, 0)$ by $L(j\omega) = kL'(j\omega)$. That is often easier to do.

Example 5.11   Plot the Nyquist locus for $L(s) = k/(s+1)^3$. For what values of k is the closed-loop system stable?

Solution   We have

$$L'(j\omega) = \frac{1}{(j\omega + 1)^3}.$$

The behavior of $L'(j\omega)$ at $\omega = 0$ and $\omega \to \infty$ is easily established:

$$L'(j0) = 1$$

$$\lim_{\omega \to \infty} |L'(j\omega)| = 0$$

$$\lim_{\omega \to \infty} \angle L'(j\omega) = -270°.$$

It will soon become apparent that the real-axis crossing points are key quantities. Clearly, $L(j\omega)$ is real if

$$\operatorname{Im} L'(j\omega) = 0.$$

Here,

$$L'(j\omega) = \frac{1}{-j\omega^3 - 3\omega^2 + 3j\omega + 1}$$

$$= \frac{(1 - 3\omega^2) + j(\omega^3 - 3\omega)}{(1 - 3\omega^2)^2 + (\omega^3 - 3\omega)^2}.$$

We see that $\operatorname{Im} L'(j\omega) = 0$ if

$$\omega^3 - 3\omega = 0$$

or

$$\omega = 0 \pm \sqrt{3}.$$

At $\omega = 0$, $L'(j\omega) = 1$, obviously a real-axis point. At $\omega = \sqrt{3}$,

$$L'(j\sqrt{3}) = \frac{(1 - 9)}{(1 - 9)^3} = -\frac{1}{8}.$$

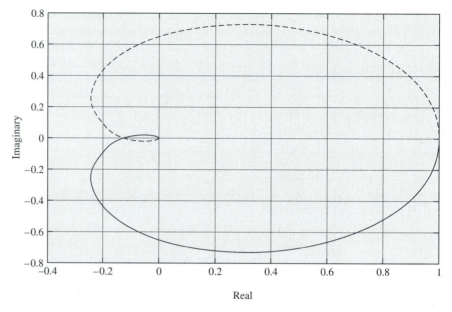

Figure 5.16 The Nyquist plot

The locus is plotted in Figure 5.16 (MATLAB nyquist). Figure 5.17 (MATLAB bode) shows the Bode plots for $L'(j\omega)$.

At $\omega = 0$, $L'(j\omega) = 1$ (0 db and $0°$ of phase). As ω increases, the phase becomes more and more negative, so the locus moves clockwise; the magnitude decreases, so the locus moves toward the origin. The negative real axis is crossed at $-\frac{1}{8}$, which occurs for $\omega = \sqrt{3}$. The locus tends to the origin as $\omega \to 0$, and approaches it at an angle of $-270°$.

To assess stability, we must count encirclements of the point $(-\frac{1}{k}, 0)$ by the locus $L'(j\omega)$. For stability, it is necessary that $N = 0$, since $P = 0$ [$L(s)$ has no RHP poles]. The point $(-\frac{1}{k}, 0)$ will not be encircled if either (i) $(-\frac{1}{k}, 0)$ lies to the *left* of $(-\frac{1}{8}, 0)$ or (ii) $(-\frac{1}{k}, 0)$ lies to the *right* of $(1, 0)$. Therefore, stability follows if either

$$-\frac{1}{k} < -\frac{1}{8} \quad \text{or} \quad k < 8, \quad k > 0$$

or

$$-\frac{1}{k} > 1 \quad \text{or} \quad k > -1, \quad k < 0$$

or, more concisely,

$$-1 < k < 8.$$

If $(-\frac{1}{k}, 0)$ lies between $(-\frac{1}{8}, 0)$ and the origin, $N = 2$ and the closed-loop system has two RHP poles. This occurs, then, for $k > 8$.

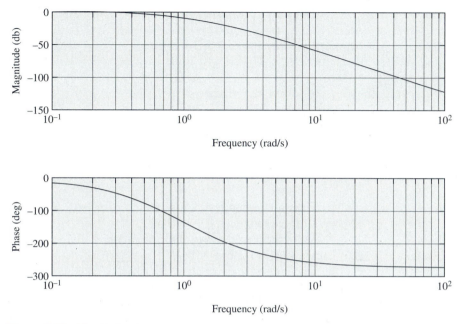

Figure 5.17   The Bode plots

If $(-\frac{1}{k}, 0)$ lies between the origin and $(1, 0)$, $N = 1$ and there is one RHP pole, for $k < -1$.

The reader is invited to verify these conclusions by applying Routh's criterion to this example.

5.5.5 The Nyquist Contour for the Case of j-Axis Poles and Zeros

There are practical instances where $L(s)$ has poles or zeros on the imaginary axis, with a pole at the origin being particularly frequent. Some modification must be made, because the principle of the argument does not allow poles or zeros on the s-plane contour.

The D-contour is modified so as to skirt around the j-axis poles and zeros. As shown in Figure 5.18, the contour avoids these poles with detours along semicircles of small radii. The mapping of the D-contour in the L-plane is the frequency response, interrupted by the mappings of the small semicircles.

Semicircles of infinitesimal radii centered on poles map into circular arcs of large radii in the L-plane. Consider a pole of multiplicity m, at $s = jy$, and write

$$L(s) = \frac{1}{(s - jy)^m} L_1(s)$$

where $L_1(s)$ has no pole or zero at $s = jy$.

Along the semicircle of radius ρ about $s = jy$, we have $s = jy + \rho e^{j\theta}$ and

$$L(s) = \frac{1}{\rho^m e^{jm\theta}} L_1(jy + \rho e^{j\theta}).$$

For very small ρ, $|L_1(jy + \rho e^{j\theta})| \approx |L_1(jy)|$ and

$$|L(s)| \approx \frac{|L_1(jy)|}{\rho^m} \tag{5.21}$$

which is a constant tending to infinity as $\rho \to 0$. Thus, the mapping is a circular arc of large radius.

As for the phase, we see that θ in Figure 5.18 *increases* from $-90°$ to $+90°$. This causes $e^{-jm\theta}$ to *decrease* by $180°m$ degrees, causing the semicircle to be traversed in the counterclockwise direction. (The indentation can also be made to the left of the pole, in which case the direction is clockwise.) It is helpful to remember that the mapping is conformal, so that a $90°$ right turn in the D contour maps into a $90°$ right turn in the L-plane.

To summarize, the infinitesimal semicircle around a pole maps into a circular arc of large radius. The arc runs counterclockwise and goes through an angle of

$$\pi m + \not\angle L_1(jy+) - \not\angle L_1(jy-). \tag{5.22}$$

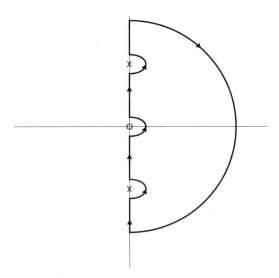

Figure 5.18   Indented Nyquist contour for the case of imaginary axis poles and zeros

As Example 5.13 will also show, the difference between the angles at $jy+$ and $jy-$ can be crucial in the determination of N.

In the neighborhood of a j-axis zero, we write

$$L(s) = (s - jy)^m L_1(s)$$

$$= \rho^m e^{jm\theta} L_1(jy + \rho e^{jm\theta}).$$

We see that a semicircle of infinitesimal radius maps into an L-plane semicircle of infinitesimal radius, which, in the limit, collapses into the point jy. The angle of L undergoes a change of $m\pi$. If m is odd, $m\pi = \pi$, and the L-contour crosses the origin; if m is even, $m\pi = 0$, so the L-contour reaches and leaves the origin at the same angle, i.e., does not go through the origin.

Example 5.12   Sketch the Nyquist plot for $L(s) = k/[s(s+1)^2]$, and determine the range of k for stability.

Solution   Figure 5.19 shows the D-contour. The Nyquist plot of $L'(s) = 1/[s(s+1)^2]$ is sketched in Figure 5.20. This plot is not to scale due to the difficulty of showing both the large semicircle and the detail near the origin.

From Figure 5.19, we see that there are no open-loop poles inside the D-contour, so $P = 0$ and we need $N = 0$ for stability. From Figure 5.20, this will be the case if $0 < k < 2$.

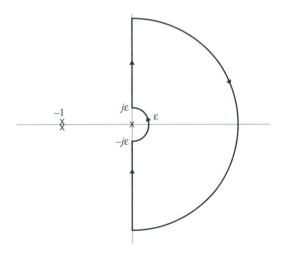

Figure 5.19   Indented Nyquist contour

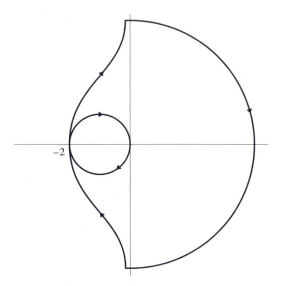

Figure 5.20 The Nyquist plot

Example 5.13 Sketch the Nyquist plot for $L(s) = k\frac{(s+1)}{s^2}$, and determine the range of k for stability.

Solution Figure 5.21 shows the D-contour and the Nyquist plot of $L'(s) = (s+1)/s^2$. The phase at $\omega = \rho$ is $-180° + \tan^{-1} \rho$, so the point $s = j\rho$ maps into 0+, slightly below

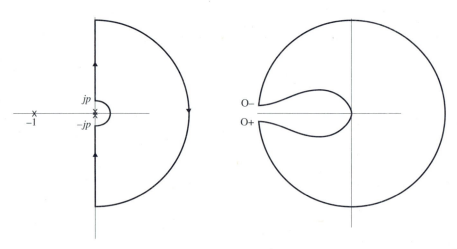

Figure 5.21 Nyquist contour and Nyquist plot, system with a double pole at $s = 0$

the negative real axis. As ω increases, the phase increases (because of the zero) and the magnitude decreases, with the given result. To map the small semicircle, we use the fact that, at $s = -j\rho$, the D-contour takes a 90° right turn and, because of the double pole, goes through nearly (but not quite) 360°.

There are no open-loop poles within the S-contour, so $N = 0$ for stability. This will occur if the $(-1/k,\ 0)$ point is located anywhere on the negative real axis—i.e., if any $k > 0$.

Example 5.14

Sketch the Nyquist plot for $L(s) = k[(s^2 + 1)/(s - 1)^3]$, and determine the range of k for stability.

Solution

Figure 5.22 shows the pole-zero plot and the D-contour. For $0 < \omega < 1$, the phase of $L'(s) = (s^2 + 1)/(s - 1)^3$ is

$$-90° + 90° - 3(180° - \theta) = -180° + 3\theta.$$

As ω approaches 1 from below, the magnitude approaches zero and the phase $-45°$ ($\theta \to 45°$). The Nyquist locus crosses the origin (odd multiplicity); the magnitude grows, then tends to zero with a phase of 90° as $\omega \to \infty$ (Fig. 5.22).

The real-axis crossing at -1 is for $\omega = 0$. The other crossing clearly occurs for $\omega > 1$, where the phase is equal to 3θ. Therefore, the crossing occurs for $\theta = 60°$,

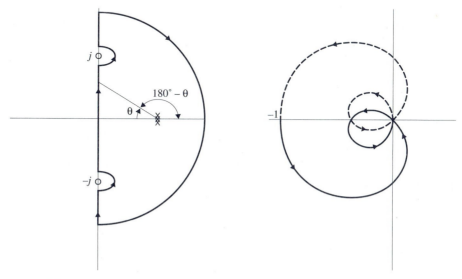

Figure 5.22 Nyquist contour and Nyquist plot, system with zeros on the imaginary axis

or $\omega = \sqrt{3}$. At that frequency,

$$|L'(j\sqrt{3})| = \frac{|-3+1|}{(3+1)^{3/2}} = \frac{1}{4}.$$

Since $P = 3$, we require $N = -3$ for stability. That will take place if $\frac{1}{4} < -\frac{1}{k} < -0$, or $4 < k < \infty$.

◆ ◆ ◆ **R E M A R K**

Useful links can be made among the Routh criterion, the Root Locus, and the Nyquist criterion. For a j-axis crossing at $s = j\omega_0$ of the Root Locus of $L(s)$, the angle condition is

$$\not{\angle}\, L(j\omega_0) = 180°.$$

The angle condition also implies that the Nyquist plot crosses the negative real-axis at $\omega = \omega_0$. Positive real-axis crossings of the Nyquist plot correspond to j-axis crossings of the Root Locus for $k < 0$.

When the Root Locus crosses the j-axis, it does so at a value of k that creates one whole row of zeros in the Routh array. Solving for the roots of the auxiliary polynomial, as shown in Section 5.3, yields the imaginary roots of the characteristic polynomial for that particular value of k—i.e., the j-axis crossing points of the Root Locus, which are also the real-axis crossing frequencies of the Nyquist plot.

◆

5.5.6 Counting Encirclements

It is sometimes tricky to count encirclements, especially if only the positive-frequency half of the Nyquist plot is given, as is the case with most software packages. The following procedure is proposed. We count the net number of real-axis crossings to the left of the point $(-\frac{1}{k}, 0)$, with clockwise (decreasing phase) crossings being positive. The number of encirclements must equal the net number of crossings, because the Nyquist locus is a closed contour. (The procedure presented here is given more ample justification in Vidyasagar et al. [7].)

Figure 5.23 shows a positive-frequency Nyquist plot, with parts of the negative-frequency portion indicated with dashed lines. We see that (i) a real-axis crossing of the positive-frequency portion for ω_0, $0 < \omega_0 < \infty$, actually corresponds to two crossings of the complete locus, and (ii) the real-axis points at $\omega = 0$ and $\omega = \infty$ contribute one crossing of the complete locus. This leads to a simple rule: to obtain N, count algebraically (positive for clockwise) the real-axis crossings left of $(-\frac{1}{k}, 0)$—once if $\omega = 0$ or ∞ or twice if $0 < \omega < \infty$. In Figure 5.23, $N = 0$ for $(-\frac{1}{k}, 0)$ in region A, -1 for $(-\frac{1}{k}, 0)$ in region B, and $1 (= 2 - 1)$ for $(-\frac{1}{k}, 0)$ in region C.

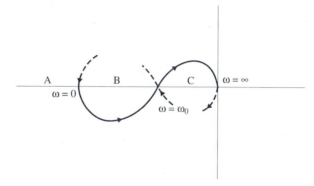

Figure 5.23 Counting encirclements from the positive frequency portion of the Nyquist plot

Quite often, the frequency response is presented as Bode plots rather than Nyquist plots. Since only the real-axis crossings are considered, it is relatively simple to obtain the necessary information from Bode plots. Bode plots offer the advantage of logarithmic scales, avoiding the scaling problems that are frequently present in Nyquist plots.

Real-axis crossings of the Nyquist plot occur at frequencies where (i) the phase is $180°$ (negative real-axis crossing), (ii) the phase is $0°$ (positive real-axis crossing), or (iii) the magnitude is zero (crossing at the origin). We can proceed from the Bode plots to a sketch of real-axis crossings. From the sketch, the Nyquist criterion is applied. Let us restrict our consideration to proper rational transfer functions, examining three types of real-axis crossings: (i) at nonzero but finite distances from the origin, (ii) at the origin, and (iii) at arbitrarily large distances from the origin (this corresponds to indented contours when there are j-axis poles).

The first set of crossings occurs at frequencies where the phase is $0°$ or $180°$; those are easily determined from the Bode phase plot. The distance of a Nyquist plot crossing from the origin is given by the magnitude Bode plot at the particular frequency. The direction of the real-axis crossing (downward or upward) is ascertained from the phase curve, by observing whether the phase crosses $0°$ or $180°$ from lower-half-plane values to upper-half-plane values, or vice versa.

Crossings at finite, nonzero distance from the origin can also occur at $\omega = 0$ and $\omega = \infty$. The crossing direction is obtained from the behavior of the phase as ω moves away from 0 or approaches ∞.

The Nyquist plot may also cross the real axis at the origin. This can take place at $\omega_0, 0 < \omega_0 < \infty$, if the transfer function has a zero at $s = j\omega_0$. The phase angle flips by $180°$ times the multiplicity of the zero as ω crosses through ω_0. If the multiplicity is even, the Nyquist locus changes by $360°$; it goes into the origin and back out along the same path, so there is no net crossing. If the multiplicity is odd, the locus angle changes by $180°$, i.e., crosses the origin. Therefore, a zero at $s = j\omega_0, 0 < \omega_0 < \infty$, contributes a crossing if its multiplicity is odd, and no crossing if it is even. The real-axis crossing, if there is one, will be downward if the phase at $\omega = \omega_{0-}$ corresponds to the top half plane; otherwise it will be upward.

Crossings at the origin can also occur for $\omega = 0$ if the transfer function has a zero at $s = 0$, and for $\omega = \infty$ if the transfer function is strictly proper. In each case, there is one crossing, regardless of multiplicity. The direction can be determined from the behavior of the phase as ω tends to 0 or ∞.

Finally, the Nyquist plot crosses the real axis arbitrarily far from the origin, as the s-plane contour follows indentations around j-axis poles. If the indentations are made to the right of the poles, the Nyquist locus follows a circular arc of large radius, clockwise because of the decreasing phase. A simple pole at $s = j\omega_0$, $\omega_0 \neq 0$, will generate one crossing—downward on the positive real axis if the phase at ω_{0-} corresponds to the top half plane, or upward on the negative real axis if the phase at ω_{0-} corresponds to a point in the lower half plane. Poles with higher multiplicity will generate several crossings, but always downward on the positive real axis and upward on the negative real axis.

The reasoning is the same for a pole at $s = 0$, even though the Bode plot does not show the phase at $\omega = 0_-$; that is of little consequence, since the phase at $\omega = 0_-$ is just the negative of the phase at $\omega = 0_+$, which is available. It is necessary to be careful in the case of poles of even multiplicity m at $s = 0$. The phase tends to $0°$ or $180°$ as ω tends to zero. The phase $\angle L[j\omega(0_+)]$ is close to the asymptotic value, and $\angle L[j\omega(0_-)] = -\angle L[j\omega(0_+)]$ is also close to the asymptotic value. The circular arc of large radius begins at $L[j\omega(0_-)]$, runs clockwise, and ends at $L[j\omega(0_+)]$, so the total angular change is slightly different from $180m$. Depending on whether $L[j\omega(0_-)]$ is above or below the real axis, there could be either m or $m - 1$ crossings of the real axis.

Example 5.15

Obtain the Bode plot for $L(s) = \frac{k}{(s-1)(s+2)(s+5)}$ with $k = 1$, and find the range of k for stability.

Solution

Figure 5.24 shows the Bode plots for $L'(j\omega)$ (MATLAB bode).[1] Since $P = 1$, we need $N = -1$ for stability. There are two crossings of the negative real axis (phase $= 180°$), at $\omega = 0$ and $\omega = 1.81$ rad/s, and one crossing of the origin, at $\omega = \infty$. The locations and directions are as follows:

- For $\omega = 0$: at -20 db, downward (increasing phase)

- For $\omega = 1.81$ rad/s: at -29.45 db, upward (decreasing phase)

- For $\omega = \infty$: at the origin, downward (the Nyquist lower approaches the origin from $-270°$, in the top half plane)

---

[1]The magnitude plot of $(-j\omega\tau + 1)$ is the same as that of $(j\omega\tau + 1)$; the phase plot of $(-j\omega\tau + 1)$ is the negative of that of $(j\omega\tau + 1)$.

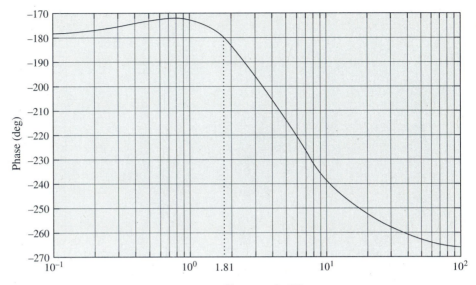

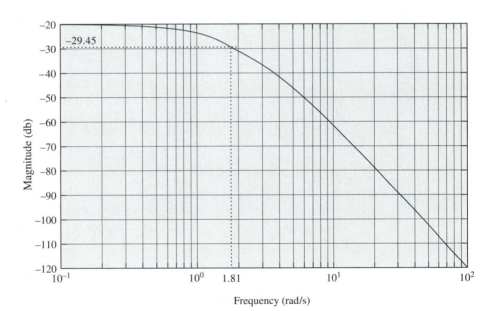

Figure 5.24 Bode plots

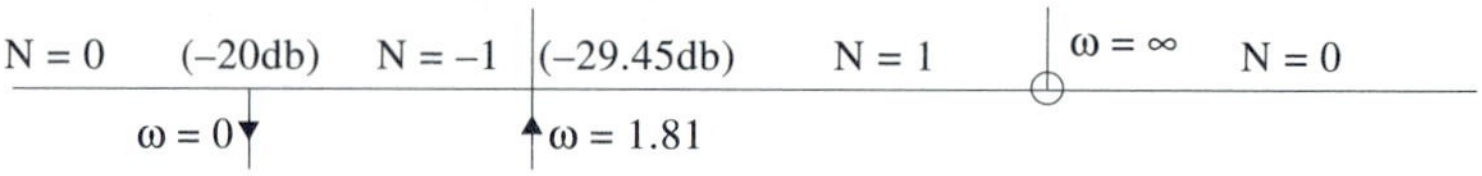

| N = 0 | (−20db) | N = −1 | (−29.45db) | N = 1 | ω = ∞ | N = 0 |

ω = 0   ω = 1.81

Figure 5.25 Real-axis crossings of the Nyquist plot

Figure 5.25 shows the real-axis crossings of the Nyquist plot, with the values of N corresponding to each interval. The stability interval, $N = -1$, is achieved for $k > 0$ and

$$-29.45 \text{ db} < \frac{1}{|k|} < -20 \text{ db}$$

or

$$20 \text{ db} < k < 29.45 \text{ db}.$$

Example 5.16   Use Bode plots to study the stability of

$$L(s) = \frac{k(s^2 + 4)}{(s + 1)^2(s^2 + 9)(s + 2)}.$$

Solution   Figure 5.26 shows the Bode plots (MATLAB bode). Since $P = 0$, we need $N = 0$ for stability. The real-axis crossings are as follows:

- For $\omega = 0$: on the positive real axis (phase $= 0°$) at a distance of -13.2 db from the origin, downward
- For $\omega = 2$: corresponding to the j-axis zero, at the origin, upward (because the phase at $\omega = 2_-$ is $-171°$, corresponding to the lower half plane)
- For $\omega = 2.22$: on the positive real axis (phase $= -360°$) at a distance of -37.7 db from the origin, downward
- For $\omega = 3$: corresponding to the j-axis pole, far from the origin on the negative real axis, upward (because the circular arc begins at a point with a phase between $-360°$ and $-400°$, in the lower half plane)
- for $\omega = \infty$; at the origin, downward (because the approach is at $-630°$, equivalent to $-270°$)

Figure 5.27 shows the real-axis crossings of the Nyquist plot. For $N = 0$, the point $-1/k$ must be located to the right of last axis crossing; k must be negative and $1/|k|$ must be greater than -13.2 db, or 0.219. In short,

$$-4.56 < k < 0.$$

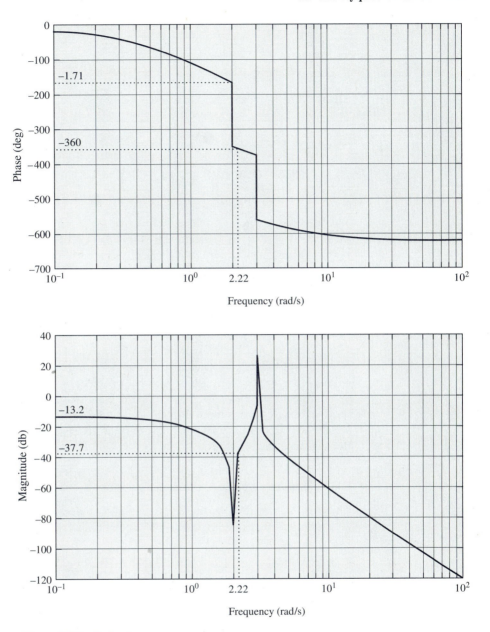

Figure 5.26   Bode plots

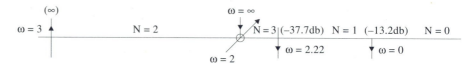

Figure 5.27   Real-axis crossings of the Nyquist plot

PERFORMANCE LIMITATIONS DUE TO PLANT UNCERTAINTY

5.6.1 Introduction

In Chapter 4, Section 4.5, three factors that limit performance were identified: the presence of RHP zeros, sensor noise, and actuator limitations. There is one more important factor: plant uncertainty. Typically, a control system is designed for a plant modeled by some transfer function or state description. We know, of course, that this model is only an approximation. It is usually possible to quantify and bound the discrepancy between the plant and the model. For example, a particular parameter p_i may be thought to lie between two limits, with a specific midrange value used in the model. (Very often, these bounds are really educated guesses. Engineering, like any profession, has its share of art.) The design model then becomes a member of a *class* of models, one member of which is believed to represent the plant.

Since we do not know which member of the class represents the plant, the safe and reasonable course is to require that the control system stabilize *all* members of the class. Since that is more difficult to do than simply stabilizing a single model, we can expect to give up something from the single-model performance to satisfy the additional requirement.

5.6.2 Stability Condition Under Uncertainty

The plant model used in the design is called the *nominal* plant. It is a member of a family of possible models for the actual plant, given the uncertainty. Let this family be denoted by a set, $\mathcal{P}$.

A control system is said to possess *robust stability* with respect to $\mathcal{P}$ if it is stable for every $P(s) \in \mathcal{P}$. Robustness implies that the control system will be stable for all models that could be expected to represent the plant.

The following assumptions are made about $\mathcal{P}$.

(i) It includes the nominal plant model $P_0(s)$.

(ii) All $P(s) \in \mathcal{P}$ have the same number of closed RHP poles (this assumption is mildly restrictive).

(iii) For any ω, the locus of the complex numbers $P(j\omega)$, $P \in \mathcal{P}$, is a connected surface.

The solid curve in Figure 5.28 is the Nyquist plot of the loop gain $F(j\omega)P_0(j\omega)$, i.e., the nominal loop gain. Consider a specific frequency, ω_0. The set $\mathcal{P}$ defines

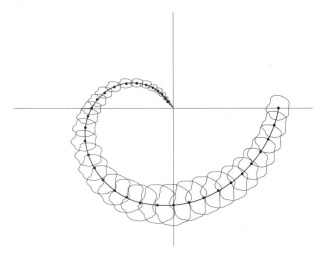

Figure 5.28   Set of Nyquist plots for a family of plants

a whole set of possible complex values of $P(j\omega_0)$, one of which is $P_0(j\omega_0)$. To each of these values there corresponds a point $F(j\omega_0)P(j\omega_0)$. As $P(j\omega_0)$ takes on all the values defined by $\mathcal{P}$, a locus of points $F(j\omega_0)P(j\omega_0)$ is generated. This locus describes a surface or curve in the complex plane, with the nominal loop gain included. Each point on the nominal Nyquist plot is replaced by a surface; the result is that the plot is surrounded by a sheath containing Nyquist plots for every member of $\mathcal{P}$.

If the closed-loop system is stable in the nominal case $P(s) = P_0(s)$, then, by the first two preceding assumptions, stability for all $P \in \mathcal{P}$ requires that the number of encirclements of the Nyquist plot of $F(s)P(s)$ be the same for all $P \in \mathcal{P}$. That will be the case if, and only if, every possible Nyquist plot "stays on the same side of" the $(-1, 0)$ point as the plot for the nominal plant. This, in turn, will be true if the Nyquist "sheath" avoids the $(-1, 0)$ point altogether. This is expressed mathematically as

$$|1 + F(j\omega)P(j\omega)| > 0, \qquad \text{all } P \in \mathcal{P} \text{ and all } \omega. \qquad (5.23)$$

5.6.3 Parametric Uncertainties: The Kharitonov Polynomials

As was pointed out in Chapter 2, Section 2.7, some of the uncertainty is *parametric*, i.e., is expressed in terms of a set of parameters. All examples in Chapter 2 contained physical parameters, so the plant models were, in fact, functions of those parameters.

A relatively new tool for the study of systems with parametric uncertainties is Kharitonov's theorem [8]. Kharitonov's work has sparked a new line of inquiry in this general area (see, for example, Barmish [9]). The theorem addresses the stability of a polynomial

$$p(s, \ \mathbf{a}) = a_0 + a_1 s + \cdots + a_m s^n$$

whose coefficients can vary over given ranges, or

$$a_i^- \leq a_i \leq a_i^+.$$

This is known as an *interval polynomial family*. It is of invariant degree if a_n is always nonzero.

We define the form *Kharitonov polynomials*:

$$k_1(s) = a_0^- + a_1^- s + a_2^+ s^2 + a_3^+ s^3 + a_4^- s^4 + a_5^- s^5 + a_6^+ s^6 + \cdots$$
$$k_2(s) = a_0^+ + a_1^+ s + a_2^- s^2 + a_3^- s^3 + a_4^+ s^4 + a_5^+ s^5 + a_6^- s^6 + \cdots$$
$$k_3(s) = a_0^+ + a_1^- s + a_2^- s^2 + a_3^+ s^3 + a_4^+ s^4 + a_5^- s^5 + a_6^- s^6 + \cdots$$
$$k_4(s) = a_0^- + a_1^+ s + a_2^+ s^2 + a_3^- s^3 + a_4^- s^4 + a_5^+ s^5 + a_6^+ s^6 + \cdots \quad \text{(5.24)}$$

Now we can state the theorem.

■ **Theorem 5.1**   **Kharitonov's Theorem** An interval polynomial family $\mathcal{P}$ with invariant degree is robustly stable if, and only if, its four Kharitonov polynomials are stable.   ■

The reader is referred to Barmish for a proof. The proof is not difficult but is too long to be included here.

■ **Example 5.17**   Assess the robust stability of the third-order interval polynomial family with $1 \leq a_0 \leq 2$, $0.5 \leq a_1 \leq 1$, $2 \leq a_2 \leq 3$, and $a_3 = 1$.

Solution   The Kharitonov polynomials are

$$k_1(s) = 1 + .5s + 3s^2 + s^3$$
$$k_2(s) = 2 + s + 2s^2 + s^3$$
$$k_3(s) = 2 + .5s + 2s^2 + s^3$$
$$k_4(s) = 1 + s + 3s^2 + s^3$$

By the Routh criterion, k_2 and k_4 are unstable, so the family is not robustly stable. In fact, k_2 and k_4 are two unstable members of $\mathcal{P}$.

In general, the coefficients of the characteristic polynomial are not the model parameters, but functions of those parameters. We may apply Kharitonov's theorem in such cases, but the result is only sufficient. The Kharitonov test may be used, but with the caveat that the results may be conservative.

Example 5.18 **(dc Servo)**

In the dc servo model of Example 2.1 (Chapter 2), some parameters may vary, as follows: $0.04 \leq K_m \leq 0.06, 6 \times 10^{-4} \leq J_m \leq 10^{-3}, 0.01 \leq J \leq 0.05$.

1. Write the Kharitonov polynomial, and give the range of k for which all four are stable.
2. Calculate the true range of k for stability, under pure-gain feedback, of this family of plants.

Solution The transfer function is

$$P(s) = \frac{\theta}{v} = \frac{NK_m/J_e L}{s^3 + (R/L)s^2 + \left(\frac{N^2 K_m^2}{J_e L}\right)s}.$$

With a control gain k, the characteristic polynomial is

$$s^3 + \frac{R}{L}s^2 + \frac{N^2 K_m^2}{J_e L}s + k\frac{NK_m}{J_e L}$$

and

$$\frac{R}{L} = 24, \qquad 23.75 \leq \frac{N^2 K_m^2}{J_e L} \leq 107.5, \qquad 49.48 \leq \frac{NK_m}{J_e L} \leq 149.4.$$

The Kharitonov polynomials are

$$k_1 = 49.48k + 23.75s + 24s^2 + s^3$$
$$k_2 = 149.4k + 107.5s + 24s^2 + s^3$$
$$k_3 = 149.4k + 23.75s + 24s^2 + s^3$$
$$k_4 = 49.48k + 107.5s + 24s^2 + s^3.$$

Let

$$k_i = A_i k + B_i s + 24s^2 + s^3.$$

Applying the Routh criterion yields $0 < k < (24B_i)/A_i$ for stability. The upper limit is 11.52 for k_1, 17.26 for k_2, 3.815 for k_3, and 52.14 for k_4. The result is therefore $0 < k < 3.815$ as a sufficient condition for stability.

This result is conservative; k_3 is obtained with a_0^+ (maximum K_m, minimum J_m and J) and a_1^- (minimum K_m, maximum J_m and J), which, of course, is impossible. Thus, k_3 is not a member of the set $\mathcal{P}$.

To calculate the true range of k for stability, we observe that the Nyquist plot for the open-loop plant has a real-axis crossing at $\omega = \omega_0$ such that

$$Im\left[-j\omega_0^3 - 24\omega_0^2 + j\frac{N^2 K_m^2}{J_e L}\omega_0\right] = 0$$

or

$$\omega_0 = \frac{NK_m}{\sqrt{J_e L}}$$

and

$$P(j\omega_0) = \frac{\frac{NK_m}{J_e L}}{\frac{-24N^2 K_m^2}{J_e L}} = -\frac{1}{24NK_m}.$$

For stability, the number of encirclements must be zero, so

$$-\frac{1}{k} < -\frac{1}{24NK_m}$$

or

$$0 < k < 24NK_m.$$

To satisfy this under all conditions requires $k < 11.52$. (Note that this result can also be obtained by applying the Routh criterion directly to the characteristic polynomial.)

5.6.4 Unstructured Uncertainties

A parametric model is based upon consideration of some physical phenomena thought to represent the salient features of the plant. In practice, a model used for control system design does not include all physical effects, for two reasons: (i) the design model is a simplified version of a more complex simulation model, and (ii) secondary physical effects are excluded because of complexity and engineering study costs. Such effects are incorporated in the *unstructured uncertainty*. We consider the model

$$P(s) = [1 + \Delta(s)]P_0(s) \tag{5.25}$$

where $P_0(s)$ is the nominal model and $\Delta(s)$ is a transfer function with a known magnitude bound; i.e.,

$$|\Delta(j\omega)| \le \ell(j\omega) \tag{5.26}$$

where $\ell(j\omega)$ is a real, nonnegative number. It is assumed that no phase information on Δ is available, so $\not< \Delta(j\omega)$ may have any value between $0°$ and $360°$.

Figure 5.29 illustrates the multiplicative term $1 + \Delta(j\omega)$, which is a complex number represented by a vector drawn from the origin to some point on or within the circle. For $\ell(j\omega) < 1$,

$$1 - \ell(j\omega) \le |1 + \Delta(j\omega)| \le 1 + \ell(j\omega) \tag{5.27}$$

and

$$-\arcsin \ell(j\omega) \le \not< [1 + \Delta(j\omega)] \le \arcsin \ell(j\omega). \tag{5.28}$$

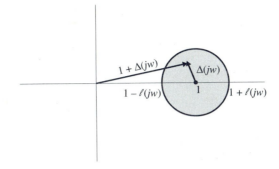

Figure 5.29 Illustration of transfer function uncertainty

For $\ell(j\omega) > 1$,

$$0 \le |1 + \Delta(j\omega)| \le 1 + \ell(j\omega) \tag{5.29}$$

and

$$0° \le \not\angle [1 + \Delta(j\omega)] \le 360°. \tag{5.30}$$

Note that $|1 + \Delta(j\omega)|$, expressed in decibels, is added to the log magnitude of $P_0(j\omega)$, and that the phase of $[1 + \Delta(j\omega)]$ is added to the phase of $P_0(j\omega)$. In particular, Equation 5.30 shows that, if $\ell(j\omega) > 1$, the phase of $P(j\omega)$ may have any value between $0°$ and $360°$, i.e., is completely uncertain.

It is often easier to include parametric uncertainties in an unstructured uncertainty because the latter is expressible by one real function, $\ell(j\omega)$. Solving Equation 5.25 yields

$$\Delta(j\omega) = \frac{P(j\omega)}{P_0(j\omega)} - 1 \tag{5.31}$$

and $\ell(j\omega)$, the upper bound on $|\Delta(j\omega)|$, is

$$\ell(j\omega) = \max_{\mathcal{P}} \left| \frac{P(j\omega)}{P_0(j\omega)} - 1 \right|. \tag{5.32}$$

Example 5.19 **(dc Servo)**

For the dc servo of Example 5.18, calculate $\ell(j\omega)$ with the given parameter variations.

Solution The computation is relatively straightforward. First, $P(j\omega)$ is calculated for 25 values of the parameter pair (K_m, J_e). Then $P_0(j\omega)$ is calculated from the nominal

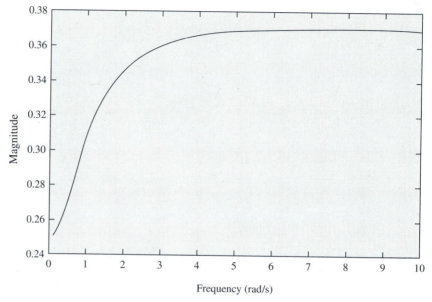

Figure 5.30   Upper bound on the uncertainty

plant, with $K_m = 0.05$, $J = 0.02$, and $J_m = 8 \times 15^{-4}$. Finally, $\Delta(j\omega)$ and $\ell(j\omega)$ are computed using Equations 5.31 and 5.32. Figure 5.30 shows $\ell(j\omega)$.

In general, $\ell(j\omega)$ is small at low frequencies and larger at high frequencies. This reflects the fact that low-frequency effects are usually easier to model. The actual $\ell(j\omega)$ used in a design is usually the product of an educated guess; uncertainty is, quite naturally, uncertain.

5.6.5 Performance Limitations

The effect of unstructured uncertainties is summarized in the following result.

■ **Theorem 5.2**   Assume that (i) the closed-loop system is stable in the nominal case $P(s) = P_0(s)$, and (ii) the unstructured uncertainty is such that all plants in the family $\mathcal{P}$ generated by the unstructured uncertainty $\Delta(s)$, $|\Delta(j\omega)| \leq \ell(j\omega)$, have the same number of RHP poles. Then the system possesses robust stability with respect to $\mathcal{P}$ (i.e., is stable for all members of $\mathcal{P}$) if, and only if,

$$|T_0(j\omega)| < \frac{1}{\ell(j\omega)} \tag{5.33}$$

where T_0 is the complementary sensitivity with $P = P_0$.

Proof: This is easily proved with reference to Figure 5.31. The loop gain is

$$L(j\omega) = [1 + \Delta(j\omega)]L_0(j\omega)$$
$$= L_0(j\omega) + \Delta(j\omega)L_0(j\omega).$$

Since $\Delta(j\omega)$ has arbitrary phase and $|\Delta(j\omega)| \le \ell(j\omega)$, the locus of $L(j\omega)$ is a circle of radius $|L_0(j\omega)|\ell(j\omega)$, centered at $L_0(j\omega)$.

From Figure 5.31, $L(j\omega)$ will not include the $(-1, 0)$ point if, and only if,

$$|L_0(j\omega)|\ell(j\omega) < |1 + L_0(j\omega)|$$

or if

$$\frac{|L_0(j\omega)|}{|1 + L_0(j\omega)|} = |T_0(j\omega)| < \frac{1}{\ell(j\omega)}. \quad \blacksquare$$

If $\ell(j\omega) < 1$, the uncertainty constraint is not serious, since $T_0(j\omega)$ can be made arbitrarily close to its ideal value of 1. It is clear that $|T_0(j\omega)|$ must be less than 1 if $\ell(j\omega) > 1$, which means that the control bandwidth cannot extend into a region where the model uncertainty is relatively high. Thus, the achievable performance is seen to be restricted by uncertainty. For the dc servo of Example 5.14, $\ell(j\omega) < 1$ for all ω, so the constraint expressed by Equation 5.34 is not severe.

A practical consequence of this theorem is that the design model must be relatively accurate over the desired system bandwidth. High bandwidth control requires not only quality sensors and actuators, but considerable modeling effort as well.

Theorem 5.2 may be employed to establish the validity of certain model simplifications that are used to make the design easier. The following example shows how this is done.

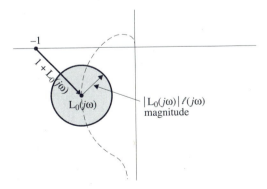

Figure 5.31 Illustration of certain quantities in the complex plane

Example 5.20

In most practical situations, the actuator dynamics are much faster than those of the plant, and are simply ignored in the design of the control system. Consider the system of Figure 5.32, where the transfer functions are known exactly. The second-order underdamped transfer function is quite common for actuators. For example, gear trains connected to motors may have some elasticity; hydraulic actuators have oscillatory high-frequency modes due to fluid compressibility, which results in springlike behavior.

Suppose the actuator is modeled by the gain k. Calculate $\ell(j\omega)$ and, hence, the design bandwidth that can be achieved with this simplification.

Solution

$$[1 + \Delta(s)]kP(s) = \frac{k}{\left(s/\omega_0\right)^2 + 2\zeta\left(s/\omega_0\right) + 1}P(s)$$

so that

$$\Delta(s) = \frac{-(s/\omega_0)^2 - 2\zeta(s/\omega_0)}{(s/\omega_0)^2 + 2\zeta(s/\omega_0) + 1}.$$

Now,

$$\ell^2(j\omega) = |\Delta(j\omega)|^2 = \frac{(\omega/\omega_0)^4 + 4\zeta^2(\omega/\omega_0)^2}{[1 - (\omega/\omega_0)^2]^2 + 4\zeta^2(\omega/\omega_0)^2}.$$

Clearly, $\ell(j\omega) = 1$ when

$$(\omega/\omega_0)^4 = [1 - (\omega/\omega_0)^2]^2$$
$$= 1 - 2(\omega/\omega_0)^2 + (\omega/\omega_0)^4$$

or

$$\omega = \frac{1}{\sqrt{2}}\omega_0.$$

The reader is invited to show that $\ell(j\omega) > 1$ for $\omega > \frac{1}{\sqrt{2}}\omega_0$.

Therefore, if the model simplification is used, we cannot guarantee stability if the bandwidth much exceeds $\omega_0/\sqrt{2}$. (This condition is conservative, of course, because Theorem 5.2 "throws away" phase information.)

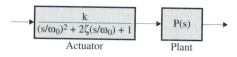

Figure 5.32 Actuator with dynamics and plant

Problems

5.1 Let $P(s) = (s - 1)/[(s + 1)(s + 2)]$ and $F(s) = k/(s - 1)$.

  a. Obtain minimal realizations for P and F.

  b. Combine the realizations obtained in part (a) to derive a third-order realization for the 1-DOF system.

  c. Show that the 1-DOF closed-loop system has an unobservable mode at $s = 1$, independent of k.

5.2 Repeat Problem 5.1 for $P(s) = 1/[(s - 1)(s + 1)]$ and $F(s) = k[(s - 1)/(s + 2)]$, replacing the word "unobservable" in part (c) with the word "uncontrollable."

5.3 Given $F(s)P(s) = [k(s - 2)]/(s^2 + 1.6s + .8)$:

  a. Find the closed-loop characteristic polynomial, and use the Routh array to calculate the range of k for which its roots are stable.

  b. With k at its upper limit for stability, calculate the two imaginary roots by solving for the roots of the auxiliary polynomial.

5.4 Repeat Problem 5.3 for $F(s)P(s) = k/[(s + 1)^2(s + .5)]$.

5.5 For $F(s)P(s) = k[(s + 1)/(s^2 + 1)(s + 4)]$, show that the closed-loop characteristic polynomial is stable for all $k > 0$.

5.6 Show that, if the loop gain is $F(s)P(s) = k/[(s + 1)^2(s - .5)]$, the closed-loop system is unstable for all values of k.

5.7 Repeat Problem 5.6 for $F(s)P(s) = [k(s - 1)]/[(s + 2)(s - 2)]$.

5.8 In Problem 4.28 (Chapter 4), an H^∞ design was carried out for the plant $P(s) = [2(-s + 1)]/[(s + 3)(s + 6)]$, with the weight function $W(s) = (s + 1)/(10s + 1)$. The compensator $F(s)$ was not proper, and we wish to design a new one, $F'(s) = [F(s)]/(Ts + 1)^2$, that attenuates high frequencies. Over what range of values of T is the closed-loop system stable?

5.9 In Problem 4.29 (Chapter 4), an H^∞ design was carried out for $P(s) = [2(-s + 1)]/[(s + 3)(-s + 6)]$, with the weight function $(0.1s + 1)/(s + 1)$. Repeat the redesign exercise of Problem 5.8.

5.10 Sketch (by hand; do not compute) the Root Locus for the system of Problem 5.3, with $k > 0$.

5.11 Repeat Problem 5.10 for the system of Problem 5.4.

5.12 Sketch the Root Locus for the system of Problem 5.5. Corroborate the truth of the assertion made there.

5.13 Repeat Problem 5.12 for the system of Problem 5.6.

5.14 Repeat Problem 5.12 for the system of Problem 5.7.

5.15 *dc servo, simplified model* For the dc servo of Problem 2.4 (Chapter 2) (with $L = 0$), obtain the Root Locus for pure-gain feedback. Calculate the feedback gain k for a damping factor $\zeta = 0.707$.

(M) **5.16** *Servo with flexible shaft* For the dc servo of Problem 2.5 (Chapter 2) (or, equivalently, Problem 3.14 in Chapter 3), obtain the Root Locus for pure-gain feedback with θ_2 as the output. Compute the value of k for which the system becomes unstable. For that value of k, what is the damping factor of the complex poles far from the origin?

(M) **5.17** *Heat exchanger* For the linearized model of Problem 2.14 (Chapter 2) (or, equivalently, Problem 3.18 in Chapter 3), obtain the Root Locus for pure-gain feedback, with ΔF_H as input and ΔT_{C3} as output. Calculate k such that all complex poles have a damping factor $\zeta \leq 0.707$. For what value of k does the system become unstable?

(M) **5.18** *Chemical reactor* For the chemical reactor transfer function $\Delta T/\Delta Q$ of Problem 3.19 (Chapter 3), obtain the Root Locus for pure-gain feedback. What is the range of gain k values that will stabilize the system?

5.19 *High-wire artist* For the linearized model of Problem 2.18 (Chapter 2) (or Problem 3.16 in Chapter 3), with τ as input and ϕ as output, show that stabilization by pure-gain feedback is not possible.

5.20 *Crane* For the linearized model of Problem 2.17 (Chapter 2) (or Problem 3.20 in Chapter 3), with F as input and x as output, show that stabilization by pure-gain feedback is not possible.

5.21 *Two-pendula problem* Show that the linearized model of Problem 2.19 (Chapter 2) (or Problem 3.21 in Chapter 3), with F as input and x as output, cannot be stabilized by pure-gain feedback.

(M) **5.22** *Servo with flexible shaft* The servo of Problem 2.5 (Chapter 2) is controlled by pure-gain feedback, with θ_2 as the output and $k = 1.0$ as the gain. We wish to study the movement of closed-loop poles in response to changes in the torsion constant K.

    a. With $v = v_{sp} - k\theta_2$, write the state equations describing the closed-loop system, with v_{sp} as the new input.

    b. Following Example 5.8, write the closed-loop system A matrix as a constant matrix plus another matrix, linear in K.

    c. Follow Example 5.8 and write the system as a new pure-gain system, where K is the gain (set v_{sp} to zero; it does not affect the poles).

    d. Obtain the Root Locus with K as the parameter.

(M) **5.23** Compute the Nyquist plot ($\omega \geq 0$) for the loop gain of Problem 5.3, with $k = 1$, and determine the range(s) of k for which the closed-loop system is stable.

(M) **5.24** Repeat Problem 5.23 for the loop gain of Problem 5.4.

M   **5.25**   Repeat Problem 5.23 for the loop gain of Problem 5.5.

       5.26   Use the Nyquist method to prove the result of Problem 5.6.

       5.27   Repeat Problem 5.26 for the loop gain of Problem 5.7.

M   **5.28**   Plot magnitude and phase Bode plots for the loop gain of Problem 5.3, with $k = 1$. From the Bode plots, identify the real-axis crossings of the Nyquist plot, and give the range of k for which the closed-loop system is stable.

M   **5.29**   Repeat Problem 5.28 for the loop gain of Problem 5.4.

       5.30   Figure 5.33 shows Bode plots for the loop gain of a system. Identify the real-axis crossings of the Nyquist plot. What is the range of stabilizing gains if the loop gain has no poles in the open RHP? What is the range if the loop gain has one RHP pole?

       5.31   Repeat Problem 5.30 for the Bode plots of Figure 5.34.

M   **5.32**   *High-wire artist*   In Problem 4.31 (Chapter 4), an H^∞ optimal control was designed, leading (as usual) to a nonproper controller. We wish to redesign the controller, taking into account actuator limitations and model uncertainty. We propose to achieve this by replacing the controller $F(s)$ with $[F(s)]/(\tau s + 1)^n$, where n is chosen to ensure that the new controller will have one more pole than zeros.

         a. Compute the range of positive values of τ for which the closed-loop system is stable.

         b. Choose a stabilizing value of τ.

         c. Using the linearized model, compute the responses $\theta(t)$ and $\phi(t)$ for $\theta(0) = 1$, $\phi(0) = -1$, and $\omega_\theta(0) = \omega_\phi(0) = 0$.

         d. Using the nonlinear model, repeat part (c) for $\theta(0) = a$, $\phi(0) = -a$, and $\omega_\theta(0) = \omega_\phi(0) = 0$. Do this for several values of a, to explore the range of validity of the control.

---

   * Hint   Since the system is stable with $\tau = 0$, and since the change introduces no RHP poles, the number of encirclements of the $(-1, 0)$ point must be the same as with $\tau = 0$. This implies that the modified loop gain may not be equal to -1, for any values of ω and τ.

M   **5.33**   *Two-pendula problem*   Repeat Problem 5.32 for the H^∞ design of Problem 4.32 (Chapter 4). For the simulations, use the initial conditions $\theta_1(0) = a$, $\theta_2(0) = -a$, and all other state variables equal to zero.

       5.34   Assess the robust stability of the interval polynomial family based on the nominal polynomial $p(s) = s^3 + 3s^2 + 2s + 3$, where all coefficients except that of s^3 may vary independently by $\pm\epsilon$ percent. For what range of values of ϵ is robust stability assured?

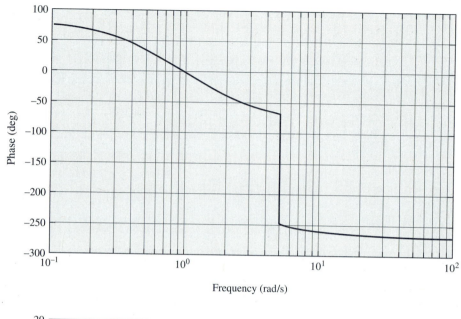

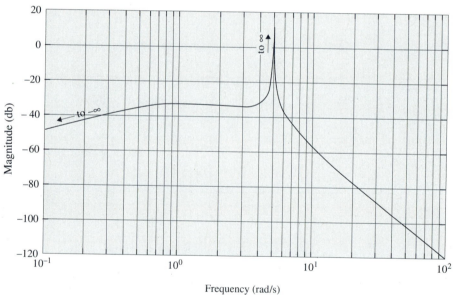

Figure 5.33 Bode plots

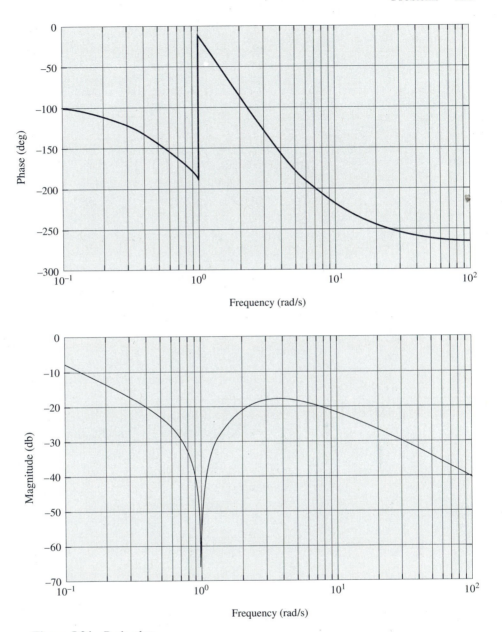

Figure 5.34   Bode plots

5.35   Repeat Problem 5.34 for the nominal polynomial $p(s) = s^4 + 3.5s^3 + 2s^2 + s + 0.5$. (Here, the coefficient of s^4 is the one that is fixed.)

(M)   **5.36**   ***Active suspension***   The suspension system of Example 2.2 (Chapter 2) is subject to parametric variations. The mass M may range between 200 and 500 kg, and the spring constant K_2 may take on values between 1×10^4 and 5×10^4 N/m.

a. Compute the ranges of values of the coefficients of the values of the transfer function y/u, $y = x_1 - x_2$. (Note: Since the extremal values of the coefficients do not necessarily occur at the extremal values of the parameters, you will need to calculate the coefficients for different values of M and K_2.)

b. For a gain feedback law $u = -k(x_1 - x_2)$, calculate the ranges of values for the coefficients of the closed-loop characteristic polynomial.

c. Calculate the range of k for which the Kharitonov polynomials are stable.

d. Using a Nyquist or Routh argument, calculate the true range of k for closed-loop stability.

(M)   **5.37**   ***Servo with flexible shaft***   The torsional spring constant K in Problem 2.5 (Chapter 2) is assumed to vary between 3 and 7 N/m.

a. Compute the ranges of values of the coefficients of θ_2/v. (See note, Problem 5.36.)

b. With θ_2 as the output and pure-gain feedback k, calculate the ranges of the values for the coefficients of the closed-loop characteristic polynomial in terms of k.

c. Obtain the Kharitonov polynomials, and compute the true range of k for which they are all stable.

d. Using a Routh or Nyquist argument, compute the true range of k for which the closed-loop stability of the family is assured.

(M)   **5.38**   In some problems, the gain is the unknown parameter.

a. Let $P(s) = kP_0(s)$, $k_{min} < k < k_{max}$. Sketch on the complex plane a particular frequency point $P_0(j\omega_0)$ and the locus of points $kP_0(j\omega_0)$ generated by k varying between its limits.

b. For a loop gain $L(s) = k[(s + 3)/s(s + 1)^2]$, determine the range of k for closed-loop stability. Let k_{min} and k_{max} be the extreme values.

c. Compute the Nyquist plots of $L(j\omega)$ for k_{min} and k_{max}, and plot them on the same plane.

(M)   **5.39**   In certain problems, the magnitude is known but the phase is not. For example, let $P(s) = e^{-s\tau}P_0(s)$, where τ, $0 \leq \tau \leq \tau_{max}$, is a time delay. Since $|e^{-j\omega\tau}| = 1$ and $\angle e^{-j\omega\tau} = -\omega\tau$, the phase of $P(j\omega)$ is uncertain.

a. Sketch, in the complex phase, a particular frequency point $P_0(j\omega_0)$ and the locus of points $e^{-j\omega_0\tau}P_0(j\omega_0)$, generated as τ varies between 0 and τ_{max}.

b. The all-pass factor $(-s\tau + 1)/(s\tau + 1)$ also has unit magnitude for $s = j\omega$ and, like the exponential, produces phase lag. For a loop gain $L(s) = [(-\tau s + 1)/(\tau s + 1)][(s+3)/s(s+1)^2]$, use the Routh criterion to determine the range of τ for closed-loop stability. Let τ_{max} be the upper limit.

c. Compute the Nyquist plots of $L(j\omega)$ for $\tau = 0$ and $\tau = \tau_{max}$. Plot.

M   **5.40**   In some situations, it is convenient to model an RHP zero as part of the uncertainty. Let

$$P(s) = \frac{-\tau s + 1}{\tau s + 1} P_0(s), \qquad 0 \le \tau \le \tau_{max}$$

and let $P_0(s)$ be the nominal plant.

a. We wish to include the Blaschke term in an unstructured uncertainty. Let $P(s) = [1 + \Delta(s)]P_0(s)$, $|\Delta(j\omega)| \le \ell(j\omega)$. Determine the minimum value of $\ell(j\omega)$ so that the family of plots generated by the unstructured uncertainty includes $P(s)$, for $0 \le \tau \le \tau_{max}$. To do this, calculate $\Delta(j\omega)$, then $|\Delta(j\omega)|$. For a given ω, calculate $\ell(j\omega)$, the maximum value taken by $|\Delta(j\omega)|$ as τ varies from 0 to τ_{max}.

b. For a loop gain $L(s) = [(-\tau s + 1)/(\tau s + 1)][(s+3)/s(s+1)^2]$, determine the complementary sensitivity $T(s)$ for $\tau = 0$. What is the allowable magnitude $\ell(j\omega)$ of an unstructured uncertainty, if stability is to be preserved?

c. From parts (a) and (b), what range of τ is guaranteed to lead to stability, according to the unstructured uncertainty analysis? How does this compare to the results of Problem 5.39(b)?

5.41   A system is constructed as a connection of two uncertain plants, $P_1(s) = [1 + \Delta_1(s)]P_{10}(s)$ and $P_2(s) = [1 + \Delta_2(s)]P_{20}(s)$.

a. Let $L(s) = P_1(s)P_2(s)$, and let $L(s) = [1 + \Delta(s)]L_0(s)$ with $L_0 = P_{10}P_{20}$. Calculate $\Delta(s)$.

b. Given $|\Delta_1(j\omega)| \le \ell_1(j\omega)$ and $|\Delta_2(j\omega)| \le \ell_2(j\omega)$, with complete phase uncertainty, calculate the upper bound on $|\Delta(j\omega)|$. Justify the statement that uncertainty is cumulative.

M   **5.42**   *dc servo, simplified model*   In Problem 2.4 (Chapter 2), a simplified model of the servo of Example 2.1 was derived, using $L = 0$. This model is to be used as a design model for a control system.

a. If $P(s)$ is the true model and $P_0(s)$ is the simplified model, calculate the multiplicative uncertainty $\Delta(s)$, where $P = (1 + \Delta)P_0$.

b. Compute $\ell(j\omega) = |\Delta(j\omega)|$, and estimate the bandwidth limit of a control system that has been designed using P_0.

M **5.43** In some robotic manipulators, feedback loops using operational amplifiers control the armature currents of the dc motors that drive the joints. Figure 5.35 has such a current-control loop as an inner loop (the constants are those of Example 2.1 in Chapter 2). Because of the relatively fast electrical phenomena, this inner loop maintains $i(t)$ near $i_d(t)$, provided $\omega(t)$ and $i_d(t)$ are relatively slow. The approximation $i \approx i_d$ is made for the design of the outer-loop controller, which is greatly simplified since the plant model becomes $\theta = (4.438/s^2)i_d$. We wish to explore the range of validity of that simplification.

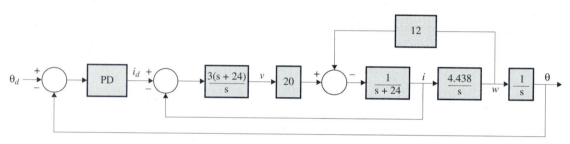

Figure 5.35 Joint control system for a robot

a. Calculate the transfer function θ/i_d.

b. The simplified model $\theta/i_d = 4.438/s^2$ is to be used. Proceed as in Example 5.20 to calculate $\Delta(s)$, where $(1+\Delta)(4.438/s^2) = $ true transfer function θ/i_d. Calculate $\ell(j\omega) = |\Delta(j\omega)|$. Estimate the bandwidth limit on the outer-loop control system if this simplification is used in the design of the outer loop.

M **5.44** *Servo with flexible shaft* In the model of Problem 2.5 (Chapter 2) for the servo with flexible shaft, the compliance parameter K has a nominal value of 500 Nm/rad. This parameter is poorly known, and we wish to estimate the bandwidth limit due to uncertainty. We assume that K may vary by $\pm x\%$. The plant input is v, and the output is θ_2.

a. Compute $\ell(j\omega)$ for $x = 1\%$, 10%, and 50%.

b. Using the results of part (a), estimate the bandwidth limit in each case. Explain why these limits are conservative.

Hint   Using a small number of points to cover the range of K, compute $|\Delta(j\omega)|$ for each point and maximize for each ω.

(M) 5.45 *Drum speed control* In Problem 3.46 (Chapter 3), the drum speed control system is divided into two independent subsystems—one with a common-mode input u_c and output ω_0, and the other with a differential-mode input u_d and output $\omega_1 - \omega_2$, each with its own transfer function. Suppose the compliance parameter K can vary between 60,000 and 100,000 Nm/rad, and the motor constant K_m between 4.0 and 8.0 Nm/A. (Note: The two K_m's, one for each motor, can vary independently.)

    a. Compute $\ell(j\omega)$ for each subsystem.

    b. Using the results of part (a), estimate for each subsystem the bandwidth limit imposed by the uncertainty. Explain why this limit is conservative.

Hint   See Problem 5.44.

(M) 5.46 *High-wire artist* For the system of Problem 2.12, linearized in Problem 2.18 (Chapter 2), the nominal values of the mass and moment of inertia of the bar are, respectively, $M_R = 2$ kg and $J_R = 1.5$ kg m^2. Suppose the range of variation for each is $\pm x\%$.

    a. Compute $\ell(j\omega)$ for the SISO plant with τ as input and θ as output, and with $x = 1\%$, 10%, and 30%.

    b. Using the results of part (a), estimate the bandwidth limit imposed by the uncertainty for each value of x. Explain why these limits are conservative.

(M) 5.47 *Crane* For the system of Problem 2.11, linearized in Problem 2.17 (Chapter 2), the mass M is assumed to vary from 500 to 2500 kg.

    a. Compute $\ell(j\omega)$ for the SISO plant with F as input and x as output.

    b. Using the results of part 9(a), estimate the bandwidth limit imposed by the uncertainty. Explain why this limit is conservative.

(M) 5.48 *Two-pendula problem* In the linearized model of Problem 2.19 (Chapter 2) for the two-pendula problem, $\ell_1 = 1$ m, $\ell_2 = 1.25$ m, and the nominal values of the cart and pendulum masses are $M = m = 1$ kg, respectively. Assume that M and m vary between .75 and 1.25 kg and that the two pendulum masses vary independently.

    a. Compute $\ell(j\omega)$ for the SISO plant with F as the input and x as the output.

    b. Using the results of part (a), estimate the bandwidth limit imposed by the uncertainty. Explain why that limit is conservative.

References

[1] Routh, E. J., *A Treatise on the Stability of a Given State of Motion*, Macmillan and Co. (1877).

[2] Evans, W. R., "Graphical Analysis of Control Systems," *Trans. A IEE Part II,* vol. 67, pp. 547–551 (1948).

[3] Shinners, S. M., *Modern Control System Theory and Applications*, 2nd ed., Addison-Wesley (1978).

[4] Dorf, R. C., *Modern Control Systems*, 5th ed., Addison-Wesley (1989).

[5] Nyquist, H., "Regeneration Theory," *Bell System Technical J.,* vol. 11, pp. 126–147, (1932).

[6] Bélanger, P. R., E. L. Adler, and N. C. Rumin, *Introduction to Circuits with Electronics: An Integrated Approach*, Holt, Rinehart and Winston (1985).

[7] Vidyasagar, M., R. K. Bertschmann, and C. S. Sallaberger, "Some Simplifications of the Graphical Nyquist Criterion," *IEEE Trans. on Automatic Control,* vol. AC-33, pp. 301–305 (1988).

[8] Kharitonov, V. L., "Asymptotic Stability of an Equilibrium Position of a Family of Systems of Linear Differential Equations," *Differensial ńye Uravneniya,* vol. 14, pp. 2086–2088 (1978).

[9] Barmish, B. R., *New Tools for Robustness of Linear Systems,* Macmillan (1994).

6 Classical Design

6.1 INTRODUCTION

The term *classical design* refers to a body of methods that is based on shaping the loop gain and is applicable to single-input, single-output (SISO) systems. Although the advent of the digital computer as a control element has made it possible to implement controllers of arbitrary structure, it has not rendered obsolete the classical methods of design, for two reasons. The first is standardization. It is convenient for a vendor to offer her customers a choice of a small number of standard software "blocks." The second reason is that, of the several hundred loops in a typical process plant, rather few are critical enough to warrant extended modeling and design efforts. Most loops are tuned "manually," and that is practically feasible only with a small number of tunable parameters (at most three). It is important to have simple controller structures that are capable of adequate control for a wide family of plant transfer functions, leaving to more complex controllers the handling of a few critical loops. Classical control is a mature subject, and the reader is referred to a number of good textbooks on the topic, including [1, 2, 3, 4].

In this chapter, design objectives will be formulated for dc steady-state time-domain and frequency-domain behavior. Several types of classical structures will be studied. A section will be devoted to systems with time delay.

6.2 DESIGN OBJECTIVES IN TERMS OF THE LOOP GAIN

6.2.1 Introduction

It is assumed that design objectives have been established for the closed-loop system in terms of (i) the dc steady-state response, (ii) the transient response, and (iii) the frequency response. In Chapter 4, the approach was to select a suitable closed-loop transfer function (e.g., S or T), subject to interpolation constraints, and to solve for the controller, $F(s)$. This almost always yields an $F(s)$ whose order is comparable with that of the plant. We cannot use that approach if the structure of $F(s)$ is predetermined (e.g., a pure gain); rather, we must work *directly* with $F(s)$.

It therefore becomes necessary to translate *closed-loop* design requirements to *open-loop* requirements, i.e., requirements on the *loop gain* $L(s) = F(s)P(s)$. Then $L(s)$ [and hence $F(s)$] is designed according to the open-loop specifications.

6.2.2 The dc Steady State

The steady-state error under constant input is perhaps the most important control specification. In regulation problems, the objective is for the output to equal a constant set point. A steady-state error appears as a bias; it is always present.

Furthermore, a regulation problem nearly always includes a constant disturbance, for the following reason. When we linearize about an operating point according to Section 2.6 (Chapter 2), we normally refer inputs, outputs, and states to their values in the desired dc steady state. The output is referred to $y^* = y_d$, so that

$$\Delta y = y - y^* = y - y_d = -e$$

which is defined in terms of quantities that are known or measured and do not depend on the model.

On the other hand, the input is referred to u^*, which is the constant value of u that results in a steady-state output of $y^* = y_d$. The value of u^* calculated by the model is not u^* but some approximation $\widehat{u}^*$. The control input u is obtained by adding the quantity Δu, the input to the linearized system, to $\widehat{u}^*$. Thus,

$$u = \Delta u + \widehat{u}^* = \Delta u + (\widehat{u}^* - u^*) + u^*.$$

Therefore, when the input is referred to the true u^*, a constant disturbance $\widehat{u}^* - u^*$, of unknown magnitude, is effectively added to the controller-generated input Δu. This constant disturbance at the input can be referred to the output, where it becomes a constant (or possibly a ramp, or higher-order integral of the step function). For a single-degree-of-freedom (1-DOF) system, recall that

$$\frac{e}{y_d} = 1 - H_d = S. \tag{6.1}$$

If $y_d(t) = Au_{-1}(t)$, a step, then

$$e(s) = S(s)\frac{A}{s}. \tag{6.2}$$

By the final value theorem, the dc steady-state error is

$$e_{ss} = \lim_{t \to \infty} e(t) = \lim_{s \to 0} se(s) = A \lim_{s \to 0} S(s). \tag{6.3}$$

Equation 6.3 assumes that the time limit exists, which is tantamount to assuming $S(s)$ to be stable. Equation 6.3 states that the steady-state error response to a set-point step of height A is $AS(0)$, if $S(0)$ exists. Since $y/d = H_{wd}(s) = S(s)$, that is also the steady-state error for a step A in $d(t)$, the disturbance applied to the output.

From Chapter 4, Equation 6.3 is written in terms of the loop gain $L(s)$ as

$$e_{ss} = A \lim_{s \to 0} \frac{1}{1 + L(s)}. \tag{6.4}$$

It is clear from Equation 6.4 that the key to a small steady-state error is to make $\lim_{s \to 0} L(s)$ large.

If $L(0)$ exists, then

$$e_{ss} = \frac{A}{1 + L(0)} = \frac{A}{1 + k_p}. \tag{6.5}$$

Here, $k_p = L(0)$ is called the *position constant*.

Example 6.1   For $L(s) = k/(s + 1)^3$, calculate the position constant. What is the lower limit on the closed-loop steady-state error for a unit-step set-point input?

Solution   Since $L(0) = k$, the position constant is $k_p = k$. The steady-state error to a unit step is given by Equation 6.5:

$$e_{ss} = \frac{1}{1 + k_p} = \frac{1}{1 + k}.$$

The value of e_{ss} is reduced by increasing k, but this is limited by stability: without stability, there is no steady state. Using Routh's criterion, $-1 < k < 8$ for stability. Therefore,

$$e_{ss} > \frac{1}{1 + 8} = \frac{1}{9}.$$

If $L(s)$ has one or more poles at $s = 0$, i.e., one or more *integrations*, then $\lim_{s \to 0} L(s) = \infty$ and

$$e_{ss} = A \lim_{s \to 0} \frac{1}{1 + L(s)} = 0. \tag{6.6}$$

That is, the steady-state error to a step is zero if the loop gain has integration.

The dc steady-state error to a ramp input is often of interest, because many signals have ramplike components. Then

$$e_{\text{ramp}} = \lim_{s \to 0} sAS(s)\frac{1}{s^2}$$

$$= A \lim_{s \to 0} \frac{S(s)}{s}. \tag{6.7}$$

In terms of the loop gain,

$$e_{\text{ramp}} = A \lim_{s \to 0} \frac{1}{s + sL(s)}$$

$$= A \lim_{s \to 0} \frac{1}{sL(s)}. \tag{6.8}$$

If $L(0)$ exists, then e_{ramp} does not; i.e., it is infinite. If $L(s)$ has one pole at $s = 0$, then $\lim_{s \to 0} sL(s)$ exists. Let that limit be k_v, known as the *velocity constant*; then

$$e_{\text{ramp}} = \frac{A}{k_v} \tag{6.9}$$

where

$$k_v = \lim_{s \to 0} sL(s). \tag{6.10}$$

If $L(s)$ has two or more poles at $s = 0$, then $\lim_{s \to 0} sL(s) = \infty$, and $e_{\text{ramp}} = 0$.

In the literature, the number of integrations in $L(s)$ is referred to as the system *type*. For a Type 0 system,

$$e_{ss} = \frac{A}{1 + k_p}, \quad k_p = L(0); \quad e_{\text{ramp}} = \infty.$$

For a Type 1 system,

$$e_{ss} = 0; \quad e_{\text{ramp}} = \frac{A}{k_v}, \quad k_v = \lim_{s \to 0} sL(s).$$

For a system of Type 2 or higher,

$$e_{ss} = e_{\text{ramp}} = 0.$$

To summarize, the dc steady-state response is determined by the limiting behavior of $L(s)$, as $s \to 0$. Small errors are achieved by large dc loop gains, with integration (infinite dc gain) actually reducing the step response error to zero.

Example 6.2   **(dc Servo)**

The dc servo of Example 2.1 (Chapter 2) is controlled by a 1-DOF control system, with a controller $F(s) = k$. Calculate the steady-state error for a unit-step reference input θ_d and for a unit step in the load torque T_L, assuming stability of the closed-loop system.

Solution   From Equation 3.21 (Chapter 3), the loop gain is

$$L(s) = \frac{88.76k}{s(s + 21.576)(s + 2.474)}.$$

The system is of Type 1, and the steady-state error to a step input is zero. From Equation 3.22 (Chapter 3),

$$\frac{\theta}{T_L} = \frac{-7.396(s + 24)}{s(s + 21.576)(s + 2.474)}.$$

For a unit step in T_L, the disturbance referred to the output—i.e., the effect on θ of T_L by itself—is

$$d(s) = \frac{-7.396(s + 24)}{s^2(s + 21.576)(s + 2.474)}$$

$$= -\left(\frac{3.325}{s^2} + \frac{A}{s} + \frac{B}{s + 21.576} + \frac{C}{s + 2.474}\right)$$

$$d(t) = -3.325tu_{-1}(t) + Au_{-1}(t) + Be^{-21.576t} + Ce^{-2.474t}.$$

In the steady state, the contributions of the two stable poles disappear. Because the system is of Type 1, the step component A/s contributes 0 to e_{ss}. That leaves the ramp component. By Equation 6.9,

$$e_{\text{ramp}} = -\frac{3.325}{k_v}$$

where

$$k_v = \lim_{s \to 0} sL(s) = \frac{88.76k}{(21.576)(2.474)}$$

$$= 1.663k.$$

Therefore,

$$e_{\text{ramp}} = \frac{-3.325}{1.663k} = \frac{-2.000}{k}.$$

That is the steady-state error for a unit step in T_L. Note that, in this example, the physical signal T_L is a step, but the disturbance referred to the output has a ramp component.

6.2.3 Transient Response

The expression relating the time-domain response of the closed loop to that of the open loop is quite complex. It cannot be used practically to infer closed-loop behavior from the open-loop transient response. Inferences about the closed-loop time response are more easily made from consideration of the pole-zero pattern.

The all-pole second-order transfer function is a convenient starting point. Let

$$\frac{y}{y_d} = H_d(s) = \frac{\omega_0{}^2}{s^2 + 2\zeta\omega_0 s + \omega_0{}^2}, \qquad \zeta \geq 0. \tag{6.11}$$

Here, ζ is the *damping factor*. For $\zeta = \sqrt{2}/2$, H_d is the second-order Butterworth low-pass filter.

The poles of $H_d(s)$ are

$$s_1, \; s_1{}^* = -\zeta\omega_0 \pm j\omega_0\sqrt{1 - \zeta^2}. \tag{6.12}$$

The poles are complex if $0 \leq \zeta < 1$. Figure 6.1 shows how the poles vary as a function of ζ. Note that $|s_1| = |s_1{}^*| = \omega_0$ if the poles are complex.

The unit-step response is calculated from

$$y(s) = \frac{1}{s} H_d(s) = \frac{\omega_0{}^2}{s(s^2 + 2\zeta\omega_0 s + \omega_0{}^2)}$$

$$= \frac{1}{s} - \frac{s + 2\zeta\omega_0}{s^2 + 2\zeta\omega_0 s + \omega_0{}^2}.$$

Using a table of Laplace transforms and assuming complex poles,

$$y(t) = 1 - e^{-\zeta\omega_0 t}\left(\cos \omega_0\sqrt{1 - \zeta^2}t - \frac{\zeta}{\sqrt{1 - \zeta^2}}\sin \omega_0\sqrt{1 - \zeta^2}t\right)$$

$$= 1 - \frac{1}{\sqrt{1 - \zeta^2}}e^{-\zeta\omega_0 t}\cos[\omega_0\sqrt{1 - \zeta^2}t + \phi] \tag{6.13}$$

where $\phi = \tan^{-1}\frac{\zeta}{\sqrt{1-\zeta^2}}$.

Step responses are plotted in Figure 6.2, for a few values of ζ. The responses are plotted against $\omega_0 t$ rather than t, to show the scaling effect of ω_0. The larger ω_0, the faster the response; what happens at $t = 1$ s for $\omega_0 = 1$ rad/s occurs at $t = 1$ ms for $\omega_0 = 1000$ rad/s.

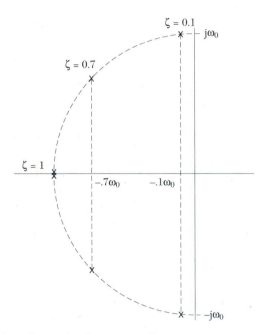

Figure 6.1 Location of second-order system poles

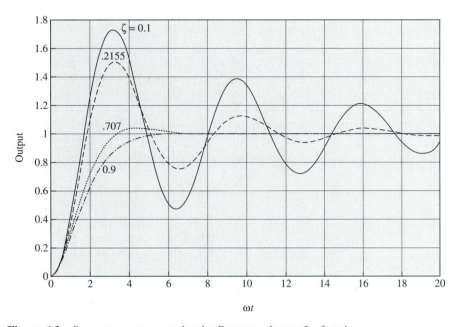

Figure 6.2 Step responses, second-order Butterworth transfer function

The shape of the response depends only on ζ. Values of ζ near zero lead to fast rise time but large overshoot and long settling time; values near 1 yield small overshoots but slow rise time. The Butterworth value $\zeta = \sqrt{2}/2$ is often used as a good compromise between speed and stability. Process control applications often use $\zeta = 0.2155$, the so-called quarter-cycle damping value (see Problem 6.5).

It can be shown [3] that the local maxima occur at times t_k given by

$$t_k = \frac{\pi + 2k\pi}{\omega_0\sqrt{1 - \zeta^2}}, \qquad k = 0, \ 1, \ 2, \ldots \tag{6.14}$$

The highest maximum is the first one, for $k = 0$; therefore, the peak time T_p is

$$T_p = \frac{\pi}{\omega_0\sqrt{1 - \zeta^2}}. \tag{6.15}$$

The peak overshoot, obtained by evaluating $y(t)$ for $t = T_p$, is

$$\text{peak overshoot} = 100e^{-\zeta\pi/\sqrt{1-\zeta^2}}. \tag{6.16}$$

As for the settling time T_s, it is approximately equal to the time at which the magnitude of the cosine in Equation 6.13 becomes 0.05; i.e., T_s satisfies

$$\frac{1}{\sqrt{1 - \zeta^2}}e^{-\zeta\omega_0 T_s} = 0.05$$

or, taking logarithms,

$$T_s = \frac{1}{\zeta\omega_0}\left(3 + \ln\frac{1}{\sqrt{1 - \zeta^2}}\right). \tag{6.17}$$

Equations 6.16 and 6.17 support the conclusion reached by examination of Figure 6.2.

Of course, the all-pole second-order system is a special case. It is a useful special case, because a practical system often has a complex pole pair relatively close to the origin, with other stable poles and zeros farther away. The transfer function, expressed in partial fractions, would have the form

$$H_d(s) = \frac{k\omega_0^2}{s^2 + 2\zeta\omega_0 s + \omega_0^2} + \sum_i \frac{A_i}{s + s_i}.$$

If $\text{Re}\ s_i \gg \zeta\omega_0$, the extra terms—which translate into terms $e^{-\text{Re}\,s_i t}e^{jI_m s_i t}$ in the time domain—go to zero much faster than the second-order transient, because $e^{-\text{Re}\,s_i t} \ll e^{-\zeta\omega_0 t}$. After a short initial period, the response is essentially that of the second-order system. In such a case, the paired complex poles near the origin are said to be *dominant poles*.

In classical design, we often try to achieve a closed-loop system with a pair of dominant poles. The prototype open-loop system is the one whose Root Locus is

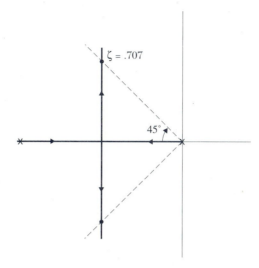

Figure 6.3   Root locus for a system with two real-axis poles

given in Figure 6.3, with two real poles. The gain is chosen so as to achieve the desired ζ for the complex closed-loop poles. In general, this idea applies to the extent that other open-loop poles are stable and that they and the open-loop zeros are much farther away from the origin than the two real poles.

The reader should guard against drawing overly strong conclusions from poles alone. The zeros are important because they help determine the coefficients of the terms in the partial fraction expansion of $y(s)$. In fact, the pole locations, taken alone, tell us only what to *avoid*, in general. For example, a system with poles in the hatched region in Figure 6.4 will have a response that may be (i) unstable (RHP poles), (ii) too oscillatory (pole pair with small ζ), (iii) slow (poles with small negative real parts). Unfortunately, avoiding the "undesirable" region does not guarantee a good response. The transfer function $(13s+5)/(s+1)(s+5)$ has two well-damped poles, but its step response, shown in Figure 6.5, still has a large overshoot.

6.2.4 Frequency Response and the Nichols Chart

Closed-loop frequency response specifications were discussed in Chapter 4. Constraints were placed on $|T(j\omega)|$ and/or $|S(j\omega)|$, depending on the frequency. This can be related to the foregoing discussion in the following way. The presence of complex poles near the j-axis (small ζ) results in sharp, high peaks in both $|T(j\omega)|$ and $|S(j\omega)|$. Figure 6.6 shows the magnitude Bode plot for the all-pole second-order system, for several values of ζ. As $\zeta \to 0$, the peak tends to infinity.

The peak value of $|T(j\omega)|$ is denoted by M_p. For the all-pole second-order system, it is given (see Problem 6.6) by

$$M_p = 1 + e^{-\zeta\pi/\sqrt{1-\zeta^2}} \tag{6.18}$$

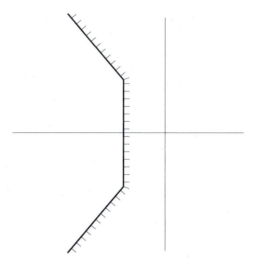

Figure 6.4 Illustration of the undesirable region (hatched) of the complex plane

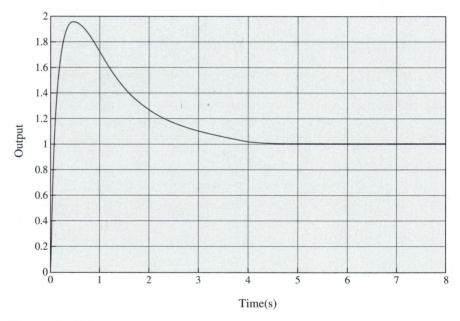

Figure 6.5 Step response

for $\zeta < 1/\sqrt{2}$. The frequency response magnitude has its maximum at $\omega = 0$ if $\zeta \geq 1/\sqrt{2}$.

For the 1-DOF system, frequency response specifications are often given in terms of $S(j\omega)$ and/or $T(j\omega)$. Let us now explore the relationship between $L(j\omega)$ and

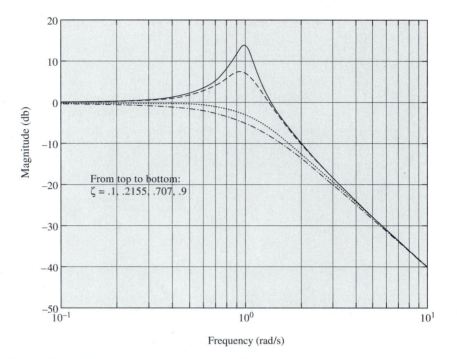

Figure 6.6   Frequency response magnitude for Butterworth second-order transfer function

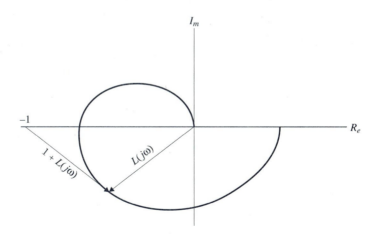

Figure 6.7   Showing $L(j\omega)$ and $1 + L(j\omega)$

$S(j\omega)$, $T(j\omega)$. Figure 6.7 illustrates the complex vectors $L(j\omega)$ and $1 + L(j\omega)$. Since

$$S(j\omega) = \frac{1}{1 + L(j\omega)}$$

it follows that $|S(j\omega)| = S_0$ for all $L(j\omega)$ such that

$$|1 + L(j\omega)| = 1/S_0. \tag{6.19}$$

That is, the locus of points in the $L(j\omega)$ plane corresponding to constant magnitude S_0 satisfies Equation 6.19. This locus is just a circle of radius $1/S_0$, centered at $(-1, 0)$. Figure 6.8 shows constant-$|S|$ circles for $|S| = .5$, 1, and 2. If, for example, $|S(j\omega)|$ is to be less than 1 at $\omega = \omega_0$, then $L(j\omega_0)$ must be outside the circle corresponding to $|S| = 1$.

The loci of constant $|T|$ are a bit more involved. Let $L(j\omega) = X + jY$. Then

$$T = \frac{L}{1 + L} = \frac{X + jY}{1 + X + jY}$$

and

$$|T|^2 = \frac{X^2 + Y^2}{(1 + X)^2 + Y^2}.$$

The locus of points (X, Y) for which $|T| = M$ satisfies

$$M^2(1 + 2X + X^2) + M^2Y^2 = X^2 + Y^2$$

or

$$(M^2 - 1)Y^2 + (M^2 - 1)X^2 + 2M^2X + M^2 = 0. \tag{6.20}$$

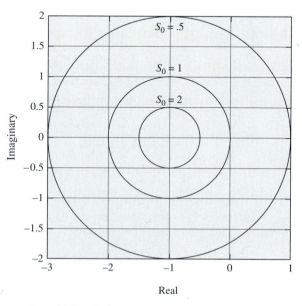

Figure 6.8 Constant sensitivity circles

If $M = 1$, then Equation 6.20 reduces to

$$X = -\frac{1}{2}. \tag{6.21}$$

That is, the vertical straight line $\text{Re } L(j\omega) = -1/2$ is the locus of points where $|T(j\omega)| = 1$.

If $M \neq 1$,

$$Y^2 + X^2 + 2\frac{M^2}{M^2 - 1}X + \frac{M^4}{(M^2 - 1)^2} + \frac{M^2}{M^2 - 1} = \frac{M^4}{(M^2 - 1)^2}$$

or

$$Y^2 + \left(X + \frac{M^2}{M^2 - 1}\right)^2 = \frac{M^2}{(M^2 - 1)^2}. \tag{6.22}$$

Equation 6.22 represents a circle of radius $M/|M^2 - 1|$, centered on the real axis ($Y = 0$) at $X = -M^2/(M^2 - 1)$. Figure 6.9 shows several of the constant-$|T|$ circles, also called M-circles. For example, if $|T(j\omega)|$ is required to be less than 2 at $\omega = \omega_0$, then $L(j\omega_0)$ must lie outside the circle corresponding to $M = 2$.

Rather than plot the imaginary part versus the real part, a *gain-phase plot* displays the gain (in decibels) versus the phase (in degrees). This was convenient in the precomputer era, because the gain and phase information could be read directly from Bode plots. Gain-phase plots do offer some advantages over Nyquist plots.

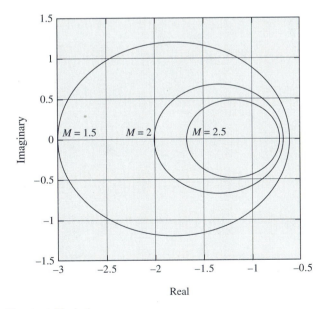

Figure 6.9   Constant-M circles

The logarithmic scales make it easier to display widely differing magnitudes; the plot for $kL(j\omega)$ is simply that for $L(j\omega)$ raised by $|k|$ db, because $20\,|kL(j\omega)| = 20\log|k| + 20\log|L(j\omega)|$.

Constant-M contours can also be plotted on the gain-phase plane, since Y and X are expressible in terms of log magnitude and phase. The contours are no longer circles, of course, but are the mapping of circles through a particular nonlinear transformation. Such constant-M contours, together with loci of constant phase of $T(j\omega)$, constitute a *Nichols chart*. From a Nichols chart and a gain-phase plot of the loop gain $L(j\omega)$, we may, for any frequency point, read the magnitude and phase of the closed-loop transmission $T(j\omega)$. In particular, we can easily ascertain the peak value of $|T(j\omega)|$ in decibels as the M-value of the contour of highest magnitude touched by the locus $L(j\omega)$.

Since $S = 1/(1 + L) = L^{-1}/(1 + L^{-1})$, the behavior of S can be studied on the Nichols chart by plotting the locus of $L^{-1}(j\omega)$ rather than $L(j\omega)$.

Example 6.3   Given $L(s) = 3/(s + 1)^3$, use the Nichols chart and gain-phase plots to obtain the peak values of $|T|$ and $|S|$ and the frequencies at which those peaks occur.

Solution   Figure 6.10 (MATLAB bode, nichols), shows the gain-phase locus and the Nichols chart. By inspection, the peak of $|T(j\omega)|$ is approximately 4.5 db and occurs at about $\omega = 1.2$ rad/s. From the gain-phase locus of $L^{-1}(j\omega)$, we infer that $|S(j\omega)|$ has a peak of approximately 6.3 db at a frequency of 1.3 rad/s. Note that

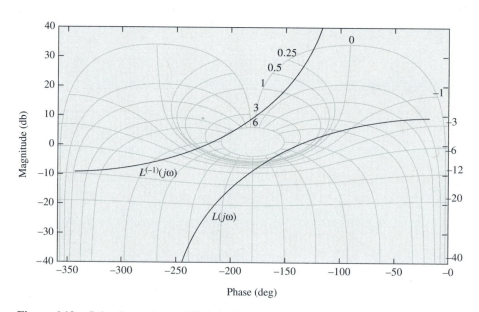

Figure 6.10   Gain-phase plot and Nichols chart

setting an upper limit on $|S|$ or $|T|$ is tantamount to surrounding the critical point $(-1, \ 0)(-180°, \ 0$ db in the gain-phase plane) by "forbidden regions" inside which L (or L^{-1}) may not be.

It is possible to design a pure-gain control system for a given maximum value of $|T(j\omega)|$ or $|S(j\omega)|$ by using the gain-phase plane. Adding k db of gain slides the whole locus of $L(j\omega)$ up by k db and the locus of $L^{-1}(j\omega)$ down by the same amount. It is usually not difficult to see how many decibels should be added or taken away in order for the locus of L or L^{-1} to just graze the M-contour corresponding to the desired peak value.

Example 6.4

(dc Servo)

For the dc servo of Example 2.1 (Chapter 2), calculate the largest value of the gain k that will satisfy the following specifications: $|T(j\omega)| \leq 2$ db, $|S(j\omega)| \leq 4$ db.

Solution

From Equation 3.21 (Chapter 3),

$$P(s) = \frac{\theta}{v} = \frac{88.76}{s(s + 2.474)(s + 21.526)}.$$

Figure 6.11 shows the gain-phase loci of $P(j\omega)$ and $P^{-1}(j\omega)$ (MATLAB bode, nichols) with the Nichols chart. With $k = 1$ (0 db), the peak value of $|T|$ is about 1 db, and that of $|S|$ is 3 db. It can be seen that a gain of about 2 db will raise the

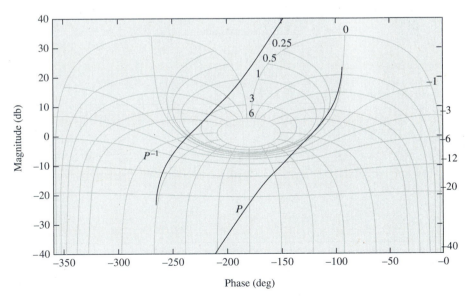

Figure 6.11 Gain-phase plots of P and P^{-1}

P-locus about enough to graze the 2-db contour and lower the P^{-1} locus to touch the 4-db contours. By trial-and-error computation of $S(j\omega)$ and $T(j\omega)$, the actual value is 2.75 db, or $k = 1.36$.

6.2.5 Stability Margins

The Nyquist plot shows the gain and the phase simultaneously. It is clear that both must be taken into account to ascertain the relationship of the Nyquist locus to the critical $(-1, 0)$ point. When working with Bode plots, in particular, it is easier to have separate gain and phase conditions. The *gain* and *phase margins*, known together as the *stability margins*, are precisely such conditions. Figure 6.12 illustrates their definitions with a Nyquist plot corresponding to a stable closed-loop system. The *gain margin* is the value of gain k that would cause the locus $kL(j\omega)$ to traverse the $(-1, 0)$ point; it is determined from the intersection of the Nyquist plot with the negative real axis. In Figure 6.12, a gain $k = 1/x_1 > 1$ would move point A to the critical point, so there is a gain margin of $1/x_1$.

The *phase margin* is the phase lag (i.e., negative angle) that, when applied to every point of the locus $L(j\omega)$, would cause its Nyquist locus to traverse the critical point. The phase margin is determined from the intersection of the Nyquist plot with the circle $|L(j\omega)| = 1$, i.e., the circle of unit radius centered at the origin. In Figure 6.12, a phase lag of θ_1 added to the locus would move point B to the critical point, so the phase margin is θ_1.

In some cases, there may be two gain or phase margins. Consider, for example, the Nyquist plot of Figure 6.13, for a plant with one RHP pole. For stability we must have $N = -1$, so the $(-1, 0)$ point must be located as shown, between $-x_1$ and $-x_2$. A gain $k > 1/x_2$ could cause instability, as would $k < 1/x_1$. Thus, there are two gain margins: an upper margin $1/x_2 > 1$, and a lower margin $1/x_1 < 1$.

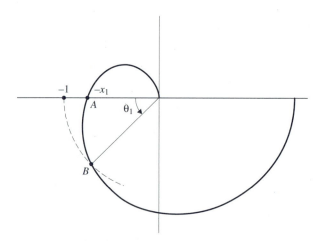

Figure 6.12 Illustrating gain and phase margins

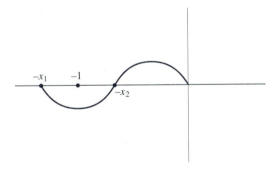

Figure 6.13   Nyquist plot of a system unit with upper and lower phase margins

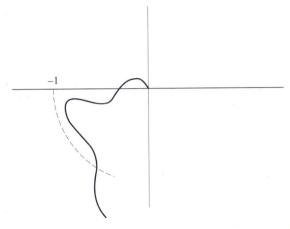

Figure 6.14   Nyquist plot showing good stability margins, but poor stability

The stability margins monitor the Nyquist plot at two points: the point where $|L(j\omega)| = 1 = 0$ db (the crossover frequency) and the point where $\angle L(j\omega) = 180°$. As Figure 6.14 shows, it is possible for those two points to be acceptably far from the critical point while the locus comes dangerously close to $(-1, 0)$. Such a situation would probably be viewed by a practitioner as pathological; it happens, but not too often. Still, this serves to show that good stability margins do not necessarily guarantee good stability properties.

Example 6.5   Given $L(s) = 6/(s + 1)^3$, obtain the magnitude and phase Bode plots and determine the stability margins.

Solution   Figure 6.15 shows the Bode plots (MATLAB bode). Since $|L(j\omega)| = 1$ (0 db) at $\omega = 1.52$ rad/s, that frequency is the crossover frequency. Because the phase at that frequency is $-169.9°$, the phase margin is $180 - 169.9° = 10.1°$.

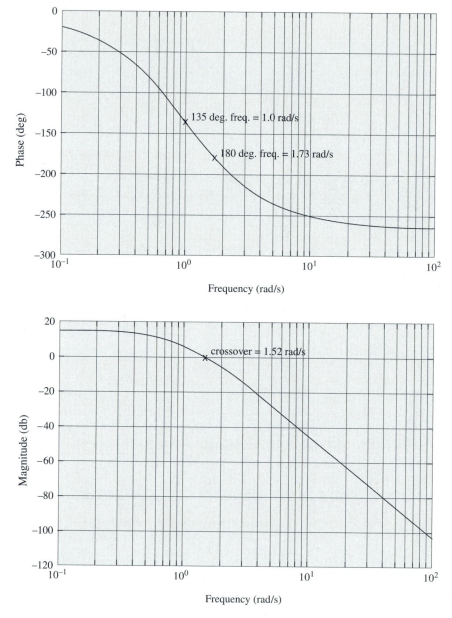

Figure 6.15   Bode plots

From the phase plot, the phase is 180° (negative real-axis crossing in the Nyquist plot) at $\omega = 1.73$ rad/s, at which frequency the magnitude is -2.47 db. The gain margin is 2.47 db. (MATLAB margin will return both the gain and phase margins.)

It is convenient to work from Bode plots to design a pure-gain controller for specified stability margins. We have, for $k > 0$,

$$20 \log |kL(j\omega)| = 20 \log k + 20 \log |L(j\omega)| \tag{6.23}$$

and

$$\not< kL(j\omega) = \not< L(j\omega). \tag{6.24}$$

From Equation 6.24, the phase curve for $kL(j\omega)$ is the same as for $L(j\omega)$. From Equation 6.23, the gain curve for $kL(j\omega)$ is just the gain curve for $|L(j\omega)|$ shifted by $20 \log k$. If $k > 1$, $\log k > 0$ and the gain curve is shifted up; the shift is downward if $k < 1$.

The following example will serve to illustrate the design procedure.

Example 6.6　For the plant of Example 6.5, determine the range of controller gain $k > 0$, so that (i) the gain margin is at least 6 db and (ii) the phase margin is at least 45°.

Solution　To adjust k for gain margin, we locate the frequency where the phase is 180°, which corresponds to a crossing of the negative real axis on the Nyquist plot. From the phase curve of Figure 6.15, that frequency is $\omega = 1.73$ rad/s. The magnitude at that frequency is -2.4 db. For a gain margin of 6 db, the loop gain on a magnitude must be at most -6 db where the phase is 180°. To achieve that, we must *lower* the gain curve by 3.6 db; therefore $k \leq -3.6$ db to satisfy the gain-margin condition.

To satisfy the phase-margin condition, we locate the frequency where the phase is $-180° +$ phase margin $= -180° + 45° = -135°$. That frequency is $\omega = 1.0$ rad/s, from the phase curve. The phase margin will be 45° if the magnitude is 0 db where the phase is $-135°$, i.e., at $\omega = 1.0$ rad/s. The magnitude is 6.51 db at $\omega = 1.0$ rad/s, so the gain curve must be lowered by 6.51 db, i.e., $k = -6.51$ db. If $k < -6.51$ db, the crossover frequency is less than 1.0 rad/s, and the phase margin ($180° +$ phase at crossover) is greater than 45°. Therefore, $k \leq -6.51$ db will satisfy the phase margin specification.

To satisfy both stability margins, we need $k \leq -6.51$ db.

Example 6.7　**(dc Servo)**

For the dc servo of Example 2.1 (Chapter 2), calculate the values of the gain k of a gain feedback controller such that the phase margin is at least 50° and the gain margin at least 6 db.

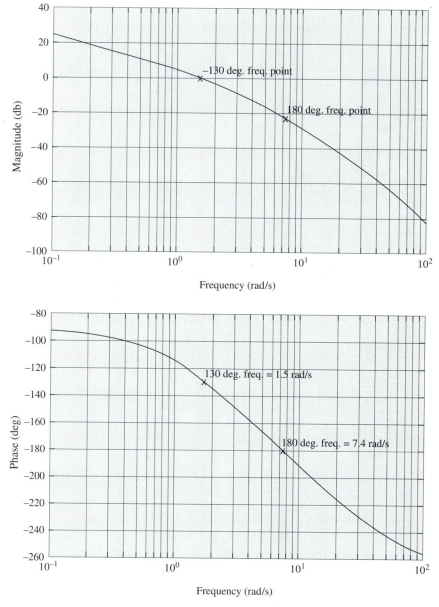

Figure 6.16 Bode plots

Solution The transfer function is given in Equation 3.21 (Chapter 3), and in Example 6.4. Figure 6.16 shows the Bode plots.

Gain margin: The phase is 180° at $\omega = 7.4$ rad/s. The magnitude at that frequency is -23.21 db. For a 6-db gain margin, the gain can be at most 17.21 db.

Phase margin: The phase is $-180° + 50° = -130°$ at $\omega = 1.5$ rad/s, and the magnitude is -0.41 db at that frequency. A gain of 0.41 db will make $\omega = 1.5$ rad/s the crossover frequency. Clearly, from Figure 6.16, a gain value less than 0.41 db will move the crossover frequency to the left, where, from Figure 6.16, the phase is greater than $-130°$. It follows that the phase margin requirement will be met with $k \le 0.41$ db. Clearly, that range of gain values will also satisfy the gain margin specifications.

6.3 DYNAMIC COMPENSATION

6.3.1 Introduction

The simplest controller design is a pure gain adjusted (as in the foregoing examples) to satisfy stability objectives. In many cases this does not satisfy the performance specifications, and it is therefore necessary to alter the shape of the loop gain frequency response by using a compensator with dynamics in cascade with the plant. This is known as *dynamic compensation.*

Bode's gain-phase relationship will give some useful insight about the shape of $L(j\omega)$ at crossover. We will then examine compensation schemes aimed at meeting steady-state specifications, followed by compensation to increase system bandwidth.

6.3.2 The Bode Gain-Phase Relationship

The Bode theorem [5] applies to systems that are stable and minimum phase, i.e., with no poles or zeros in the closed RHP. For such systems,

$$\measuredangle\, G(j\omega) = \frac{\pi}{2} \frac{d}{du} (\ln |G(j\omega)|) \Big|_{\omega_0}$$

$$+ \frac{1}{\pi} \int_{-\infty}^{\infty} \left[\frac{d}{du} (\ln |G(j\omega)|) - \frac{d}{du} (\ln |G(j\omega)|) \Big|_{\omega_0} \right] W(u)\,du \qquad (6.25)$$

where

$$u = \ln (\omega/\omega_0)$$

$$W(u) = \ln \coth \left| \frac{u}{2} \right|.$$

Since $u = \ln \omega - \ln \omega_0$, the plot of $\ln |G(j\omega)|$ versus u is just the plot of $\ln |G(j\omega)|$ versus $\ln \omega$ shifted to a new abscissa origin, $\ln \omega_0$. The slope of $\ln |G(j\omega)|$ plotted against $\ln \omega$ is the same as that of $\log |G(j\omega)|$ versus $\log \omega$, and is the slope of the Bode magnitude plot (in decibels) divided by 20.

Figure 6.17 shows the weighting function $W(u)$. It weighs the contributions of frequencies near ω_0 heavily, with much less emphasis on those frequencies relatively far from ω_0. If the slope of the Bode magnitude plot is nearly constant in the vicinity of ω_0, the bracketed term in the integral will be small near ω_0 and the

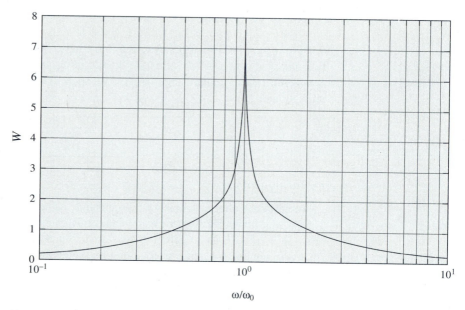

Figure 6.17   Weighting function for Bode's theorem

integral itself will be small. The integral will contribute significantly to the phase at frequencies where the slope of the magnitude curve is changing. Because the Bode magnitude plot is basically a set of straight lines, the first term in Equation 6.25 dominates, so

$$\angle\, G(j\omega_0) \sim 90^\circ \times \frac{\text{slope of Bode magnitude curve at } \omega_0 \text{ (db/decade)}}{20}. \qquad (6.26)$$

The practical consequence is this: since the phase must be greater (i.e., less negative) than -180° at crossover, the slope of the magnitude curve must be greater than -40 db/decade (less than 40 db/decade in magnitude) for a relatively wide range of frequencies around ω_0. If the crossover frequency is located in a region of approximately constant slope, that slope should be -20 db/decade.

6.3.3 Lag and Proportional-Integral Compensation

Lag Compensation in the Frequency Domain
The value of gain that is used in gain compensation is usually dictated by stability considerations. It is often the case that this gain value is not high enough to satisfy a given steady-state specification. The pure-gain design must be modified so as to achieve the steady-state specification *and* maintain the required degree of stability (i.e., the stability margins, peak frequency response, or pole damping).

We assume that a design is available (e.g., a pure-gain design) that provides satisfactory stability but not enough dc gain to give acceptable position or velocity constants. Because stability depends on the frequency behavior in the neighborhood of crossover, it is best to leave the design essentially undisturbed in that frequency range. We modify the design at and near zero frequency, because the steady-state behavior depends on the behavior of the loop gain near $\omega = 0$.

Figure 6.18 shows the structure of the lag compensator. The controller transfer function is

$$\frac{u}{e} = G_c(s) = \left(1 + \frac{k_1}{Ts + 1}\right). \tag{6.27}$$

An equivalent expression is

$$G_c(s) = \frac{Ts + 1 + k_1}{Ts + 1}$$
$$= (1 + k_1)\frac{[T/(1 + k_1)]s + 1}{Ts + 1}. \tag{6.28}$$

This form is the one seen most often in control texts.

At dc, $G_c(0) = (1 + k_1)$. From Equation 6.27, we see that at high frequencies, where $|k_1/(j\omega T + 1)| \ll 1$, $G_c \approx 1$. The basic idea is to choose k_1 such that $G_c(0) = (1 + k_1)$ yields a satisfactory position or velocity constant, and to choose T such that $G_c \approx 1$ in the critical frequencies region where the stability margins are determined. This ensures satisfactory dc gain without disturbing the stability.

Suppose the dc specification is $G_c(0) = k_0$. Then k_1 is chosen as follows:

$$(1 + k_1) = k_0$$

or

$$k_1 = (k_0 - 1). \tag{6.29}$$

It is assumed that $k_0 > 1$, so $k_1 > 0$; otherwise, the pure-gain control yields satisfactory dc performance.

Let ω_c be some frequency in the critical region. We shall take ω_c to be the crossover frequency, but it could also be, for example, the frequency at which the phase is $180°$.

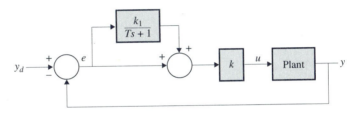

Figure 6.18   The lag compensator

To ensure that $G_c(j\omega_c) \approx 1$, we refer to Equation 6.27 and impose

$$\left| \frac{k_1}{j\omega_c T + 1} \right| = \frac{k_1}{\sqrt{\omega_c^2 T^2 + 1}} = a \ll 1$$

or, solving for $\omega_c T$,

$$\omega_c T = \sqrt{\frac{k_1^2}{a^2} - 1} \qquad (6.30)$$

where a is such that $k_1/a > 1$.

To summarize the procedure for lag compensation:

(i) Carry out a pure-gain design, observing the specified stability indicators. Identify ω_c, the crossover frequency.

(ii) Calculate the compensator dc gain required to meet the dc steady-state specifications. If the gain of step (i) exceeds this, stop; if not, go on.

(iii) Calculate k_1 so that the gain of step (i) multiplied by $(1 + k_1)$ equals the required compensator dc gain.

(iv) With $a \sim 0.1$ or less, compute the time constant T from Equation 6.30.

(v) Form the complete compensator by solving Equation 6.28 for $G_c(s)$ and multiplying by the gain of step (i).

Example 6.8   (dc Servo)

In Example 6.2, the dc servo of Example 2.1 (Chapter 2) was shown to have a dc steady-state error to a unit load torque of $-2.000/k$, with pure-gain feedback. The Root Locus design of Example 5.6 sets k at 0.657 to satisfy dynamics specifications. Modify the design, using a lag compensator, so that the dc steady-state error to a load torque of 0.01 Nm is $0.5°(= .00875$ rad) in magnitude.

Solution   A Bode plot of the loop gain with $k = 0.657$ shows that the crossover frequency is $\omega_c = 1.030$ rad/s, with a phase margin of 64.2°. To meet the specification,

$$|e_{ss}| = \frac{2.00}{k_0}(.01) = .00875$$

or

$$k_0 = 2.28.$$

Clearly, the pure-gain controller with $k = 0.657$ cannot meet the requirement. If the dc gain is to be 2.28, the lag compensator dc gain must be

$$k_0 = G_c(j0) = 2.28/.657 = 3.47.$$

From Equation 6.29,

$$k_1 = k_0 - 1$$

$$= 3.47 - 1 = 2.47.$$

The value of $a \ll 1$ is to be selected so as not to cause appreciable changes in the stability margin. Some trial and error may be necessary. With $a = 0.1$, from Equation 6.30,

$$T = \frac{1}{1.030}\sqrt{\frac{2.47^2}{.01} - 1}$$

$$\approx \frac{1}{1.030}\frac{2.47}{.1} = 24.0 \text{ s}.$$

From Equation 6.28, the complete compensator, including the original gain of 0.657, is

$$G_c(s) = 2.28\frac{6.92s + 1}{24.0s + 1}.$$

The new phase margin is 58.4°, and the crossover is 1.039 rad/s. Figure 6.19 (MATLAB step) shows closed-loop responses to a step load torque of -0.01 Nm with and without the lag compensator, and Figure 6.20 shows the response to a unit step in the set point θ_d for the compensated system.

Lag Compensation: Root Locus Viewpoint

As shown by Equation 6.28, the lag compensator is of the form

$$G_c(s) = k\frac{\alpha Ts + 1}{Ts + 1}, \qquad \alpha = 1/(1 + k_1) < 1.$$

When $G_c(s)$ is cascaded with the plant, the loop gain $L(s)$ acquires a real pole at $-1/T$ and a real zero at $-1/\alpha T$. Now T is relatively large compared to the plant time constants, because the lag is meant to act at frequencies well below the plant bandwidth. This means that the pole-zero pair introduced by G_c will be relatively close to the origin.

Figure 6.21 shows the pole-zero pair and a piece of the root locus *before* the introduction of the lag compensator. Let s^* be a point on the original locus for the plant $P(s)$, and suppose that it is a closed-loop pole for some gain k^*. Then, by Root Locus definitions,

$$\angle P(s^*) = 180° \tag{6.31}$$

$$k^* = -\frac{1}{P(s^*)}. \tag{6.32}$$

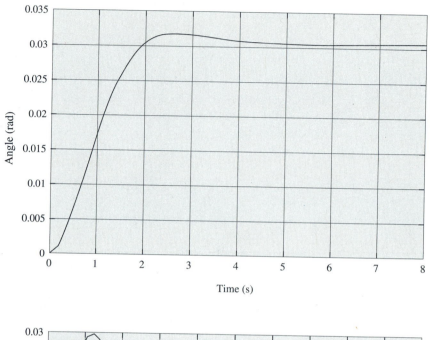

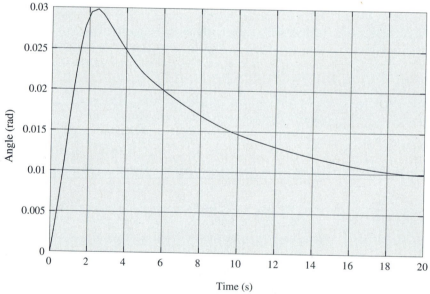

Figure 6.19 Disturbance response, pure gain, and lag compensation, dc servo

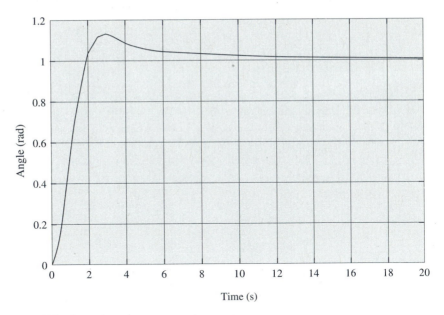

Figure 6.20   Set-point step response, dc servo

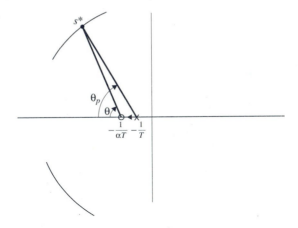

Figure 6.21   Root locus view of lag compensation

With the compensator in place, s^* will be a point on the new Root Locus if

$$\measuredangle G_c(s^*)P(s^*) = \measuredangle G_c(s^*) + \measuredangle P(s^*) = 180°$$

or, since

$$\measuredangle P(s^*) = 180°,$$

$$\measuredangle G_c(s^*) = 0. \qquad (6.33)$$

But, from Figure 6.21,

$$\measuredangle \, G_c(s^*) = \theta_z - \theta_p.$$

If T is large enough to ensure that $1/\alpha T \ll |s^*|$, then $\theta_z \approx \theta_p$, and Equation 6.33 is approximately satisfied. Therefore, some point near s^* will in fact satisfy the angle condition, and the Root Locus will be only slightly perturbed by the lag compensator, for $|s| \gg 1/T$.

With the lag, the Root Locus gain condition is

$$k \frac{\alpha T s^* + 1}{T s^* + 1} P(s^*) = -1.$$

By Equation 6.32,

$$k = k^* \frac{T s^* + 1}{\alpha T s^* + 1}.$$

If $|s^*| \gg 1/\alpha T$, then $|\alpha T s^*| \gg 1$ and

$$k \approx k^* \frac{T s^*}{\alpha T s^*} = \frac{k^*}{\alpha}.$$

Since $\alpha < 1$, it follows that $k > k^*$. In essence, the lag compensator allows a higher gain to be used for the same closed-loop poles as in the pure-gain case. The same stability properties are thus achieved with a higher gain, improving the dc steady-state behavior.

There is one important difference in the Root Locus. As Figure 6.21 shows, there is a new real closed-loop pole between $-1/T$ and $-1/\alpha T$; it is relatively near the origin, and hence relatively slow. This pole accounts for the slow tail in the load torque response of Figure 6.19. The tail is also present in the set-point response of Figure 6.20, since the output must eventually go to 1, but is much less obvious then in Figure 6.19. This is because $y/y_d = T(s)$ has the zeros of $L(s)$ as its zeros, so that $T(s)$ has a zero at $-1/\alpha T$ that almost cancels the slow closed-loop pole. On the other hand, the closed-loop transmission from T_L to y has the same poles but does not have a zero that would cause near cancellation of the slow lag pole; hence the sizeable contributions of that pole to the response. One conclusion can be drawn from this discussion: we should not make T larger than necessary, because the slow closed-loop pole created by the lag may appear in some of the transients.

Proportional-Integral Control

Proportional-integral (PI) control is essentially a limiting case of the lag compensation. It is obtained by letting T go to infinity while keeping the ratio k_1/T finite.

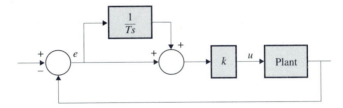

Figure 6.22   Proportional-integral (PI) control

As shown in Figure 6.22, the PI controller has the transfer function

$$G_c(s) = k\left(1 + \frac{1}{Ts}\right)$$

$$= k\frac{Ts+1}{Ts}. \qquad (6.34)$$

In this equation, k is the gain that achieves the desired degree of stability in the pure-gain design. The integration increases the type number of the system by one—from Type 0 to Type 1, from Type 1 to Type 2, and so on. The design approach is similar to that used with lag compensation except that there is no counterpart of the gain k_1, because integration provides infinite gain at dc. The time constant T is chosen to make $G_c \approx k$ in the neighborhood of the crossover frequency ω_c. That will be true if

$$\frac{1}{\omega_c T} = a \ll 1. \qquad (6.35)$$

Example 6.9   **(Active Suspension)**

For the suspension system of Example 2.2 (Chapter 2), feedback is used from the output $y = x_1 - x_2$ to the force input u. We wish to design a controller that will return the distance y between the two masses to some desired set point in the dc steady state, so as to maintain that distance in spite of load changes. We require gain and phase margins 6 db and 60°, respectively.

Solution   From Equation 3.23 (Chapter 3),

$$\frac{y}{u} = \frac{0.02334(s^2 + 85.861)}{(s^2 + 12.305s + 639.76)(s^2 + 1.6954s + 9.3787)}.$$

First we design a pure-gain controller. Figure 6.23 shows the Bode plots. Since the loop gain magnitude is $-\infty$ db where the phase is 180° ($\omega = 9.26$ rad/s), a pure-gain controller with any positive value of k achieves the gain margin. The phase is $-180° + 60° = -120°$ at $\omega = 3.51$ rad/s, where the magnitude is -67.7 db.

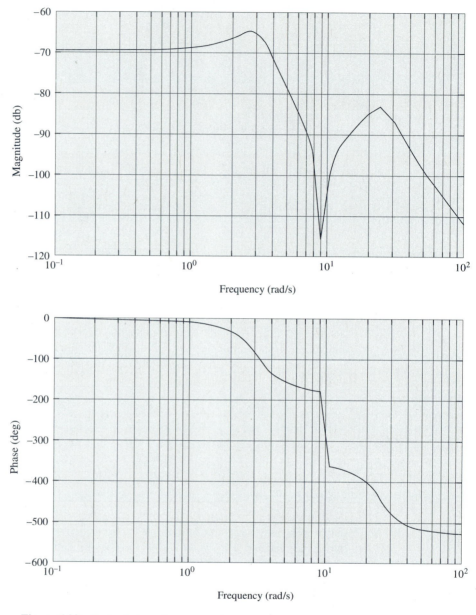

Figure 6.23   Bode plots, active suspension

A gain of 67.7 db makes that frequency the crossover frequency, and the phase margin specification is satisfied for $k \leq 67.7$ db, for which the gain margin also holds.

   To meet the dc requirement of zero error, we rely on PI control. We use the maximum value of k consistent with the stability margins because this also maximizes the crossover frequency and hence the bandwidth; so $k = 67.7$ db ($k = 2427$).

With $a = 0.1$, Equation 6.35 yields

$$T = \frac{1}{a\omega_c} = \frac{1}{(.1)(3.51)} = 2.85 \text{ s.}$$

The PI controller is

$$G_c(s) = 2427\left(1 + \frac{1}{2.85s}\right).$$

6.3.4 Lead and Proportional-Derivative Compensation

Lead Compensation in the Frequency Domain
Lead compensation is used to increase the bandwidth of the closed-loop system. This is done by raising the crossover frequency, which (as the reader will recall) is roughly equal to the bandwidth.

To understand the principle, refer to Figure 6.15. It is easy to see that, for a phase margin of 45°, the crossover frequency must be set at $\omega = 1.0$ rad/s; the reason is that $\angle L(j\omega) = -135°$ at $\omega = 1.0$. Suppose that the crossover frequency is to be doubled, to 2.0 rad/s *while the 45° phase margin is maintained*. Since the phase at $\omega = 2.0$ must be $-135°$, it will be necessary to boost the phase of $L(j2)$ from $-190.3°$ to $-135°$, a difference of 55.3°. The purpose of lead compensation is to provide this phase boost, i.e., to provide phase lead near crossover.

A lead compensator has a transfer function

$$G_c(s) = k\frac{\alpha Ts + 1}{Ts + 1}, \qquad \alpha > 1. \tag{6.36}$$

As shown in Figure 6.24, (for $\alpha = 10$), G_c provides phase lead in the midfrequency range. In that range, the compensator adds roughly 20 db/decade to the slope of the magnitude plot of the plant. This is consistent with Bode's theorem, insofar as it raises the slope in a region where it would normally be too negative to allow placement of the crossover frequency.

To calculate the maximum phase lead and the frequency where it occurs, we write

$$\phi = \angle G_c(j\omega) = \tan^{-1}\alpha\omega T - \tan^{-1}\omega T$$

and differentiate with respect to ω. This yields

$$\frac{d}{d\omega}\phi = \frac{\alpha T}{1 + (\alpha\omega T)^2} - \frac{T}{1 + (\omega T)^2}.$$

To maximize, set the RHS to zero. This yields $\omega T = 1/\sqrt{\alpha}$. To calculate the maximum phase, we compute $\angle G_c(j\omega)$ with $\omega T = 1/\sqrt{\alpha}$. After some trigonometric manipulations, this turns out to be $\angle G_c = \sin^{-1}(\alpha - 1)/(\alpha + 1)$.

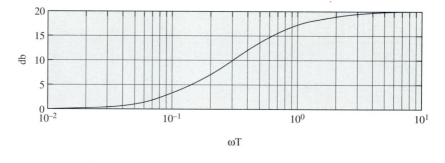

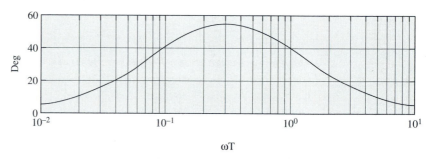

Figure 6.24 Bode plots for a lead compensator

To summarize,

$$\phi_{\max} = \sin^{-1} \frac{\alpha - 1}{\alpha + 1} \qquad (6.37)$$

which occurs at

$$\omega_{\max} = \frac{1}{T\sqrt{\alpha}}. \qquad (6.38)$$

We shall need one more result. We compute

$$\begin{aligned}
|G_c(j\omega_{\max})| &= k \left[\frac{\alpha^2 \omega_{\max}^2 T^2 + 1}{\omega_{\max}^2 T^2 + 1} \right]^{1/2} \\
&= k \left[\frac{\alpha + 1}{(1/\alpha) + 1} \right]^{1/2} \\
&= k\sqrt{\alpha}
\end{aligned}$$

We redefine the gain in Equation 6.36 and use

$$G_c(s) = \frac{k_1}{\sqrt{\alpha}} \frac{\alpha T s + 1}{T s + 1}. \qquad (6.39)$$

This normalization has the advantage that $|G_c(j\omega_{\max})| = k_1$.

To design a lead compensator, we follow these steps:

(i) Ascertain the phase to be added at the desired crossover frequency in order that a desired phase margin be achieved. Let this phase be ϕ_{max}.

(ii) Calculate α from Equation 6.37, given ϕ_{max}.

(iii) With ω_{max} equal to the desired crossover frequency, calculate T from Equation 6.38.

(iv) Since $|G_c(j\omega_{max})| = k_1$, k_1 is just that gain required to make $|k_1 L(j\omega_{max})| = 1$; in decibels, $k_1 = -|L(j\omega_{max})|$ db.

Example 6.10   **(dc Servo)**

The dc servo pure-gain design of Example 5.6 (Chapter 5) ($k = 0.657$) turns out to have a phase margin of 64.2° and a crossover frequency of 1.030 rad/s. Design a lead compensator such that the same phase margin is obtained, but at a crossover frequency of 3 rad/s.

Solution   At $\omega = 3$ rad/s, the magnitude of the loop gain is -12.6 db and the phase is $-150.0°$. To preserve the phase margin of 64.2° with a crossover at 3 rad/s, we must have

$$\measuredangle L(j3) = -(180° - 64.2°) = -115.8°.$$

Therefore, we need to add $150.0° - 115.8° = 34.2°$ of phase lead at $\omega = 3$. From Equation 6.37,

$$\sin 34.2° = 0.562 = \frac{\alpha - 1}{\alpha + 1}$$

which, solving for α, yields $\alpha = 3.56$. Using Equation 6.38,

$$T = \frac{1}{3\sqrt{3.56}} = 0.177 \text{ s}.$$

Finally, the gain k_1 is given by

$$k_1 = -|L(j3)| = 12.6 \text{ db, or } 4.26.$$

The lead compensator to be added is

$$G_c(s) = \frac{4.26}{\sqrt{3.56}} \frac{(3.56)(.177)s + 1}{0.177s + 1}$$

$$= 2.258 \frac{0.630s + 1}{0.177s + 1}.$$

This compensator is cascaded with the pure-gain controller ($k = 0.657$). The complete controller is

$$G_c(s) = 1.483 \frac{0.630s + 1}{0.177s + 1}.$$

Figure 6.25 shows the Bode plots with and without lead compensation.

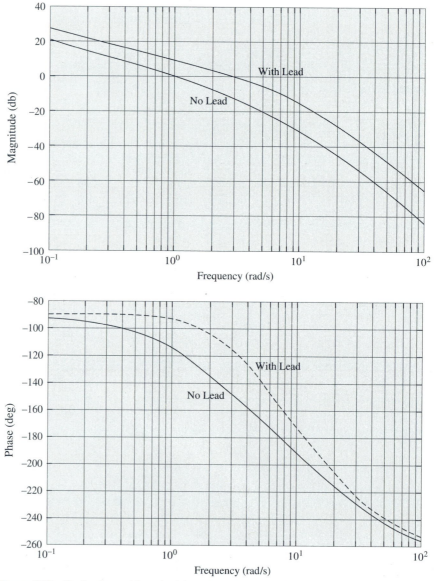

Figure 6.25   Bode plots with and without lead compensation, dc servo

Proportional-Derivative Compensation

A proportional-derivative (PD) compensator has the transfer function

$$G_c(s) = k(Ts + 1). \tag{6.40}$$

It may be viewed as a limiting form of the lead compensator, as T tends to 0 and αT tends to a finite value. Since G_c is not proper, it cannot be realized unless the derivative of the error can be measured as well as the error.

The phase of the compensator at the desired crossover frequency ω_c is

$$\measuredangle G_c(j\omega_c) = \tan^{-1} \omega_c T. \tag{6.41}$$

To normalize the magnitude at ω_c, we write

$$G_c(s) = \frac{k_1}{\sqrt{\omega_c^2 T^2 + 1}}(Ts + 1) \tag{6.42}$$

so that $|G_c(j\omega_c)| = k_1$.

The design procedure parallels that for the lead. Using Equation 6.42, T is chosen to obtain a desired phase lead at $\omega = \omega_c$. The gain k_1 is the gain needed to make $k_1|L(j\omega_c)| = 1$; i.e., $k_1 = -|G_c(j\omega_c)|$ db.

6.3.5 Lead-Lag and Proportional-Integral-Derivative Control

Lead-lag compensation is simply a combination of both types of compensation. First a lead compensator is designed to push the bandwidth to some desired value. Then a lag compensator is added to the design to shape the low-frequency portion of the loop gain. The result is a compensator of the form

$$G_c(s) = k\frac{\alpha T_1 s + 1}{T_1 s + 1}\frac{\beta T_2 s + 1}{T_2 s + 1} \tag{6.43}$$

where $\alpha < 1$ and $\beta > 1$.

Example 6.11   **(dc Servo)**

Design a lead-lag compensator for a dc servo that satisfies the crossover-frequency and phase-margin specifications of Example 6.10 and the dc specifications of Example 6.8.

Solution   The first part of the design is exactly as in Example 6.10. The resulting compensator is

$$G_{c1}(s) = 1.483\frac{0.630s + 1}{0.177s + 1}.$$

In Example 6.8, it was determined that the dc gain of the compensator had to be at least 2.28. The dc gain of $G_{c1}(s)$ is 1.483, so the lag compensator must contribute

$$k_0 = G_{c2}(j0) = \frac{2.28}{1.483} = 1.537.$$

From Equation 6.29,

$$k_1 = 1.537 - 1 = 0.537.$$

Note that the process of increasing the crossover frequency has increased the gain at all frequencies, as Figure 6.25 shows. This means that the lag is not called upon to deliver as much additional gain as in Example 6.5.

From Equation 6.30, with $a = 0.1$ and $\omega_c = 3$ rad/s,

$$T = \frac{1}{3}\sqrt{\frac{.537^2}{.1^2} - 1}$$

$$= 1.76 \text{ s}.$$

Because this time constant is much smaller than that in Example 6.8, it should avoid the long, slow tail in the transients. The lag portion is

$$G_{c2}(s) = 1.537\frac{1.14s + 1}{1.76s + 1}.$$

The complete compensator is

$$G_c(s) = G_{c2}(s)G_{c1}(s)$$

$$= 2.28\frac{(1.14s + 1)\,(.630s + 1)}{(1.76s + 1)\,(.177s + 1)}.$$

Figures 6.26 and 6.27 show, respectively, the responses to a unit step in the set-point θ_d and to a step load torque of -0.01 Nm. These should be compared to the responses of Figures 6.19 and 6.20.

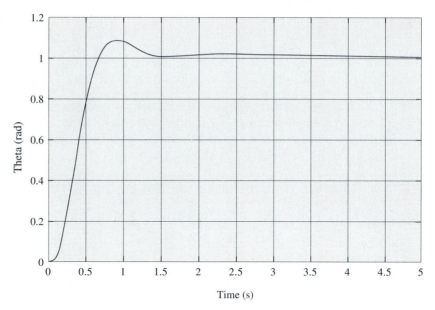

Figure 6.26   Step response, dc servo with lead-lag compensation

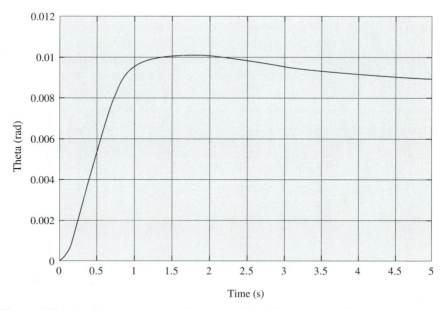

Figure 6.27   Load torque response, dc servo with lead-lag compensation

◆ ◆ ◆ **R E M A R K**

The two steps in the procedure should be performed as in Example 6.11. The lead is designed first and generally affects the loop gain at all frequencies. The lag follows and, because of its minimal effect on the loop gain in the neighborhood of crossover, does not undo the effect of the lead. ◆

Proportional-Integral-Derivative Compensation

The proportional-integral-derivative controller is the workhorse of the process industry, where it is often called a "three-term controller." If we cascade a PI controller with a PD controller, we obtain

$$G_c(s) = k\left(\frac{1}{T_1 s} + 1\right)(T_2 s + 1)$$

$$= k\left(1 + \frac{T_2}{T_1}\right) + \frac{k}{T_1 s} + kT_2 s$$

$$= \kappa\left(1 + \frac{1}{T_I s} + T_D s\right) \tag{6.44}$$

where

$$\kappa = k(1 + T_2/T_1) \text{ is the } \textit{proportional gain}$$

$$T_I = T_1 + T_2 \text{ is the } \textit{integral, or reset, time}$$

$$T_D = T_1 T_2/(T_1 + T_2) \text{ is the } \textit{derivative, or rate, time.}$$

In the form of Equation 6.44, the PID controller is seen to provide separate adjustments of low frequencies (through the integral term), midfrequencies (through the proportional term), and high frequencies (through the derivative term). This accounts for its great flexibility and for its widespread use in practice.

Example 6.12   **(dc Servo)**

Design a PID controller for the dc servo that satisfies the crossover-frequency and phase-margin specifications of Example 6.10.

Solution   We begin with a PD controller. As in Example 6.10, we must provide 34.2° of phase lead at $\omega = 3$.

$$\angle(j3T_2 + 1) = \tan^{-1} 3T_2 = 34.2°$$

$$T_2 = 0.226 \text{ s.}$$

The PI term is such that $\frac{1}{j\omega T_1} + 1 \approx 1$ at crossover, i.e.,

$$\frac{1}{3T_1} = \alpha \ll 1.$$

Take $\alpha = 0.1$, so that $T_1 = 3.33$ s.

We need a loop gain of unit magnitude at $\omega = 3$; since $|L(j3)| = -12.6$ db, the controller must contribute 12.6 db, or 4.26, at $\omega = 3$. Thus,

$$k \left| \frac{1}{j(3.33)(3)} + 1 \right| |j(3)(.226) + 1| = 4.26$$

or

$$k = \frac{4.26}{(1.005)(1.208)} = 3.51.$$

Therefore,

$$G_c(s) = 3.51 \left(\frac{1}{3.33s} + 1 \right)(0.226s + 1)$$

$$= 3.75 \left(1 + \frac{1}{3.556s} + 0.211s \right).$$

6.4 FEEDFORWARD CONTROL

When feedforward is used, it is normally in addition to feedback, as in Figure 6.28. The transfer function from disturbance to output is

$$\frac{y}{w} = P_w S - QPS = (P_w - QP)S. \tag{6.45}$$

The feedforward design problem is to choose a stable Q such that $(P_w - QP)S$ is close to zero. Since $|S(j\omega)|$ is usually small at low frequency, the designer can focus on a frequency range where $|S(j\omega)|$ becomes appreciable; that will take place near crossover. In other words, feedback is used to handle the low frequencies, and feedforward to extend the bandwidth beyond the crossover frequency. It is understood that the transfer functions P_w and P must be relatively accurate in the frequency range where feedforward is to act.

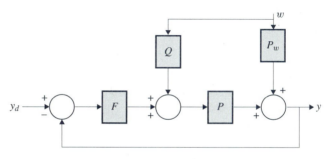

Figure 6.28 Feedforward compensation

We want $Q(j\omega) \sim [P_w(j\omega)]/[P(j\omega)]$ in a frequency region near crossover. It is often sufficient to approximate this with a simple lead or lag [6], or in some cases with a constant.

Example 6.13 (dc Servo)

In Example 6.8, a lag compensator was designed for the dc servo, with a resulting closed-loop bandwidth of about 1 rad/s. The load torque is assumed available for measurement, to be used as an input for feedforward control. Design such a control to improve the response to disturbances up to 10 rad/s.

Solution From Equations 3.21 and 3.22,

$$P(s) = \frac{\theta}{v} = \frac{88.76}{s(s + 21.526)(s + 2.474)}$$

$$P_w(s) = \frac{\theta}{T_L} = \frac{-7.396(s + 24)}{s(s + 21.526)(s + 2.474)}$$

so that

$$\frac{P_w}{P} = \frac{-7.396}{88.76}(s + 24) = -2.00\left(\frac{s}{24} + 1\right).$$

Over the specified frequency range, the approximation $Q = -2.00$ is valid, so the feedforward control is

$$v_{ff} = -2.00 T_L.$$

Figure 6.29 shows the magnitude of the frequency response from disturbance to output, with and without feedforward.

6.5 2-DOF CONTROL

The two-degrees-of-freedom (2-DOF) configuration of Chapter 4 is repeated in Figure 6.30. In Chapter 4, the transfer function from set point to output was calculated to be

$$\frac{y}{y_d} = RPS. \tag{6.46}$$

Recall that R may have RHP poles to cancel any RHP zeros of PS *except* those that are also RHP zeros of P.

Since $F = T/SP$, we write Equation 6.46 as

$$\frac{y}{y_d} = \frac{R}{F}T. \tag{6.47}$$

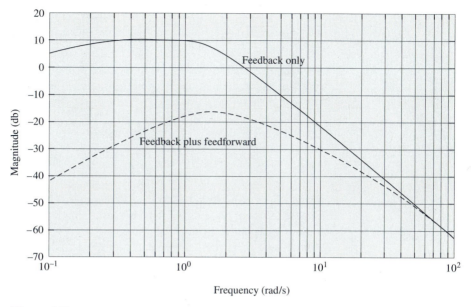

Figure 6.29 Disturbance frequency response with and without feedforward, dc servo

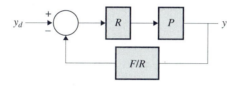

Figure 6.30 A 2-DOF feedback system

We choose $R(s) = F(s)R'(s)$, so that

$$\frac{y}{y_d} = R'(s)T(s). \tag{6.48}$$

Assuming that there are no unstable pole-zero cancellations between F and P, the unstable poles of F are zeros of S and are thus allowable. It will suffice to make R' stable to satisfy the conditions for internal stability.

There are two cases of interest, depending on whether the bandwidth of the response to y_d is to be less or greater than that of $T(j\omega)$:

Case 1: Decreased bandwidth. We may wish to do this to avoid saturating the actuators in response to sudden changes in the set point, or perhaps to meet a requirement for a smooth, overdamped response. Since $T(j\omega) \sim 1$ in the frequency band under consideration, $R'(s)$ may be any low-pass function.

Case 2: Increased bandwidth. Here, we wish to have $R'T \sim 1$ over a wider range of frequencies than the bandwidth of T. Since $T \sim 1$ up to about crossover, we must choose $R'(s)$ to be approximately the inverse of T near crossover and beyond. If the frequency range is extended only moderately, T^{-1} may be approximated by a simple transfer function.

An instance of Case 1 often found in practice is the modification of a PID design to lower the bandwidth of the set-point response. The PID controller is

$$F(s) = k\left(\frac{1}{T_1 s} + 1\right)(T_2 s + 1)$$

$$= \frac{k}{T_1}\frac{(T_1 s + 1)(T_2 s + 1)}{s}.$$

Typically, T_1 will be larger than T_2. If the bandwidth is greater than $1/T_1$, it is reasonable to use $R'(s) = \frac{1}{T_1 s + 1}$, which, from Equation 6.48, will be approximately the set-point response.

From Figure 6.30, the two compensator blocks will be

$$\text{Forward path:} \quad R = FR' = \frac{k}{T_1}\frac{T_2 s + 1}{s}$$

$$\text{Feedback path:} \quad \frac{F}{R} = \frac{1}{R'} = T_1 s + 1.$$

The effect is to differentiate only the measurement and not the reference, so that step changes in the latter do not produce "spikes" on the control. Note that this design has a PI control in the forward path and a *PD* in the feedback path.

Example 6.14   **(Active Suspension)**

In Example 6.9, a PI controller was designed for the active suspensions. The bandwidth of 3.51 rad/s would suggest a time constant of the order $0.3s$, probably too fast to lead to comfortable set-point changes. A response to set point with a time constant of the order of $3s$ is desired. Use a 2-DOF design to achieve this.

Solution   The PI controller is

$$F(s) = G_c(s) = 2427\frac{(2.85s + 1)}{2.85s}.$$

Since $1/2.85$ is well below the bandwidth of 3.51 rad/s, we choose $R'(s) = 1/(2.85s + 1)$. The time constant of $2.85s$ is close enough to $3s$, and leads to simplification.

The two compensators are:

$$\text{Forward path:} \quad R = FR' = \frac{851.6}{s}$$

$$\text{Feedback path:} \quad \frac{F}{R} = \frac{1}{R'} = 2.85s + 1.$$

Example 6.15   (dc Servo)

In Example 6.11, a lead-lag compensator was used to achieve a bandwidth of 3 rad/s, with desirable low-frequency properties. Design a 2-DOF system to extend the bandwidth to 10 rad/s, for the set-point response.

Solution   The compensator from Example 6.11 is

$$F(s) = G_c(s) = 2.28 \frac{(1.14s + 1)(0.630s + 1)}{(1.76s + 1)(0.177s + 1)}.$$

Figure 6.31 shows the inverse complementary sensitivity $T(j\omega)$. A rough approximation is given by $s/4 + 1$. To obtain a proper compensator, we use a lead:

$$R' = \frac{(s/4) + 1}{(s/40) + 1}.$$

The two compensators are:

Forward path: $F = 2.28 \dfrac{(1.14s + 1)(.630s + 1)(.25s + 1)}{(1.76s + 1)(.177s + 1)(.025s + 1)}$

Feedback path: $\dfrac{1}{R'} = \dfrac{0.25s + 1}{.25s + 1}.$

Figure 6.32 shows the responses for the 1-DOF and 2-DOF designs.

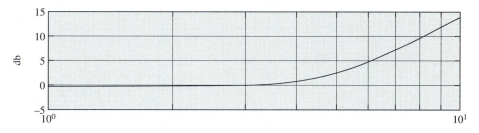

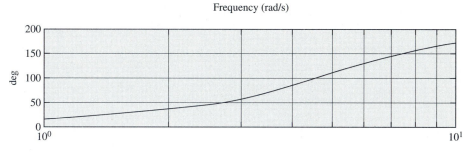

Figure 6.31   Inverse of complementary sensitivity

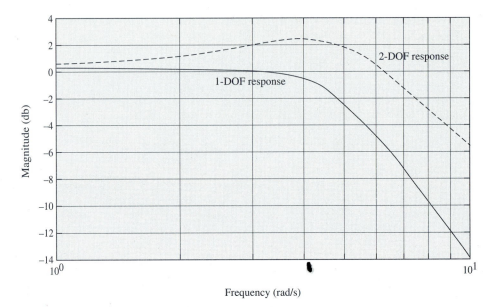

Figure 6.32   Frequency responses, 1-DOF and 2-DOF designs

6.6 SYSTEMS WITH DELAY

System descriptions often include pure delays. For example, in a system that includes a communication link, the finite speed of light causes delay; this is particularly important if part of the system is on another planet. Another, more down-to-earth example of delay is the time taken to transport materials on a conveyor belt or via a moving web.

A pure delay has the input–output description

$$y(t) = u(t - T), \qquad T > 0. \tag{6.49}$$

This element does not have a finite-state description. From Equation 6.49, it is seen that the output $y(t), t > 0$, depends not only on future inputs but also on the immediate past inputs, i.e., on $u(t), -T < t \le 0$. We may think of this as a delay line, where what first comes out is what is initially stored in the line. Thus, the state is actually a function, $u(t), -T < t \le 0$. Since this function has an infinite number of points, it cannot be represented by a finite set of numbers. (Approximations are possible, of course.)

Taking Laplace transforms of Equation 6.49 yields

$$y(s) = e^{-sT} u(s). \tag{6.50}$$

The transfer function e^{-sT} is analytic everywhere in the complex plane, including the j-axis. Since

$$|e^{-j\omega T}| = 1 \qquad (6.51)$$

$$\measuredangle\, e^{-j\omega T} = -\omega T \qquad (6.52)$$

the delay element is "all-pass" and has phase lag increasing linearly with frequency.

As we will show presently, frequency-response methods are applicable to systems with delay.

Example 6.16   For $P(s) = ke^{-sT}/(\tau s + 1)$, compute the range of k for stability, given unity feedback.

Solution   With $P'(s) = e^{-sT}/(\tau s + 1)$, it is easy to see that

$$|P'(j\omega)| = \frac{|e^{-j\omega T}|}{|j\omega\tau + 1|} = \frac{1}{(\tau^2\omega^2 + 1)^{1/2}}$$

$$\measuredangle\, P'(j\omega) = -\omega T - \tan^{-1}\omega\tau.$$

As ω increases, the magnitude decreases monotonically while the phase becomes indefinitely more negative. The result, shown in Figure 6.33, is a spiral Nyquist plot.

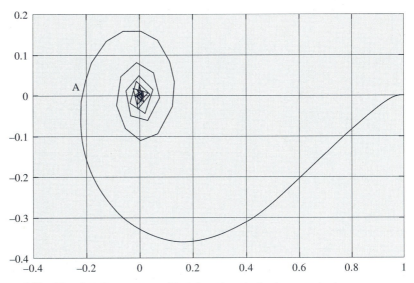

Figure 6.33   Nyquist plot, system with delay plus single time constant

Because there are no RHP poles, we require zero encirclements, so the point $(-1/k, 0)$ must be to the left of point A. The system is simple enough to allow some analytical headway. Since the phase is $180°$ at A,

$$-\omega T - \tan^{-1}\omega\tau = -\pi - 2\pi n$$

where $n =$ a nonnegative integer. Each value of n corresponds to a negative-real-axis crossing. Point A corresponds to the lowest-frequency crossing, i.e., to $n = 0$. Therefore, the frequency satisfies

$$\tan^{-1}\omega\tau = \pi - \omega T$$

or

$$\omega\tau = \tan(\pi - \omega T)$$
$$\frac{\tau}{T}(\omega T) = -\tan\omega T.$$

This transcendental equation has no analytic solution. Figure 6.34 shows the functions $-\tan\omega T$ and $\frac{\tau}{T}(\omega T)$ versus ωT. The intersection point ω_0 (the lowest-frequency crossing of the negative real axis) is near $\omega T = \pi/2$ for large τ/T, and nearer $\omega T = \pi$ for small τ/T.

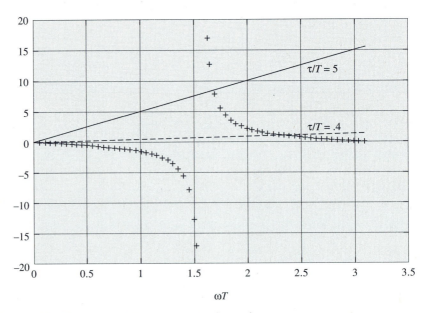

Figure 6.34 Illustration of the transcendental equation

For stability, the point $(-1/k, \ 0)$ must be to the left of A, so

$$-\frac{1}{k} < -|P'(j\omega_0)|$$

$$k < \frac{1}{|P'(j\omega_0)|} = (\tau^2\omega_0^2 + 1)^{1/2} = \left(\frac{\tau^2}{T^2}\omega_0^2 T^2 + 1\right)^{1/2}.$$

If τ/T is large (the delay is small with respect to the time constant), then the upper limit for stable k is also large. This is what we expect, because, with no delay at all $(T = 0)$, the first-order system is stable for all k. As τ/T decreases (delay becomes more important), the stability limit also decreases. The following table lists a few values.

τ/T	$\omega_0 T$	Limit on k
.1	2.86	1.04
.5	2.29	1.52
1	2.03	2.26
2	1.84	3.81
5	1.69	8.50
10	1.63	16.35

Unfortunately, many of the techniques we have studied so far (such as the Routh criterion and the Root Locus) and the state methods still to come require finite-state representation. This motivates us to approximate the delay with a rational transfer function called a Padé approximation. The idea is to match the leading terms of the series for e^{-sT}. Since e^{-sT} is all-pass, we try the all-pass transfer function

$$H(s) = \frac{-as + 1}{as + 1}, \qquad a > 0.$$

The reader is invited to show, by long division, that

$$H(s) = 1 - 2as + 2a^2s^2 - 2a^3s^3 + \cdots.$$

Compare this to

$$e^{-sT} = 1 - sT + \frac{1}{2}s^2 T^2 - \frac{1}{6}s^3 T^3 + \cdots.$$

With $a = T/2$, the two series match up to and including the term in s^2. The *first-order* Padé approximation is

$$H(s) = \frac{(-T/2)s + 1}{(T/2)s + 1}. \tag{6.53}$$

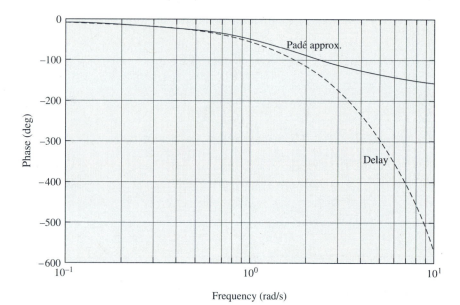

Figure 6.35 Phase curves for the pure delay, Padé delay, and Padé approximation

For comparison, the phases for e^{-sT} and for the Padé approximation are shown in Figure 6.35 for $T = 1$. (In both cases the magnitudes are unity at all frequencies.) In those (frequent) cases where the delay multiplies a rational transfer function, the replacement of e^{-sT} by the Padé approximation can be assessed by the method of Chapter 5, Section 5.6, using the concept of an unstructured uncertainty.

The plant is $e^{-sT} P(s)$, and the model is $\frac{(-T/2)s+1}{(T/2)s+1} P(s)$. With $\Delta(s)$ as the multiplicative uncertainty, we have

$$[1 + \Delta(s)]\frac{(-T/2)s + 1}{(T/2)s + 1} P(s) = e^{-sT} P(s)$$

and therefore

$$\Delta(s) = \frac{(T/2)s + 1}{(-T/2)s + 1} e^{-sT} - 1.$$

The magnitude bound is

$$\ell(j\omega) = \left| \frac{\frac{1}{2} j\omega T + 1}{-\frac{1}{2} j\omega T + 1} e^{-j\omega T} - 1 \right|.$$

This was computed and is plotted in Figure 6.36. We see that $\ell(j\omega)$ is less than 1 for $\omega T < 0.614$. According to the results of Section 5.6, that is the bandwidth limit of a design that uses this Padé approximation.

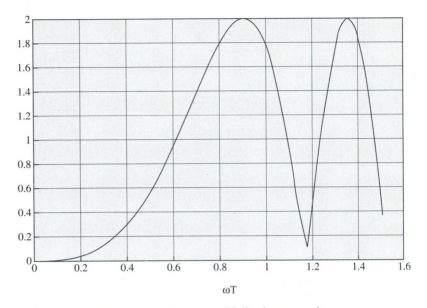

Figure 6.36 The Padé approximation as a multiplicative uncertainty

Example 6.17 Replace the delay in Example 6.16 with a first-order Padé approximation, and use the Routh criterion to predict the stability range for k.

Solution With the approximation, we obtain the closed-loop characteristic equation from

$$1 + k\frac{(-T/2)s + 1}{(T/2)s + 1} \cdot \frac{1}{\tau s + 1} = 0$$

$$\frac{1}{2}\tau T s^2 + \left(\frac{1}{2}T + \tau - \frac{1}{2}kT\right)s + k = 0.$$

Assuming τ, $T > 0$, it follows easily that, for stability,

$$\frac{1}{2}T + \tau - \frac{1}{2}kT > 0, \qquad k > 0$$

or

$$0 < k < 1 + 2\frac{\tau}{T}.$$

The correspondence with the results of Example 6.16 is rather coarse. Even for $\tau/T = 0.1$, the upper limit is 15% higher than the true value.

The standard control structures may be applied to systems with delay, of course, but a controller structure exists that is especially adapted to such systems. The *Smith predictor* [7] uses the structure shown in Figure 6.37, where the transfer function $P_m(s)$ is a model of $P(s)$, the plant, with the delay removed. The controller, which is the portion in the dotted line, requires a delay line for implementation. In a digital control system, that is accomplished by storing past values of the signal to be delayed.

The controller transfer function is

$$\frac{u}{e} = \frac{F}{1 + P_m F(1 - e^{-sT})}$$

and the overall transmission is

$$\frac{y}{y_d} = \frac{\frac{FPe^{-sT}}{1 + P_m F(1 - e^{-sT})}}{1 + \frac{FPe^{-sT}}{1 + P_m F(1 - e^{-sT})}}$$

$$= \frac{FPe^{-sT}}{1 + FP_m - FP_m e^{-sT} + FPe^{-sT}}.$$

If $P_m = P$, then

$$\frac{y}{y_d} = \frac{FP}{1 + FP} e^{-sT}. \tag{6.54}$$

The transfer function $FP/(1 + FP)$ is the closed-loop transmission obtained by using the controller F with the plant, with the delay removed. That suggests the following design procedure for a Smith predictor: design a compensator $F(s)$ for the plant without the delay, and use F in the controller of Figure 6.37. The resulting dynamic behavior is simply that of the control loop designed without the delay, except that all responses are delayed by T.

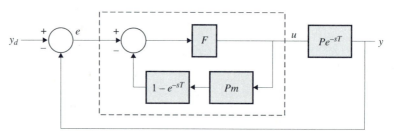

Figure 6.37 The Smith predictor

Problems

6.1   Given $P(s) = [-(s-1)]/[(s+1)^2(s^2+s+1)]$ and pure-gain control (gain $k > 0$):

    a. Calculate the range of values of gain k for which the controller yields a gain margin of at least 6 db and a phase margin of at least $50°$.

    b. Calculate, as a function of k, the steady-state error for a unit-step set-point input, assuming stability.

6.2   Repeat Problem 6.1 for $P(s) = [(s+1)]/[s(s^2+s+1)(s+4)]$, and calculate the steady-state error for a unit-ramp set-point input.

6.3   To investigate the concept of dominant complex poles, consider

$$H_d(s) = \frac{1}{(Ts+1)(s^2+2\zeta s+1)}.$$

    a. With $\zeta = 0.7$, compute the step response and obtain the peak overshoot, rise time, and settling time for $T = 0$. Repeat for increasing values of T. For what values of T (approximately) do the overshoot, use time, and settling time, respectively, change by more than 20% from their values of $T = 0$?

    b. Repeat part (a) for $\zeta = .4$ and $\zeta = .1$. Describe the influence of the pole at $-1/T$ as a function of ζ.

6.4   Repeat Problem 6.3 for $H_d(s) = (Ts+1)/(s^2+2\zeta s+1)$.

6.5   Quarter-cycle damping is defined as the case where the overshoot of each peak of the step response is $1/4$ that of the previous peak. Calculate the value of ζ that will lead to this.

6.6   Prove Equation 6.18.

6.7   *Servo with flexible shaft*   The dc servo of Problem 2.5 (Chapter 2) (or, equivalently, of Problem 3.14 in Chapter 3) has two pairs of complex, underdamped poles. One way of controlling such poles is by placing LHP complex zeros relatively near; since closed-loop poles migrate to zeros as the gain is increased, zeros "attract" the underdamped poles away from the j-axis.

    Consider the compensator

$$F(s) = k\frac{(s+100\pm j100)(s+5\pm j5)}{(s+200)^4}.$$

    a. Obtain the Root Locus.

    b. Calculate the range of k for stability.

c. Calculate (roughly) the value of k for which the lowest of the damping factors associated with any pair of complex poles is maximized; i.e., maximize the smallest damping factor.

d. For k as in part (c), compute the closed-loop step response.

6.8 ***Drum speed control*** In Problem 3.46 (Chapter 3), the drum speed control system is divided into two independent subsystems: one with a common-mode input u_c and an output ω_0, and the other with a differential-mode input u_d and an output $\omega_1 - \omega_2$. Each system is characterized by a pair of underdamped poles.

In line with the idea outlined in Problem 6.7 for dealing with underdamped poles, the following compensators are proposed:

$$F_1(s) = k_1 \frac{(-s + 250 \pm j250)}{(s + 800)^2}$$

for the subsystem with u_c as input, and

$$F_2(s) = k_2 \frac{(s + 250 \pm j250)}{(s + 800)^2}$$

for the other subsystems. For each subsystem:

a. Obtain the Root Locus.

b. Select the value of k_i, $i = 1$ or 2, for which the lowest of the damping factors associated with any pair of complex poles is maximized (approximately).

c. Sketch the block diagram for the complete (multivariable) control system.

d. Obtain (closed-loop) responses for unit-step changes in the references to ω_0 and $\omega_1 - \omega_2$.

6.9 ***Maglev*** In Problem 3.22 (Chapter 3), the inputs to the Maglev vehicle were redefined in terms of three new inputs: the levitation common-mode (Δu_{LC}) and differential-mode (Δu_{LD}) inputs, and the guidance differential-mode (Δu_{GD}) input. It was shown that the heave mode is affected only by Δu_{LC}, and that the transfer function has two real poles located symmetrically with respect to the origin.

a. Use a compensator $k[(s + p)/(s + 2p)]$ of the plant, where $-p$ is the stable pole, and select k so that the closed-loop poles have $\zeta = \sqrt{2}/2$.

b. For an initial state $\Delta z(0) = a$, $\Delta v_z(0) = 0$, what is the largest value of a such that (i) $|\Delta u_{LC}| \le 600$ and (ii) $|\Delta z(t)| \le .014$?

6.10 ***Maglev*** Following upon Problem 6.9, we wish to control the other two degrees of freedom. It makes physical sense to control Δy (lateral motion) with Δu_{GD} and $\Delta \theta$ (roll) with Δu_{LD}. This time, however, it is not true that Δu_{GD} affects only Δy and Δu_{LD} affects only $\Delta \theta$, but we ignore this cross-coupling in the first instance.

a. Proceed as in part (a) of Problem 6.9 to design each of the two loops.

b. Using the full, interacting model for an initial state $\Delta y(0) = a$, $\Delta v_y(0) = 0$, what is the largest value of a such that (i) $|\Delta u_{L1}|, |\Delta u_{L2}| \le 600$; (ii) $|\Delta u_{G1}|, |\Delta u_{G2}| \le 150$; (iii) $|\Delta y(t)| \le .014$; and (iv) $|\Delta\theta(t)| \le .014/1.2$.

c. Repeat part (b) for $\Delta\theta = a$, $\Delta\omega = 0$.

6.11 The step response of $H_d(s) = (13s + 5)/[(s + 1)(s + 5)]$ is shown in Figure 6.5.

a. Obtain M_p from a Bode magnitude plot.

b. Use Equation 6.18 to calculate ζ for an all-pole second-order system with the same M_p.

c. What would be the percent overshoot of the step response of an all-pole second-order system with that same ζ? Compare with the overshoot of H_d.

6.12 *Servo with flexible shaft*   The transfer function θ_2/v of the servo with flexible shaft was computed in Problem 3.14 (Chapter 3). From the Bode plot, calculate the range of values, if any, of the gain k of a pure-gain controller for which the closed-loop system is stable.

6.13 *Heat exchanger*   For the heat exchanger transfer function $\Delta T_{c3}/\Delta F_H$ of Problem 3.18 (Chapter 3):

a. Calculate, from the Bode plot, the range of values of gain k in a pure-gain controller for which the phase margin is at least $40°$.

b. For the maximum value of k allowable in part (a), use the Nichols chart to compute the maximum values of $|T(j\omega)|$ and $|S(j\omega)|$, and the frequencies at which these maxima occur. Compare with the crossover frequency.

c. With k as in part (a), calculate the dc steady-state error resulting from a step change of $1°$ in ΔT_{c0}.

6.14 *Chemical reactor*   For the transfer function $\Delta T/\Delta Q$ of Problem 3.19 (Chapter 3):

a. Compute the Bode plot and calculate the range of values of the gain k of a pure-gain controller for which the system is stable.

b. Using the Nichols chart, compute the stabilizing value of k such that the maximum value of $|T(j\omega)|$ is 3 db.

6.15 a. Design a lag compensator for the plant of Problem 6.1. The gain margin should be 6 db (nearly), but the steady-state error to a unit step should be 0.15.

b. Compute the Root Loci, with and without the lag.

c. Compute the error response to a unit-step set point, with and without the compensator.

6.16 **a.** Design a lag compensator for the plant of Problem 6.2. The phase margin should be 50° (nearly), but the steady-state error to a unit ramp should be 0.1.

b. Compute the Root Loci, with and without the lag.

c. Compute the error response to a unit-ramp input with and without the compensator.

6.17 *Heat exchanger*

a. Modify the design of Problem 6.13 by changing from proportional to PI control. Keep the same stability margins (approximately).

b. Obtain, for both the proportional and PI designs, the response ΔT_{c3} to unit step changes in the set point to ΔT_{c3} and in the disturbance ΔT_{c0}.

6.18 *Flow control*   The flow control example of Problem 2.10 (Chapter 2) has a model without dynamics. An integral control $\Delta u = (1/\tau s)(F_d - F)$ is to be used.

a. Using the results of Problem 2.16 (Chapter 2), calculate the range spanned by the small-signal gain $\Delta F/\Delta u$ as P_1^* ranges from 2 to 5 kpa and P_2^* lies between 0.5 and 1 kpa.

b. Calculate τ so that the crossover frequency of the linearized loop gain is at least 60 rad/s for all conditions in part (a). What is the maximum crossover frequency?

6.19 *Flow control*   The dynamics of the valve stem are actually not instantaneous, but are modeled as $u/u_d = \omega_0^2/(s^2 + \sqrt{2}\omega_0 s + \omega_0^2)$. The fact that these dynamics were not taken into account in Problem 6.18 raises the question of stability of the design. For the value of τ obtained in Problem 6.18, what condition must ω_0 satisfy in order that the small-signal closed-loop system be stable for all small-signal gains in the range obtained from Problem 6.18(a)? (Note: Stability of all possible small-signal models is necessary to the stability of a nonlinear system but is *not* sufficient.)

6.20 *Flow control*   It is often practical in process control to design flow control loops as a first step in the design process. Figure 6.38 shows a structure, known as *cascade control* in the process industries, for the level-control problem of Example 2.4 (Chapter 2). The level ℓ appears as an input to the flow, because $F_{\text{out}} = cu\sqrt{\ell}$. The flow loop is the *inner loop*, and the level loop is the *outer loop*.

If the flow-loop dynamics are much faster than the dynamics of the level process, then: (i) flow loop transients can be considered to take place at constant ℓ, and the inner loop is designed as an LTI system; (ii) transients in the error $F_{\text{out}} - F_d$ are so fast that they do not affect the level, and the outer loop is designed with F_d ($\approx F_{\text{out}}$) as its input. In effect, the flow loop takes care of the valve nonlinearity, which does not need to be considered in the design of the outer loop.

a. Design the flow control loop with $G_{c1} = 1/\tau s$. Choose τ such that the 3-db bandwidth is at least 30 rad/s for all values of ℓ between 0.5 m and 1.5 m. (Use $c = 2.0$ m$^{3/2}$/s.)

b. Design the level control loop with $G_{c2} = k$. Assume that $F_{out} \approx F_d$, i.e., the flow loop has a transmission of 1. (See Equation 2.28 for the model, with $A = 1$ m^2.) The 3-db bandwidth should be 1 rad/s.

c. Simulate the system (including the valve) and compute the responses, from equilibrium at $\ell_d = 1$ m to step changes $\ell_d = 1.2$ m, 1.5 m, 0.8 m, and 0.5 m.

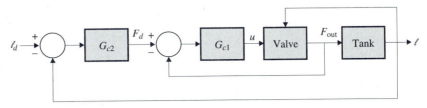

Figure 6.38 Cascaded control loops

6.21 It is often necessary to tune a controller experimentally, and several rule-of-thumb methods have been developed for that purpose. As an example, let us go through the following steps, assuming a stable plant:

a. Use pure-gain control and increase the gain to k_0, where the step response is on the verge of instability. Argue from Root Locus or Nyquist ideas that the oscillation frequency ω_0 is such that $\angle P(j\omega_0) = 180°$, and that the gain k_0 is such that $k_0|P(j\omega_0)| = 1$.

b. Use the information obtained in (a) to design a PI controller with a 6-db gain margin where the integral action has negligible effect at ω_0.

c. Try this on the plant $P(s)$ of Problem 6.1, with $k = 1$, using the model for simulation only (i.e., perform the "experiments" using the model). Compute the step response.

6.22 Industrial controllers are usually provided with various "fixes" to deal with the real, nonlinear world. One effect to be avoided is *reset windup*. Figure 6.39 shows a pure-integral controller feeding an actuator that is subject to saturation. If the actuator saturates, the input u cannot increase to help reduce the error as rapidly as we would wish. The error is integrated, and the variable v grows still more. When the error finally changes sign, some time must pass before the integrator is "discharged" and its output falls within the linear range.

a. With $K = 4$, $M = 1$, and $P(s) = 1/(s + 1)$, simulate the system for a unit step in y_d. Show $y(t)$, $v(t)$, $e(t)$, and $u(t)$.

b. Repeat part (a) *without* the saturation.

c. One "fix" is to introduce some logic and clamp v at $\pm M$, as follows:

$$\text{If } v(t) = M \text{ and } e(t) > 0, \text{ then } v(t) \text{ is fixed at } M.$$

$$\text{If } v(t) = -M \text{ and } e(t) < 0, \text{ then } v(t) \text{ is fixed at } -M.$$

Otherwise, v is K times the integral of e.

Repeat part (a) with this modification.

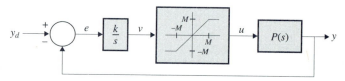

Figure 6.39 Model used to study reset windup

6.23 **a.** Design a lead compensator for the system of Problem 6.1, for a phase margin of 50° and a crossover frequency of 0.9 rad/s.

b. To show the effect of an increased bandwidth on the control amplitude, calculate the plant input in response to a unit-step set point, for the pure-gain design of Problem 6.1 and for the design of part (a).

c. Add lag compensation to the design of part (a) so as to obtain a steady-state error of 0.15 to a unit step. Compute the step response.

6.24 **a.** Design a lead compensator for the system of Problem 6.2. Keep the same phase margin, but increase the crossover frequency to 1.5 rad/s.

b. Calculate the error and plant input response to a unit ramp, for the pure-gain design of Problem 6.2 and for the design of part (a).

c. Add lag compensation to the design of part (a) so as to obtain a steady-state error of 0.1 to a unit ramp. Compute the error response to a unit ramp in this case.

6.25 *dc servo, simplified model* For the dc servo of Problem 2.4 (chapter 2), design a PD compensator to obtain a phase margin of 60° and a crossover frequency of 4 rad/s. Obtain the closed-loop unit-step response.

6.26 *Servo with flexible shaft* In Problem 6.7, a Root-Locus-based design was carried out to move the underdamped poles away from the imaginary axis. Consider this as the initial phase of a design, and use the closed-loop system as the plant for the second phase. For the second phase, design a lead-lag compensator to achieve: (i) a phase margin of 50°, (ii) a crossover frequency of 10 rad/s; (iii) a steady-state error to a unit step of 0.05 rad.

6.27 *Chemical reactor* In Problem 6.14, a pure-gain compensator was designed to stabilize the reactor and achieve a peak value of 3 db for $|T(j\omega)|$.

a. Using this closed-loop system as the plant, design a pure-gain controller that will achieve a 60° phase margin.

b. Design a PI controller that retains (approximately) the phase margin of part (a).

c. Obtain the closed-loop step response, using the linearized model of the plant.

d. Obtain the step responses for increasing amplitudes, using the nonlinear model of the plant. What can you conclude about the range of validity of the linear model?

6.28 *Heat exchanger*

a. Modify the proportional-control design of Problem 6.13 to PID control. Retain the same phase margin, but increase the crossover frequency to 2.8 rad/s.

b. Compute the closed-loop step response, using the linearized model.

c. Compute the closed-loop responses to steps of increasing amplitudes, using the full nonlinear model. What can you conclude about the range of validity of the linear model?

6.29 *Crane* The plant with the transfer function $\Delta x/F$ of Problem 3.20 (Chapter 3) cannot be stabilized by pure-gain feedback.

a. Design a lead compensator for a phase margin of $50°$ and a crossover frequency of 0.6 rad/s.

b. With the lead compensator of part (a) but letting the gain vary, compute the Root Locus.

c. Compute the unit-step and unit-ramp responses for the compensator of part (a), using the linear model.

d. Compute the responses to steps and ramps of increasing amplitudes, using the full nonlinear model. What can you conclude about the range of validity of the linear model?

6.30 In Problem 6.23, a lead-lag compensator was designed for the system of Problem 6.1. Suppose the disturbance-to-output transmission is given as

$$\frac{y}{w} = P_w(s) = \frac{.5(s+2)}{(s+1)^2(s^2+s+1)}.$$

a. Add a feedforward control such that the magnitude of the disturbance frequency response is at most -10 db to 4 rad/s.

b. Compute the responses to a unit-step disturbance with and without feedforward.

6.31 In Problem 6.24, a lead-lag compensator was designed for the system of Problem 6.2. Suppose the disturbance-to-output transmissions is given as

$$\frac{y}{w} = P_w(s) = \frac{.5(s+2)^2}{s(s^2+s+1)(s+4)}.$$

   a. Add a feedforward control such that the magnitude of the disturbance frequency reponse is at most -10 db to 6 rad/s.

   b. Compute the responses to a unit-step disturbance with and without feedforward.

6.32 *Heat exchanger* In a heat exchanger, the objective is to control the outlet temperature of the cold stream by varying the flow rate of the hot stream. Suppose the flow rate of the cold stream is subject to variations, ΔF_C. A PID control was designed in Problem 6.28.

   a. Add a feedforward control such that, up to 4 rad/s, the magnitude of the disturbance frequency reponse is at most -10 db. (The pertinent transfer functions were calculated in Problem 3.18, Chapter 3.)

   b. Obtain the disturbance step responses with and without feedforward, using the linearized model.

6.33 *Blending tank* In Problem 3.47 (Chapter 3), the linearized equations of Problem 2.13 (Chapter 2) were transformed to diagonal form, with $\Delta \ell$ and ΔC as the state variables. Use the numerical values of Problem 2.13.

   a. Design a PID control for the composition, with ΔF_A as the input. The bandwidth is to be 4 rad/min with a phase margin of $50°$.

   b. For the level loop, use ΔF_0 as the control variable and ΔF_A as the disturbance input. Design a pure-gain feedback for a closed-loop time constant of 5 s, and add a feedforward to cancel out the disturbance.

   c. Using the linearized model, simulate to obtain the closed-loop responses to unit steps in the set points to $\Delta \ell$ and ΔC.

   d. Using the full nonlinear model, simulate as in part (c) for steps of increasing amplitudes. What can you conclude about the range of validity of the linear model?

6.34 *Chemical reactor* In Problem 6.27, a PI controller was designed for the linearized model of the chemical reactor. The major disturbance in this application is ΔF, the change in flow rate.

   a. Add a feedforward control to the design to increase the disturbance response bandwidth to 2.5 rad/s.

   b. For the linearized model, simulate the response to a step disturbance.

6.35 *Maglev* In Problem 6.10, two SISO loops were designed to control a 2×2 system with interacting variables. It may be asserted that the influence of Δu_{LD} on Δy is minimal, but that may not be said of the influence of Δu_{GD} on $\Delta \theta$. One way to take care of this is to consider Δu_{GD} as a disturbance input affecting $\Delta \theta$.

   a. Design a feedforward control to nullify some of the effect of Δu_{GD} on $\Delta \theta$.

   b. Repeat parts (b) and (c) of Problem 6.10.

6.36 In Problem 6.23, a lead-lag compensator was designed for the system of Problem 6.1. Modify the design to obtain a 2-DOF system such that (i) both the forward-path and feedback-path compensators are proper or strictly proper and (ii) the frequency response to the reference input has a bandwidth of 3 rad/s and a peak response of no more than 3 db.

6.37 In Problem 6.24, a lead-lag compensator was designed for the system of Problem 6.2. Repeat Problem 6.36 for this system, but with a bandwidth of 6 rad/s.

6.38 *dc servo, simplified model*   Modify the design of Problem 6.25 to a 2-DOF system such that the voltage input required by the response to a reference step of 2 rad does not exceed 3 V.

6.39 *Chemical reactor*   In a chemical reactor, it is often desirable to avoid excessive speed in the response to a step change in the set point. Modify the design of Problem 6.27 to a 2-DOF design such that (i) both the forward-path and feedback-path compensators are proper or strictly proper and (ii) the step response is overdamped, with a dominant time constant of 10 s.

6.40 A second-order Padé approximation of the form $H(s) = (as^2 - bs + 1)/(as^2 + bs + 1)$ may be developed.

a. Use long division to obtain the series for $H(s)$.

b. Choose a, and b so as to match the series for e^{-sT} to as many terms as possible.

c. Obtain $\ell(j\omega)$, following the procedure set out in Section 6.6, for the first-order approximation. For what value of ωT does ℓ first reach the value of 1?

6.41 For $P(s) = e^{-.5s}/[(.5s + 1)(s + 1)]$:

a. Compute the Nyquist plot and calculate the range of gain for which a pure-gain feedback system is stable.

b. Replace the delay with a first-order Padé approximation, and use the Routh criterion to calculate the range of stable gains.

6.42 In process control, the step response is sometimes used to generate an approximate model of the form $\kappa e^{-sT}/(\tau s + 1)$. For $\kappa = 2, \tau = .5$ s and $T = 1$ s:

a. Design a proportional controller with a phase margin of $50°$.

b. Add integral action such that the phase at crossover is almost as in part (a).

c. Outline an algorithm that would accomplish this tuning for any given κ, τ, and T.

6.43 Repeat Problem 6.42 for $\kappa = 2, \tau = 1$ s, and $T = 0.5$ s.

6.44 If the delay is the dominant dynamic element, lead compensation is not effective as a means of increasing bandwidth. For the model of Problem 6.42, with $\kappa = 1$, $\tau = 0.1$ s, and $T = 1$ s,

    a. Design a proportional controller with a phase margin of $50°$. Note the crossover frequency.

    b. Given that the phase contribution of a lead compensator cannot exceed $+90°$, by how much can the crossover frequency be increased if the $50°$ phase margin is to be retained?

6.45 *Heat exchanger* Calculate the step response of the transfer function $\Delta T_{c3}/\Delta F_H$ in Problem 3.18 (Chapter 3), and estimate "by eye" the dc gain κ, the delay time T, and the time constant τ. Use the method of Problem 6.42 to design a PI controller, and obtain the closed-loop step response.

6.46   **a.** Repeat the first two steps of Problem 6.42 for the same plant, but without the delay.

    b. Design a Smith predictor based on the result of part (a). How does the closed-loop bandwidth of this design compare to that of Problem 6.42?

References

[1] Dorf, Richard C., *Modern Control Systems*, 6th Edition, Addison-Wesley (1992).
[2] Franklin, G. F., J. D. Powell, and A. Emami-Naeini, *Feedback Control of Dynamic Systems*, 2nd Edition, Addison-Wesley (1991).
[3] Kuo, B. C., *Automatic Control Systems*, 4th Edition, Prentice-Hall (1982).
[4] Hostetter, G. H., C. J. Savant, Jr. and R. T. Stefani, *Design of Feedback Control Systems*, 2nd Edition, Saunders (1989).
[5] Bode, H. W., *Network Analysis and Feedback Amplifier Design*, Van Nostrand (1945).
[6] Shinskey, F. G., *Process-Control Systems*, McGraw-Hill (1967).
[7] Smith, O. J. M., *Feedback Control Systems,* McGraw-Hill (1958).

Chapter

7

State Feedback

7.1 INTRODUCTION

In this chapter, we study a class of design methods that are fundamentally different from the classical methods presented in Chapter 5. In *state feedback*, the controls are generated as linear combinations of the state variables. State methods appeared in the late 1950s, although the basic ideas behind them are much older. State feedback focuses on time-domain features of the system responses. The mathematical tools are drawn from linear algebra and optimization theory; since operations on matrices larger than 3×3 are difficult to do by hand, state methods had to wait for the advent of the digital computer to become practical; hence their rather late entry in the fifties. The sixties and seventies saw a great flurry of research activity on state feedback methods. During that period many researchers tended to regard the frequency-based methods as passé; fortunately, practitioners did not share that view. Nowadays, it is generally recognized that both state and frequency methods have their places in the control engineer's design kit. Much good research work has been directed at showing linkages between the two. In most cases, the frequency domain is the better way to define specifications. On the other hand, it appears that state methods are better suited to numerical work. What is emerging are methods where problems are cast in the frequency domains and translated into a state formulation for algorithmic solutions, thus combining the best of both worlds.

This chapter is organized as follows. We begin with a discussion of some properties of state feedback. We then describe pole placement methods, particularly for single-input systems. We go on to the so-called *linear–quadratic* (LQ) regulator, probably the centerpiece of so-called modern control. Both pole placement and the LQ regulator require that the whole state vector be available, a condition that often is not met in practice. We devise dynamic systems called *observers* to generate estimates of the state. The final section discusses control systems designed by using a state feedback control law, but with the state replaced by its estimate.

291

PROPERTIES OF STATE FEEDBACK

An LTI system, with the state description

$$\dot{\mathbf{x}} = A\mathbf{x} + B\mathbf{u}$$

$$\mathbf{y} = C\mathbf{x} \tag{7.1}$$

is given, where the state vector $\mathbf{x}$ is of dimension n, the input $\mathbf{u}$ is of dimension r, and the output $\mathbf{y}$ is of dimension m. A *state feedback law* is an expression for the input $\mathbf{u}$ in terms of the state $\mathbf{x}$. A *linear* state feedback law has the form

$$\mathbf{u} = -K\mathbf{x} \tag{7.2}$$

where K, the state gain matrix, is of dimensions $r \times n$. With $\mathbf{u}$ given by Equation 7.2, the *closed-loop* state description becomes

$$\dot{\mathbf{x}} = (A - BK)\mathbf{x}$$

$$\mathbf{y} = C\mathbf{x}. \tag{7.3}$$

Equation 7.3 describes an autonomous linear, time-invariant (LTI) system. If K can be chosen such that $A - BK$ has all its eigenvalues in the open LHP, then $\lim_{t \to \infty} \mathbf{x}(t) = 0$, and $\lim_{t \to 0} \mathbf{y}(t) = 0$. As it stands, this control law performs a regulation function and drives the state and the output to zero from any initial state.

If we wish to regulate to some constant output set point $\mathbf{y}_d \neq 0$, we must first calculate the dc steady state for which $\mathbf{y}_d$ is the output. From Chapter 2, the steady-state input $\mathbf{u}^*$ and state $\mathbf{x}^*$ satisfy

$$0 = A\mathbf{x}^* + B\mathbf{u}^*$$

$$\mathbf{y}_d = C\mathbf{x}^*. \tag{7.4}$$

A unique solution to Equation 7.4 exists if, and only if, the matrix $\left[\begin{smallmatrix} A & B \\ C & 0 \end{smallmatrix} \right]$ is non-singular. From Chapter 3, that means that there is no transmission zero at $s = 0$. We shall assume this to be true, in which case $\mathbf{x}^*$ and $\mathbf{u}^*$ are linear in $\mathbf{y}_d$ and given by expressions of the form

$$\mathbf{x}^* = M_x \mathbf{y}_d$$

$$\mathbf{u}^* = M_u \mathbf{y}_d. \tag{7.5}$$

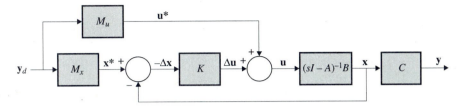

Figure 7.1 Steady-state and incremental signals, state feedback

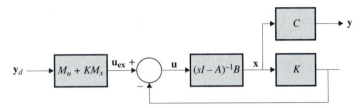

Figure 7.2 State feedback in terms of full signals (incremental plus steady state)

With $\Delta\mathbf{x} = \mathbf{x} - \mathbf{x}^*$, $\Delta\mathbf{y} = \mathbf{y} - \mathbf{y}_d$, and $\Delta\mathbf{u} = \mathbf{u} - \mathbf{u}^*$, it was established in Chapter 2 that

$$\Delta\dot{\mathbf{x}} = A\,\Delta\mathbf{x} + B\,\Delta\mathbf{u}$$

$$\Delta\mathbf{y} = C\Delta\mathbf{x}. \tag{7.6}$$

The state feedback law $\Delta\mathbf{u} = -K\,\Delta\mathbf{x}$, with K appropriately selected, will drive $\Delta\mathbf{x}$ and $\Delta\mathbf{y}$ to zero, thus achieving the regulation objective.

Since $\mathbf{u} = \mathbf{u}^* + \Delta\mathbf{u}$, the control system is as shown in Figure 7.1. Figure 7.2 follows from expanding $\mathbf{u}$ as follows:

$$\mathbf{u} = \mathbf{u}^* + \Delta\mathbf{u} = \mathbf{u}^* - K\,\Delta\mathbf{x}$$

$$= \mathbf{u}^* - K(\mathbf{x} - \mathbf{x}^*)$$

$$= (M_u + KM_x)\mathbf{y}_d - K\mathbf{x}$$

$$= \mathbf{u}_{ex} - K\mathbf{x} \tag{7.7}$$

where $\mathbf{u}_{ex} = (M_u + KM_x)\mathbf{y}_d$. The response for a given initial state $\mathbf{x}(0)$ is calculated by solving

$$\Delta\dot{\mathbf{x}} = (A - BK)\,\Delta\mathbf{x} \tag{7.8}$$

for $\Delta \mathbf{x}(0) = \mathbf{x}(0) - \mathbf{x}^*$. The state, control, and output are then obtained as

$$\mathbf{x}(t) = \mathbf{x}^* + \Delta \mathbf{x}(t)$$

$$\mathbf{u}(t) = \mathbf{u}^* + \Delta \mathbf{u}(t)$$

$$\mathbf{y}(t) = \mathbf{y}_d + C \, \Delta \mathbf{x}(t). \tag{7.9}$$

As we shall establish in the next section, the gain K may, under certain conditions, be selected to place the closed-loop eigenvalues at selected locations in complex plane. For the moment, we present a simple example to show the pole-shifting property.

Example 7.1   Let $A = \begin{bmatrix} 0 & 1 \\ 0 & 0 \end{bmatrix}$ and $B = \begin{bmatrix} -1 \\ 1 \end{bmatrix}$. Calculate the closed-loop eigenvalues for the state feedback gain $K = [k_1 \quad k_2]$.

Solution   We have

$$A - BK = \begin{bmatrix} 0 & 1 \\ 0 & 0 \end{bmatrix} - \begin{bmatrix} -1 \\ 1 \end{bmatrix} [k_1 \quad k_2]$$

$$= \begin{bmatrix} 0 & 1 \\ 0 & 0 \end{bmatrix} - \begin{bmatrix} -k_1 & -k_2 \\ k_1 & k_2 \end{bmatrix}$$

$$= \begin{bmatrix} k_1 & 1+k_2 \\ -k_1 & -k_2 \end{bmatrix}.$$

The characteristic equation is

$$\det(sI - A + BK) = 0$$

or

$$\det \begin{bmatrix} s - k_1 & -1 - k_2 \\ k_1 & s + k_2 \end{bmatrix} = 0$$

$$s^2 + (k_2 - k_1)s + k_1 = 0.$$

The closed-loop eigenvalues are

$$s_1, s_2 = \frac{k_1 - k_2 \pm \sqrt{k_1^2 - 2k_1 k_2 + k_2^2 - 4k_1}}{2}.$$

Clearly, they change with k_1 and k_2.

By contrast, the zeros are unchanged by state feedback.

■ **Theorem 7.1**   Let z be a transmission zero of the system $\dot{\mathbf{x}} = A\mathbf{x} + B\mathbf{u}$, $\mathbf{y} = C\mathbf{x} + D\mathbf{u}$, and let the closed-loop system be defined by the state feedback law $\mathbf{u} = -K\mathbf{x} + \mathbf{u}_{ex}$. Then z is also a transmission zero of the closed-loop matrix transfer function relating $\mathbf{y}$ to $\mathbf{u}_{ex}$.

Proof: From Chapter 3, z satisfies

$$\text{rank} \begin{bmatrix} -zI + A & B \\ C & D \end{bmatrix} < n + r.$$

Since the above matrix is not of full rank, there exists a vector $[\begin{smallmatrix} \mathbf{v}_1 \\ \mathbf{v}_2 \end{smallmatrix}]$ such that

$$\begin{bmatrix} -zI + A & B \\ C & D \end{bmatrix} \begin{bmatrix} \mathbf{v}_1 \\ \mathbf{v}_2 \end{bmatrix} = 0. \tag{7.10}$$

With state feedback, the system equations are

$$\dot{\mathbf{x}} = A\mathbf{x} - BK\mathbf{x} + B\mathbf{u}_{ex}$$
$$\mathbf{y} = C\mathbf{x} - DK\mathbf{x} + D\mathbf{u}_{ex}.$$

A closed-loop zero ζ satisfies

$$\text{rank} \begin{bmatrix} -\zeta I + A - BK & B \\ C - DK & D \end{bmatrix} < n + r. \tag{7.11}$$

We claim that this holds with $\zeta = z$. To see this, form

$$\begin{bmatrix} -zI + A - BK & B \\ C - DK & D \end{bmatrix} \begin{bmatrix} \mathbf{v}_1 \\ \mathbf{v}_2 + K\mathbf{v}_1 \end{bmatrix} = \begin{bmatrix} (-zI + A)\mathbf{v}_1 + B - BK\mathbf{v}_1 + BK\mathbf{v}_1 \\ C\mathbf{v}_1 + D\mathbf{v}_2 - DK\mathbf{v}_1 + DK\mathbf{v}_1 \end{bmatrix}$$
$$= \begin{bmatrix} (-zI + A)\mathbf{v}_1 + B\mathbf{v}_2 \\ C\mathbf{v}_1 + D\mathbf{v}_2 \end{bmatrix} = 0$$

in view of Equation 7.10. The matrix in Equation 7.11 is rank-deficient with $\zeta = z$, and z is a zero. ■

An important property of state feedback is that it preserves controllability. The following theorem states this precisely.

■ **Theorem 7.2**   A system controlled by state feedback is controllable with $\mathbf{u}_{ex}$ as input if, and only if, the open-loop system is controllable.

Proof: Refer to Figure 7.2. Suppose the open-loop system is controllable. Then, for any state $\mathbf{x}_f$ and time $T > 0$, there exists an input $\mathbf{u}^*(t)$ that takes the state from $\mathbf{x} = \mathbf{0}$ at $t = 0$ to $\mathbf{x} = \mathbf{x}_f$ at $t = T$. Let the state evolution under the action of $\mathbf{u}^*(t)$ be $\mathbf{x}^*(t)$ [where, of course, $\mathbf{x}^*(T) = \mathbf{x}_f$]. Now, let $\mathbf{u}_{ex}(t) = \mathbf{u}^*(t) + K\mathbf{x}^*(t)$. Clearly,

$\mathbf{u}(t) = \mathbf{u}^*(t)$ and $\mathbf{x}(t) = \mathbf{x}^*(t)$, so that the state of the closed-loop system at time T is $\mathbf{x}_f$. Thus, there exists an input $\mathbf{u}_{ex}(t)$ that takes the state of the closed-loop system from 0 to an arbitrary $\mathbf{x}_f$ at an arbitrary time T, i.e., the closed-loop system is controllable.

If the open-loop system is not controllable, the state can never have a component along an uncontrollable state when driven from $\mathbf{x} = \mathbf{0}$, regardless of what input $\mathbf{u}(t)$ is applied. This is true no matter how that input is generated, including by state feedback. Therefore, the closed-loop system is also uncontrollable.   ∎

From the modal viewpoint, an uncontrollable mode remains uncontrollable with state feedback. Recall that, if s_i is an uncontrollable mode, the corresponding left eigenvector $\mathbf{w}_i^T$ of the system A matrix satisfies $\mathbf{w}_i^T B = 0$. Now, in such a case,

$$\mathbf{w}_i^T (A - BK) = s_i \mathbf{w}_i^T - \mathbf{w}_i^T BK = s_i \mathbf{w}_i^T$$

so s_i is also an eigenvalue of the closed-loop system, with the same left eigenvector as before. Since the closed-loop B matrix is still B, and since $\mathbf{w}_i^T B = 0$, the result follows.

In fact, uncontrollable modes are unchanged not only by state feedback, but by any other means of generating $\mathbf{u}$.

7.3 POLE PLACEMENT

A look back at Example 7.1 will show that the two gains k_1 and k_2 can be chosen to achieve any desired closed-loop characteristic polynomial and, hence, any real or complex conjugate eigenvalue pair. The reason is that k_1 and k_2 allow independent adjustment of the coefficients of the first and zero powers of s. That is true in general for controllable systems.

We need the following result:

Lemma 7.1   Let $(A, \mathbf{b})$ be a controllable single-input, single-output (SISO) system of order n. There exists a nonsingular $n \times n$ matrix T such that, with $A_c = T^{-1}AT$ and $\mathbf{b}_c = T^{-1}\mathbf{b}$, $(A_c, \mathbf{b}_c)$ is the controllable canonical form.

Proof: Consider the system

$$\dot{\mathbf{x}} = A\mathbf{x} + \mathbf{b}u$$

$$\mathbf{y} = I\mathbf{x} \tag{7.12}$$

where I is the $n \times n$ identity matrix. This realization is minimal because (i) $(A, \mathbf{b})$ is controllable and (ii) (I, A) is obviously observable since all state variables are measured. That being the case, the vector transfer function $\mathbf{h}(s) = \mathbf{y}(s)/u(s)$ is of order n. In particular, a controllable canonical form can be written for each element of $\mathbf{h}(s)$. The A matrix is common to all, because the modes are common, and the rows of the C matrix are written directly from the numerator polynomials of $\mathbf{h}(s)$. We write

$$\dot{\mathbf{z}} = A_c \mathbf{z} + \mathbf{b}_c u$$

$$\mathbf{y} = T\mathbf{z} \tag{7.13}$$

as the desired realization.

Since Equations 7.12 and 7.13 are realizations of the same transfer function, the corresponding impulse responses are the same. Thus,

$$e^{At}\mathbf{b} = Te^{A_c t}\mathbf{b}_c, \qquad \text{for all } t.$$

Taking time derivatives and evaluating at $t = 0$ yields

$$\mathbf{b} = T\mathbf{b}_c$$

$$A\mathbf{b} = TA_c\mathbf{b}_s$$

$$A^2\mathbf{b} = TA_c^2\mathbf{b}_c$$

$$\vdots = \vdots$$

$$A^{n-1}\mathbf{b} = TA_c^{n-1}\mathbf{b}_c$$

or

$$[\mathbf{b} \quad A\mathbf{b} \quad \ldots \quad A^{n-1}\mathbf{b}] = T[\mathbf{b}_c \quad A_c\mathbf{b}_c \quad \ldots \quad A_c^{n-1}\mathbf{b}_c].$$

We identify the two controllability matrices as $\mathcal{C}$ and $\mathcal{C}_c$, and write

$$T = \mathcal{C}\mathcal{C}_c^{-1} \tag{7.14}$$

which is justified because $\mathcal{C}_c$ is nonsingular (the controllable canonical form is controllable). The matrix T is nonsingular since it is the product of two $n \times n$ nonsingular matrices.

From Equations 7.12 and 7.13,

$$\mathbf{y} = \mathbf{x} = T\mathbf{z}$$

$$\mathbf{z} = T^{-1}\mathbf{x}$$

and

$$\dot{\mathbf{z}} = T^{-1}AT\mathbf{z} + T^{-1}\mathbf{b}u$$

Comparison of this with Equation 7.13 yields the desired result.   ∎

■ **Theorem 7.3**   (Pole-placement theorem) Given an *n*th-order, controllable system (A, **b**) and a monic *n*th-order polynomial $p(s)$, there exists a unique gain vector $\mathbf{k}^T$ such that the characteristic polynomial of $A - \mathbf{b}\mathbf{k}^T$ is $p(s)$.

Proof: We begin with the controllable canonical form,

$$A_c = \begin{bmatrix} 0 & 1 & 0 & \dots & 0 \\ 0 & 0 & 1 & \dots & 0 \\ \hdotsfor{5} \\ 0 & 0 & \dots & 0 & 1 \\ -a_0 & -a_1 & \dots & \dots & -a_{n-1} \end{bmatrix}, \qquad \mathbf{b}_c = \begin{bmatrix} 0 \\ 0 \\ \vdots \\ 0 \\ 1 \end{bmatrix}.$$

Let κ^T be a row vector of dimension *n*. Then

$$A_c - \mathbf{b}_c \kappa^T = \begin{bmatrix} 0 & 1 & \dots & 0 \\ 0 & 0 & \dots & 0 \\ \hdotsfor{4} \\ 0 & 0 & \dots & 1 \\ -a_0 - \kappa_1 & -a_1 - \kappa_2 & \dots & -a_{n-1} - \kappa_n \end{bmatrix}.$$

Since this matrix is in companion form, its characteristic polynomial is written directly as

$$p_c(s) = s^n + (a_{n-1} + \kappa_n)s^{n-1} + \dots + (a_1 + \kappa_2)s + (a_0 + \kappa_1). \qquad (7.15)$$

It is clear that the components of κ can be picked to obtain any desired *n*th-order, monic polynomial.

By Lemma 7.1, there exists a nonsingular $n \times n$ matrix T such that $A = TA_cT^{-1}$, $\mathbf{b} = T\mathbf{b}_c$. Thus,

$$T(A_c - \mathbf{b}_c\kappa^T)T^{-1} = A - \mathbf{b}\mathbf{k}^T$$

where

$$\mathbf{k}^T = \kappa^T T^{-1}. \qquad (7.16)$$

The characteristic polynomials of $A_c - \mathbf{b}_c\kappa^T$ and $A - \mathbf{b}\kappa^T$ are the same, because similarity transformations preserve eigenvalues. Therefore, since there is a unique κ^T that results in $p_c(s) = p(s)$, $\mathbf{k}^T$ is also unique.   ■

We now present an algorithm for pole placement, based on Lemma 7.1 and Theorem 7.3. From Equation 7.16,

$$\mathbf{k}^T T = \kappa^T.$$

Let us multiply each side by a vector as follows:

$$\mathbf{k}^T T \begin{bmatrix} 1 \\ s \\ \vdots \\ s^{n-2} \\ s^{n-1} \end{bmatrix} = \kappa^T \begin{bmatrix} 1 \\ s \\ \vdots \\ s^{n-2} \\ s^{n-1} \end{bmatrix}. \qquad (7.17)$$

From Equation 7.15, we see that the right-hand side (RHS) is

$$\kappa_1 + \kappa_2 s + \cdots + \kappa_n s^{n-1} = p_c(s) - p_0(s)$$

where $p_0(s)$ is the open-loop characteristic polynomial; i.e., $p_0(s) = \det(sI - A)$. As for the left-hand side (LHS) of Equation 7.17, recall that, in Lemma 7.1, T was formed by extracting each row from the corresponding row of the numerator of the transfer function. For example, given a fourth-order system, the row $2s^3 + 3s^2 + s + 1$ would, according to the rule for the controllable canonical form, correspond to the C vector $[1 \quad 1 \quad 3 \quad 2]$. The multiplication on the left-hand side (LHS) of Equation 7.17 simply restores the transfer-function numerator. Then

$$T \begin{bmatrix} 1 \\ s \\ \vdots \\ s^{n-1} \\ s^{n-1} \end{bmatrix} = \det(sI - A)\mathbf{h}(s) = Adj(sI - A)\mathbf{b}.$$

We may therefore write Equation 7.17 as

$$p_c(s) = \det(sI - A) + \mathbf{k}^T Adj(sI - A)\mathbf{b}. \qquad (7.18)$$

◆ ◆ ◆ **R E M A R K**
Different algorithms are used in numerical computation (MATLAB commands acker and place). ◆

To summarize the procedure for pole-placement design:

1. Calculate the determinant and the adjoint of $(sI - A)$.

2. Compute the desired closed-loop characteristic polynomial.

3. Apply Equation 7.18 to generate linear equations for the gain components.

4. Calculate the dc steady state ($\mathbf{x}^*$ and u^*) in terms of the desired output.

5. Obtain the control law $u = u^* - \mathbf{k}^T(\mathbf{x} - \mathbf{x}^*)$.

7. If the transfer function y/y_d is desired, it has the same zeros as the plant, and the specified closed-loop poles. Furthermore, since the state tends to the dc steady

state that yields y_d as the output, y tends to y_d for constant y_d, so the transfer function $y/y_d = 1$ at $s = 0$. This fact is used to calculate the gain.

8. If the time response is desired, use the procedure indicated by Equations 7.8 and 7.9.

Example 7.2   (dc Servo)

For the dc servo of Example 2.1 (Chapter 2), compute the state feedback gain that places the closed-loop poles at $-3 \pm 3j$ and -24. Give the control law for regulation to a set $\theta = \theta_d$, and write the closed-loop transfer function θ/θ_d.

Solution   The state equation is Equation 2.19, where

$$A = \begin{bmatrix} 0 & 1 & 0 \\ 0 & 0 & 4.438 \\ 0 & -12 & -24 \end{bmatrix}; \qquad b = \begin{bmatrix} 0 \\ 0 \\ 20 \end{bmatrix}.$$

The open-loop eigenvalues are $0, -2.424$, and -21.576, so the open-loop characteristic polynomial is

$$\det(sI - A) = s(s + 2.424)(s + 21.576)$$
$$= s^3 + 24s^2 + 53.37s.$$

To apply Equation 7.18, we must calculate

$$Adj(sI - A)b = Adj \begin{bmatrix} s & -1 & 0 \\ 0 & s & -4.438 \\ 0 & 12 & s+24 \end{bmatrix} \begin{bmatrix} 0 \\ 0 \\ 20 \end{bmatrix}$$

$$= \begin{bmatrix} x & x & 4.438 \\ x & x & 4.438s \\ x & x & s^2 \end{bmatrix} \begin{bmatrix} 0 \\ 0 \\ 20 \end{bmatrix}$$

$$= \begin{bmatrix} 88.76 \\ 88.76s \\ 20s^2 \end{bmatrix}.$$

(Note that, by exploiting the structure of **b**, we save work by not calculating elements of the adjoint that will be multiplied by 0.)

We are now ready to apply Equation 7.18. The desired characteristic polynomial is

$$p(s) = (s + 24)(s + 3 + 3j)(s + 3 - 3j)$$
$$= (s^3 + 30s^2 + 162s + 432).$$

We write Equation 7.18 as

$$s^3 + 30s^2 + 162s + 432 = s^3 + 24s^2 + 53.37s + [k_1 \quad k_2 \quad k_3] \begin{bmatrix} 88.76 \\ 88.76s \\ 20s^2 \end{bmatrix}$$

$$= s^3 + (24 + 20k_3)s^2 + (52.058 + 88.76k_2)s + 88.76k_1.$$

Equating coefficients yields

$$24 + 20k_3 = 30 \quad \text{or} \quad k_3 = 0.300$$

$$53.37 + 88.76k_2 = 162 \quad \text{or} \quad k_2 = 1.23$$

$$88.76k_1 = 432 \quad \text{or} \quad k_1 = 4.86.$$

To calculate the control law, we need to compute the dc steady state corresponding to $\theta = \theta_d$. Applying Equation 7.4,

$$\begin{bmatrix} 0 \\ 0 \\ 0 \end{bmatrix} = \begin{bmatrix} 0 & 1 & 0 \\ 0 & 0 & 4.438 \\ 0 & -12 & -24 \end{bmatrix} \begin{bmatrix} \theta^* \\ \omega^* \\ i^* \end{bmatrix} + \begin{bmatrix} 0 \\ 0 \\ 20 \end{bmatrix} v^*$$

$$\theta_d = \theta^*.$$

The result is $\theta^* = \theta_d$, $\omega^* = i^* = v^* = 0$. The control law, illustrated in Figure 7.3, is

$$v = v^* - K\,\Delta\mathbf{x}$$

$$= 0 - [k_1 \quad k_2 \quad k_3]\left(\begin{bmatrix} \theta \\ \omega \\ i \end{bmatrix} - \begin{bmatrix} \theta_d \\ 0 \\ 0 \end{bmatrix}\right)$$

$$v = -4.86(\theta - \theta_d) - 1.24\omega - 0.300i.$$

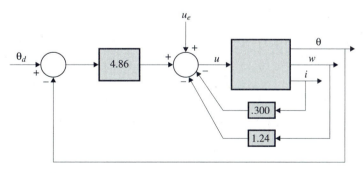

Figure 7.3 State feedback for pole placement, dc servo

To write the closed-loop transfer function, we observe that the plant has no zeros. By Theorem 7.1, the closed-loop transfer function has no zeros, so

$$\frac{\theta}{\theta_d} = \frac{k}{(s+24)(s+3+3j)(s+3-3j)}$$

$$= \frac{k}{(s+24)(s^2+6s+18)}.$$

Since $\theta \to \theta_d$ as $t \to \infty$, the transfer function must go to 1 as $s \to 0$; hence, $k = (24)(18) = 432$ and

$$\frac{\theta}{\theta_d} = \frac{432}{(s+24)(s^2+6s+18)}.$$

Example 7.3   (Pendulum on a Cart)

Design a state feedback gain to place the closed-loop poles of the inverted pendulum system of Example 2.9, with the numerical values used in Example 2.5 (Chapter 2) at the locations -4.43, -4.43, and $-2 \pm 2j$. Give the control law for regulation to the vertical equilibrium state.

Solution   The linearized state space representation is

$$\frac{d}{dt}\begin{bmatrix} x \\ v \\ \theta \\ \omega \end{bmatrix} = \begin{bmatrix} 0 & 1 & 0 & 0 \\ 0 & 0 & -9.8 & 0 \\ 0 & 0 & 0 & 1 \\ 0 & 0 & 19.6 & 0 \end{bmatrix}\begin{bmatrix} x \\ v \\ \theta \\ \omega \end{bmatrix} + \begin{bmatrix} 0 \\ 1 \\ 0 \\ -1 \end{bmatrix} u.$$

With

$$sI - A = \begin{bmatrix} s & -1 & 0 & 0 \\ 0 & s & 9.8 & 0 \\ 0 & 0 & s & -1 \\ 0 & 0 & -19.6 & s \end{bmatrix}$$

we compute

$$\det(sI - A) = s^2(s^2 - 19.6).$$

The open-loop poles are at 0, 0, and ± 4.43. We now calculate

$$Adj(sI - A) = \begin{bmatrix} x & s^2 - 19.6 & x & -9.8 \\ x & s(s^2 - 19.6) & x & -9.8s \\ x & 0 & x & s^2 \\ x & 0 & x & s^3 \end{bmatrix}$$

and

$$\mathbf{k}^T Adj(sI - A)\mathbf{b} = [k_1 \quad k_2 \quad k_3 \quad k_4] \begin{bmatrix} s^2 - 9.8 \\ s^3 - 9.8s \\ -s^2 \\ -s^3 \end{bmatrix}$$

$$= k_1(s^2 - 9.8) + k_2(s^3 - 9.8s) - k_3 s^2 - k_4 s^3.$$

The desired characteristic polynomial is

$$p(s) = (s + 4.43)^2 (s + 2 + 2j)(s + 2 - 2j)$$
$$= s^4 + 12.86 s^3 + 63.04 s^2 + 149.3 s + 157.8.$$

Applying Equation 7.18,

$$p(s) = s^4 - 19.6 s^2 + k_1(s^2 - 9.8) + k_2(s^3 - 9.8s) - k_3 s^2 - k_4 s^3$$
$$= s^4 + (k_2 - k_4)s^3 + (-19.6 + k_1 - k_3)s^2 - 9.8 k_2 s - 9.8 k_1.$$

Matching coefficients yields

$$k_2 - k_4 = 12.86$$
$$-19.6 + k_1 - k_3 = 63.04$$
$$-9.8 k_2 = 149.3$$
$$-9.8 k_1 = 157.8$$

which leads to

$$k_1 = -16, \qquad k_2 = -15.2, \qquad k_3 = -98.6, \qquad k_4 = -28.1.$$

Here, the desired dc steady state is the origin; i.e., $x^* = u^* = 0$.
   The control law is

$$u = 16x + 15.2v + 98.6\theta + 28.1\omega.$$

Figure 7.4 shows the state and input response for an initial state $\theta(0) = 15° = 0.26$ rad, $v(0) = x(0) = \omega(0) = 0$.

In the case of a multivariable system, there are more free parameters in the state feedback gain matrix than there are eigenvalues. That freedom can be used either to achieve desirable eigenvectors [1, 2] or to achieve other desirable properties [3].

Since closed-loop poles can be chosen arbitrarily, one might wonder why they should not be placed so far into the LHP that transient times are negligible. The reason is that it takes large gain values to move poles far away from their open-loop positions.

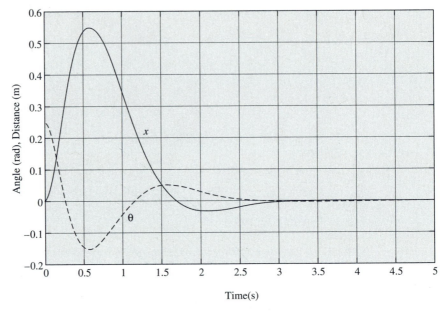

Figure 7.4 Time responses for pole-placement state feedback, pendulum-on-cart

Figure 7.2 shows the situation in closed-loop unit-feedback form. The loop gain is $L(s) = K(sI - A)^{-1}B$. If the elements of K are large, $L(j\omega)$ will tend to be large over a relatively large range of ω, and the complementary sensitivity $T(j\omega)$ for the closed loop will be near unity over a large frequency range. From the results of Chapter 5, Section 5.6, the system will not be robust unless the model is relatively good to frequencies extending beyond the bandwidth. It follows that the closed-loop poles chosen should not produce bandwidths extending beyond the range of validity of the model.

7.4 THE LINEAR–QUADRATIC REGULATOR

7.4.1 Introduction

The rather puzzling term "linear–quadratic" (LQ) is shorthand for "optimal linear regulator for a quadratic performance index." The LQ regulator is possibly the most important result of modern control. Many practical problems can be cast in the LQ framework, and there are excellent numerical algorithms for their solution.

In the LQ regulator, performance is assessed by a scalar *performance index*, of the form

$$J = \int_0^\infty [\mathbf{x}^T(t)Q\mathbf{x}(t) + \mathbf{u}^T(t)R\mathbf{u}(t)]dt \tag{7.19}$$

where Q and R, both symmetric matrices, are positive semidefinite and positive definite, respectively.

The expressions $\mathbf{x}^T Q \mathbf{x}$ and $\mathbf{u}^T R \mathbf{u}$ are quadratic forms. It is useful to recall that

$$\mathbf{x}^T Q \mathbf{x} = Q_{11} x_1^2 + 2 Q_{12} x_1 x_2 + \cdots + 2 Q_{1n} x_1 x_n$$
$$+ Q_{22} x_2^2 + \cdots + 2 Q_{2n} x_2 x_n$$
$$+ \cdots$$
$$+ Q_{nn} x_n^2.$$

If Q is positive semidefinite (in shorthand, $Q \geq 0$) then $\mathbf{x}^T Q \mathbf{x} \geq 0$ for all $\mathbf{x}$. If Q is positive definite ($Q > 0$), then $\mathbf{x}^T Q \mathbf{x} > 0$ for all $\mathbf{x} \neq 0$.

The problem to be solved is assumed to be a regulation problem where the objective is to drive the state from some initial value to zero. The term $\mathbf{x}^T(t) Q \mathbf{x}(t)$ penalizes departures of $\mathbf{x}(t)$ from zero. The penalty is quadratic, so large errors are penalized much more than small ones; for example, $Q_{11}(.1)^2 = 0.01 Q_{11}$, whereas $Q_{11}(10)^2 = 1000 Q_{11}$. The integral accumulates $\mathbf{x}^T(t) Q \mathbf{x}(t)$ over time, so long-lasting departures from $\mathbf{x}(t) = 0$ are penalized more than short-lived ones. It is desirable to have a small value for the term $\int_0^\infty \mathbf{x}^T(t) Q \mathbf{x}(t) dt$, because that implies that, on average, $\mathbf{x}(t)$ is relatively near zero.

An attempt to design a control system that minimizes $\int_0^\infty \mathbf{x}^T(t) Q \mathbf{x}(t) dt$ will almost invariably lead to a design that uses large input values. The reason is that large inputs (e.g., impulses) cause the state to change rapidly and thus drive $\mathbf{x}(t)$ to zero faster than inputs of moderate magnitude. The term $\int_0^\infty \mathbf{u}^T(t) R \mathbf{u}(t) dt$ is included in Equation 7.19 to keep tabs on the input magnitude. A design that uses large input values will yield a large value for J, because the second term in Equation 7.19 will be large; on the other hand, a design that generates small inputs will not do a good job of driving $\mathbf{x}(t)$ to zero, so the first term in Equation 7.19 will be relatively large. A design that minimizes J represents a compromise between the conflicting desiderata of good regulation and reasonably sized inputs.

The actual value of the performance index J is not usually meaningful; control system performance is really too complex and multifaceted to be reducible to a single number. We cast the control problem as an optimization problem not because we particularly wish to minimize J, but because the optimal solution is mathematically tractable and has desirable properties. The optimization framework also offers a convenient "handle" on the properties of the solution. For example, suppose the solution that minimizes $J = \int_0^\infty [Q_{11} x_1^2(t) + R_{11} u_1^2(t)] dt$ turns out to yield excessive values of $u_1(t)$ under typical simulation runs. We can increase R_{11} and solve again; the increased penalty on u_1 will yield a new solution that leads to smaller values of u_1 and somewhat larger values of x_1. We keep adjusting the weights until we reach a suitable compromise.

7.4.2 The Matrix Lyapunov Equation

If the system $\dot{\mathbf{x}} = A \mathbf{x} + B \mathbf{u}$ is controlled under the state feedback law $\mathbf{u} = -K \mathbf{x}$, the closed-loop system is given by

$$\dot{\mathbf{x}} = (A - BK) \mathbf{x} \qquad (7.20)$$

and the performance index of Equation 7.19 is

$$J = \int_0^\infty [\mathbf{x}^T(t)Q\mathbf{x}(t) + \mathbf{u}^T(t)R\mathbf{u}(t)]dt$$

$$= \int_0^\infty [\mathbf{x}^T(t)Q\mathbf{x}(t) + \mathbf{x}^T(t)K^T R K\mathbf{x}(t)]dt$$

$$= \int_0^\infty \mathbf{x}^T(t)[Q + K^T R K]\mathbf{x}(t)dt. \tag{7.21}$$

In view of Equations 7.20 and 7.21, we shall solve the following problem. Given

$$\dot{\mathbf{x}} = A\mathbf{x} \tag{7.22}$$

where A is stable, calculate

$$J = \int_0^\infty \mathbf{x}^T(t)S\mathbf{x}(t)dt \tag{7.23}$$

where S is symmetric and positive semidefinite or definite. If we can solve this problem, then we merely substitute $A - BK$ for A, and $Q + K^T R K$ for S, to solve the problem defined by Equations 7.20 and 7.21. The solution is given by the following theorem.

■ **Theorem 7.4**   Let A have all its eigenvalues in the open LHP and let S be a symmetric, positive definite or semidefinite matrix. Then, for $\mathbf{x}(0) = \mathbf{x}_0$,

$$J = \int_0^\infty \mathbf{x}^T(t)S\mathbf{x}(t)dt = \mathbf{x}_0^T P\mathbf{x}_0$$

where P is a symmetric, positive definite or semidefinite matrix satisfying the *matrix Lyapunov equation,*

$$A^T P + PA = -S. \tag{7.24}$$

Proof: With $\mathbf{x}(0) = \mathbf{x}_0$, we have $\mathbf{x}(t) = e^{At}\mathbf{x}_0$ and

$$J = \int_0^\infty \mathbf{x}_0^T e^{A^T t} S e^{At} \mathbf{x}_0 dt$$

$$= \mathbf{x}_0^T \left[\int_0^\infty e^{A^T t} S e^{At} dt \right] \mathbf{x}_0$$

$$= \mathbf{x}_0^T P\mathbf{x}_0$$

with

$$P = \int_0^\infty e^{A^T t} S e^{At} dt. \tag{7.25}$$

Since A is stable, e^{At} has only negative exponential terms, so the integral of Equation 7.25 exists. To see that P is symmetric, we calculate

$$P^T = \int_0^\infty (e^{At})^T S(e^{At})dt = \int_0^\infty e^{A^T t} Se^{At} dt = P.$$

The matrix P is at least semidefinite, because

$$\mathbf{x}_0{}^T P\mathbf{x}_0 = J = \int_0^\infty \mathbf{x}^T(t) S\mathbf{x}(t)dt \geq 0$$

since the integrand is nonnegative. (S is positive semidefinite or definite.)

We now show that P satisfies the matrix Lyapunov equation. Using Equation 7.25, we write

$$A^T P + PA = \int_0^\infty \left[A^T e^{A^T t} Se^{At} + e^{A^T t} Se^{At} A \right] dt.$$

Recall that $(d/dt)e^{At} = Ae^{At} = e^{At}A$, since A and e^{At} commute. Then

$$\frac{d}{dt}\left[e^{A^T t} Se^{At} \right] = A^T e^{A^T t} Se^{At} + e^{A^T t} Se^{At} A$$

so that

$$A^T P + PA = \int_0^\infty \frac{d}{dt}\left[e^{A^T t} Se^{At} \right] dt$$

$$= e^{A^T t} Se^{At} \Big|_0^\infty$$

$$= -S$$

because $\lim_{t\to\infty} e^{At} = 0$ by stability, and $e^{A0} = I$. That establishes the desired result. ∎

Under certain conditions, the existence of a positive definite or semidefinite P that satisfies the matrix Lyapunov equation guarantees internal stability. That is the matter addressed by the following theorem. Before presenting the theorem, let us state one mathematical fact: a symmetric positive definite or positive semidefinite matrix S can always be written as a matrix product $S = (S^{1/2})^T S^{1/2}$, where $S^{1/2}$ is called the square root of S. The matrix $S^{1/2}$ is not unique, and good algorithms to generate a square root of a particular structure exist, e.g., Cholesky decomposition algorithms [4].

■ **Theorem 7.5**   Suppose a $P \geq 0$ exists that satisfies the matrix Lyapunov equation, with $S \geq 0$. Also suppose that the pair $(A, S^{1/2})$ is detectable, from the output $\mathbf{y} = S^{1/2}\mathbf{x}$. Then the system $\dot{\mathbf{x}} = A\mathbf{x}$ is internally stable.

Proof: We define the *Lyapunov function*,

$$V = \mathbf{x}^T P \mathbf{x}.$$

We calculate

$$\dot{V} = \dot{\mathbf{x}}^T P \mathbf{x} + \mathbf{x}^T P \dot{\mathbf{x}}$$

which, since $\dot{\mathbf{x}} = A\mathbf{x}$, becomes

$$\dot{V} = \mathbf{x}^T A^T P \mathbf{x} + \mathbf{x}^T P A \mathbf{x}$$

$$= \mathbf{x}^T (A^T P + PA)\mathbf{x} = -\mathbf{x}^T S \mathbf{x}$$

$$= -\mathbf{x}^T (S^{1/2})^T S^{1/2} \mathbf{x} = -(S^{1/2}\mathbf{x})^T (S^{1/2}\mathbf{x}) = -\| S^{1/2}\mathbf{x}\|^2. \qquad (7.26)$$

By Equation 7.26, $\dot{V} \leq 0$, so $V(t)$ is nonincreasing. Also, since $P \geq 0$, we have $V \geq 0$; $V(t)$ is bounded from below by zero. The shape of $V(t)$ must therefore be as shown in Figure 7.5. $V(t)$ cannot oscillate, because that would require $\dot{V} > 0$ part of the time, and V cannot cross the real axis, because $V \geq 0$. Thus, V must tend to a constant, so $\dot{V} \to 0$. From Equation 7.26, if $\dot{V} \to 0$, then $S^{1/2}\mathbf{x} \to 0$.

Consider the system

$$\dot{\mathbf{x}} = A\mathbf{x}$$

$$\mathbf{y} = S^{1/2}\mathbf{x}.$$

If $\mathbf{y} \to 0$ as $t \to \infty$, then either (i) all modes are stable, in which case $\mathbf{x} \to 0$, or (ii) there are unstable modes, but they are not observable. However, the latter case is excluded by assumption, so A must be stable. ∎

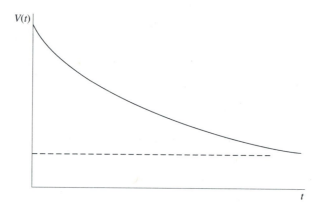

Figure 7.5 Time plot for a Lyapunov function

◆ ◆ ◆ R E M A R K

In the special case where S is positive definite, $S^{1/2}$ is nonsingular; this guarantees that the pair $(A, S^{1/2})$ is observable, since $\mathbf{x}$ is immediately recoverable from $\mathbf{x} = (S^{1/2})^{-1}\mathbf{y}$. ◆

As we shall see presently, the matrix Lyapunov equation is, in fact, a set of linear equations for the elements of P.

Example 7.4

The system $\dot{\mathbf{x}} = \begin{bmatrix} 0 & 1 \\ 0 & -1 \end{bmatrix}\mathbf{x} + \begin{bmatrix} 0 \\ 1 \end{bmatrix}u$ is to be controlled with the state feedback law $u = -kx_1$. For $\mathbf{x}(0) = \mathbf{x}_0$, calculate

$$J = \int_0^\infty [\mathbf{x}^T(t)Q\mathbf{x}(t) + ru^2(t)]dt$$

with $Q = I$.

Solution

From Equations 7.20 and 7.21, with $K = [k \ \ 0]$,

$$\dot{\mathbf{x}} = \left(\begin{bmatrix} 0 & 1 \\ 0 & -1 \end{bmatrix} - \begin{bmatrix} 0 \\ 1 \end{bmatrix}[k \ \ 0] \right)\mathbf{x}$$

$$= \begin{bmatrix} 0 & 1 \\ -k & -1 \end{bmatrix}\mathbf{x}$$

and

$$J = \int_0^\infty \mathbf{x}^T(t) \left\{ \begin{bmatrix} 1 & 0 \\ 0 & 1 \end{bmatrix} + \begin{bmatrix} k \\ 0 \end{bmatrix}r[k \ \ 0] \right\}\mathbf{x}(t)dt$$

$$= \int_0^\infty \mathbf{x}^T(t) \begin{bmatrix} 1 + rk^2 & 0 \\ 0 & 1 \end{bmatrix}\mathbf{x}(t)dt.$$

We then apply the matrix Lyapunov equation,

$$\begin{bmatrix} 0 & -k \\ 1 & -1 \end{bmatrix} \begin{bmatrix} P_{11} & P_{12} \\ P_{12} & P_{22} \end{bmatrix} + \begin{bmatrix} P_{11} & P_{12} \\ P_{12} & P_{22} \end{bmatrix} \begin{bmatrix} 0 & 1 \\ -k & -1 \end{bmatrix} = -\begin{bmatrix} 1 + rk^2 & 0 \\ 0 & 1 \end{bmatrix}$$

where the symmetry of P has been invoked, to make $P_{21} = P_{12}$. This is a matrix equation, so we must equate element by element. For the 11 element,

$$-kP_{12} - kP_{12} = -1 - rk^2.$$

For the 12 element,

$$-kP_{22} + P_{11} - P_{12} = 0.$$

For the 21 element,

$$P_{11} - P_{12} - kP_{22} = 0.$$

For the 22 element,

$$P_{12} - P_{22} + P_{12} - P_{22} = -1.$$

Note that the "12" and "21" equations are identical. The reason is that both the LHS and RHS are symmetric matrices. In any case, since P has only three distinct elements, we only need three equations. Summarized, they are

$$2kP_{12} = 1 + rk^2$$
$$P_{11} - P_{12} - kP_{22} = 0$$
$$2P_{12} - 2P_{22} = -1.$$

The solution is

$$P_{11} = \frac{1}{2k} + \frac{1}{2} + (r+1)\frac{k}{2} + \frac{rk^2}{2}$$

$$P_{12} = \frac{1}{2k} + \frac{rk}{2}$$

$$P_{22} = \frac{1}{2k} + \frac{1}{2} + \frac{rk}{2}.$$

Since $J = \mathbf{x}^T(0)P\mathbf{x}(0)$,

$$J = P_{11}x_1{}^2(0) + 2P_{12}x_1(0)x_2(0) + P_{22}x_2{}^2(0).$$

In Figure 7.6, J is plotted against k, for $\mathbf{x}(0) = \begin{bmatrix} 1 \\ 0 \end{bmatrix}$ and $r = 0, 1, 2$. As r increases, the optimal value of k decreases, reflecting the fact that increasing the penalty on the control effort tends to decrease the gain in order to reduce the control.

In Figure 7.7 we plot J versus k for $r = 1$ and two initial states, $\mathbf{x}(0) = \begin{bmatrix} 1 \\ 0 \end{bmatrix}$ and $\begin{bmatrix} 0 \\ 1 \end{bmatrix}$. It is to be noted that the minimizing value of k differs for different initial states; there does not exist a value of k that is best for all initial states.

◆ ◆ ◆ **R E M A R K**
Theorem 7.5 can be used to check stability in this example. Since $S = Q + K^T R K$ and $Q > 0$, it follows that

$$\mathbf{x}^T S\mathbf{x} = \mathbf{x}^T Q\mathbf{x} + \mathbf{x}^T K^T R K\mathbf{x} = \mathbf{x}^T Q\mathbf{x} + (K\mathbf{x})^T R(K\mathbf{x}) > 0$$

if $\mathbf{x} \neq 0$; as a result, S is positive definite, and detectability of the pair $(A, S^{1/2})$ is assured. To check stability, we must verify that $P \geq 0$. One way to do this is to

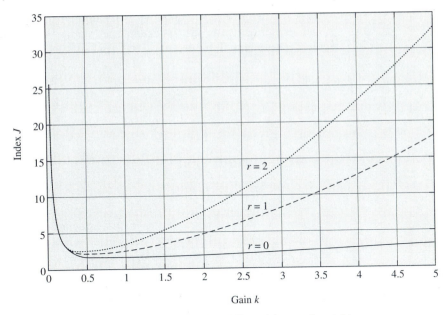

Figure 7.6 Performance index vs. gain for differential control weights

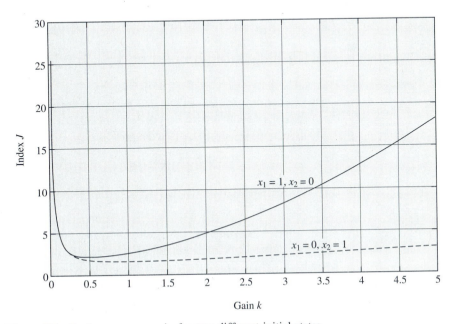

Figure 7.7 Performance vs. gain for two different initial states

verify that the principal minors of P (determinants of the $1 \times 1, 2 \times 2, 3 \times 3, \ldots$ matrices formed from the upper left-hand corner) are all positive or zero. In this case, the principal minors are

$$P_{11} = \frac{1}{2k} + \frac{1}{2} + (r+1)\frac{k}{2} + r\frac{k^2}{2}$$

and

$$P_{11}P_{22} - P_{12}{}^2 = \frac{1}{2k} + \frac{1}{2} + \frac{(3r+s)}{4}k + \frac{r}{2}k^2 + \frac{s^2k^3}{4}.$$

It is clear that both are positive if $k > 0$. ◆

The numerical solution of the matrix Lyapunov equation does not follow the by-hand procedure. The most widely used procedure makes use of orthogonal transformations to reduce the problem to a sequence of solutions of linear equations of order 1 or 2 [5].

7.4.3 The Optimal LQ Regulator

The optimal linear–quadratic regulator problem is stated as follows. Given the LTI system

$$\dot{\mathbf{x}} = A\mathbf{x} + B\mathbf{u}$$

and the initial state $\mathbf{x}(0)$, calculate the state feedback gain matrix K such that the control law $\mathbf{u}(t) = -K\mathbf{x}(t)$ minimizes the performance index

$$J = \int_0^\infty \left[\mathbf{x}^T(t)Q\mathbf{x}(t) + \mathbf{u}^T(t)R\mathbf{u}(t) \right] dt$$

where Q and R are symmetric and $Q \geq 0, R > 0$.

The closed-loop system and performance index are described by Equations 7.20 and 7.21, repeated here for convenience:

$$\dot{\mathbf{x}} = (A - BK)\mathbf{x}$$
$$J = \int_0^\infty \mathbf{x}^T(t)(Q + K^T R K)\mathbf{x}(t)dt.$$

The solution begins with the observation that, in order for the problem to be of practical interest, it must be possible to obtain internal stability with linear state feedback. Thus, the system must be *stabilizable*; i.e., any unstable modes must be controllable. Controllability of the system guarantees stabilizability, of course, because in that case *all* modes are controllable.

We now derive a necessary condition to be satisfied by a stabilizing gain to lead to a minimum of J.

■ **Theorem 7.6**   For a stabilizing gain K to minimize J for all $\mathbf{x}(0)$, it must satisfy the necessary condition

$$K = R^{-1}B^T P \tag{7.27}$$

where P satisfies the *matrix Riccati equation,*

$$A^T P + PA - PBR^{-1}B^T P + Q = 0. \tag{7.28}$$

Proof: From the preceding section and Equations 7.20 and 7.21, we have, for any stabilizing K,

$$J(K) = \mathbf{x}^T(0) P \mathbf{x}(0) \tag{7.29}$$

where P satisfies the matrix Lyapunov equation,

$$(A - BK)^T P + P(A - BK) = -Q - K^T RK. \tag{7.30}$$

Let us assume that there exists a gain that minimizes J, and let K^* be that gain. Let P^* be the corresponding P matrix; i.e., let

$$(A - BK^*)^T P^* + P^*(A - BK^*) = -Q - K^{*^T} RK^*. \tag{7.31}$$

To derive the necessary condition, we examine the effect on J of small perturbations of K from K^*. We argue that, if J is indeed minimized by K^*, any perturbation of gain must lead to an increase in J; otherwise, some value of K would result in a lower value of J than K^*. The fact that infinitesimal perturbations are used means that we are checking $J(K^*)$ against values of $J(K)$ in a neighborhood of K^*; i.e., we verify that K^* yields a *local* minimum. But, since the *global* minimum we are seeking must also be a minimum in its own neighborhood, it must also satisfy the condition for a local minimum. We let

$$K = K^* + \epsilon \, \delta K$$

where ϵ is infinitesimally small. The matrix P will change. It can be shown that P is an analytic function of the elements of K, so that

$$P = P^* + \epsilon \, \delta P_1 + \epsilon^2 \, \delta P_2 + \cdots.$$

We insert this in Equation 7.30:

$$(A - BK^* - \epsilon B \, \delta K)^T (P^* + \epsilon \, \delta P_1 + \epsilon^2 \, \delta P_2)$$
$$+ (P^* + \epsilon \, \delta P_1 + \epsilon^2 \, \delta P_2)(A - BK^* - \epsilon B \, \delta K)$$
$$= -Q - (K^* + \epsilon \, \delta K)^T R(K^* + \epsilon \, \delta K)$$

or

$$(A - BK^*)^T P^*(A - BK^*) + \epsilon[-\delta K^T B^T P^* + (A - BK^*)^T \delta P_1$$
$$+ \delta P_1(A - BK^*)^T - P^* B \, \delta K] + 0(\epsilon^2)$$
$$= -Q - K^{*^T} RK^* + \epsilon[-\delta K^T RK^* - K^{*^T} R \, \delta K] + 0(\epsilon^2).$$

Here, the notation $0(\epsilon^2)$ refers to terms of order two or more in ϵ. Matching powers of order zero gives back Equation 7.31. Matching terms in ϵ gives

$$\delta K^T (B^T P^* - RK^*) + (P^* B - K^{*^T} R) \, \delta K$$
$$+ (A - BK^*)^T \delta P_1 + \delta P_1(A - BK^*) = 0. \qquad (7.32)$$

We go back to Equation 7.29 and write

$$J(K^* + \epsilon \, \delta K) = \mathbf{x}^T(0) P \mathbf{x}(0)$$
$$= \mathbf{x}^T(0) P^* \mathbf{x}(0) + \epsilon \mathbf{x}^T(0) \, \delta P_1 \mathbf{x}(0) + \epsilon^2 \mathbf{x}^T(0) \, \delta P_2 \mathbf{x}(0) + \cdots.$$

We now argue that δP_1 must be zero. If K^* is indeed optimal, the minimum value of J is $\mathbf{x}^T(0) P^* \mathbf{x}(0)$, and therefore

$$\epsilon \mathbf{x}^T(0) \, \delta P_1 \mathbf{x}(0) + \epsilon^2 \mathbf{x}^T(0) \, \delta P_2 \mathbf{x}(0) + \cdots \geq 0 \qquad (7.33)$$

for all $\mathbf{x}(0)$.

Suppose $\delta P_1 \neq 0$. Then, for some $\mathbf{x}(0)$, $\mathbf{x}^T(0) \, \delta P_1 \mathbf{x}(0) \neq 0$. Choose ϵ small enough that the first term on the LHS of Equation 7.33 dominates and the sign of the LHS is that of the term in ϵ. By choosing the sign of ϵ, we can make $\epsilon \mathbf{x}^T(0) \, \delta P_1 \mathbf{x}(0)$ negative, thus violating Equation 7.33. Hence, it is necessary that $\delta P_1 = 0$.

If that is so, then, from Equation 7.32,

$$\delta K^T (B^T P^* - RK^*) + (P^* B - K^{*^T} R) \, \delta K = 0$$

or

$$\delta K^T (B^T P^* - RK^*) + [\delta K^T (B^T P^* - RK^*]^T = 0.$$

Since this must hold for all perturbations δK, it is necessary that

$$B^T P^* - RK^* = 0$$

or

$$K^* = R^{-1} B^T P^*.$$

Inserting this in Equation 7.30 yields

$$A^T P^* - P^* B R^{-1} B^T P^* + P^* A - P^* B R^{-1} B^T P^* = -Q - P^* B R^{-1} R R^{-1} B^T P^*$$

or

$$A^T P^* + P A^* - P^* B R^{-1} B^T P^* + Q = 0$$

where the fact that P^* is symmetric ($= P^{*^T}$) has been used. This completes the proof. ∎

As things stand, we need to solve the matrix Riccati equation, Equation 7.28, for P; calculate K from Equation 7.27; and check $A - BK$ for stability. Since the Riccati equation is nonlinear, it generally has many solutions, so we must pick out the stabilizing ones and compare values of J. Fortunately, we can avoid most of that process.

First, we impose the requirement that $(A, Q^{1/2})$ be detectable. Suppose it is not. Then there exists an unstable, unobservable mode with eigenvalue s_i and eigenvector $\mathbf{v}_i$. Then, for $\mathbf{x}(0) = \mathbf{v}_i$, $\mathbf{x}(t) = \mathbf{v}_i e^{s_i t}$ and $\mathbf{x}^T Q \mathbf{x} = \mathbf{v}_i^T (Q^{1/2})^T Q^{1/2} \mathbf{v}_i e^{2s_i t} = 0$. The performance index does not penalize this mode, so the mathematics of the optimization leave it unchanged, and the closed-loop system is not internally stable.

Given detectability of $(A, Q^{1/2})$, it can be shown [1, 7] that a positive definite or semidefinite P will yield a stabilizing K that will simultaneously be at least a local minimum. It can be shown [1, 7] that the matrix Riccati equation has *one*, and *only one*, solution $P \geq 0$. This means that there always exists one, and only one, local minimum, which is also the global minimum.

To summarize:

1. To set up a regulation problem, the state equations are written in incremental form, as in Equation 7.6. The performance index is defined with respect to $\Delta \mathbf{x}$ and $\Delta \mathbf{u}$.

2. The Riccati equation has a unique positive definite or semidefinite solution. That is the one that must be found, among all possible solutions of Equation 7.28.

3. Once P is known, the optimum gain is calculated from Equation 7.27.

4. Given the gain K, the response from a given initial state is computed as shown in Equations 7.8 and 7.9.

Example 7.5   Calculate the optimal state feedback gain for the plant

$$\dot{\mathbf{x}} = \begin{bmatrix} 0 & 1 \\ 0 & 0 \end{bmatrix} \mathbf{x} + \begin{bmatrix} 0 \\ 1 \end{bmatrix} u \quad \text{and} \quad J = \int_0^\infty [x_1^2(t) + \rho u^2(t)] dt, \qquad \rho > 0.$$

Solution   With

$$\mathbf{x}^T Q \mathbf{x} = q_{11} x_1{}^2 + 2 q_{12} x_1 x_2 + q_{22} x_2{}^2$$

the first term in the integrand, $x_1{}^2$, implies that $q_{11} = 1$, $q_{12} = q_{22} = 0$; i.e.,

$$Q = \begin{bmatrix} 1 & 0 \\ 0 & 0 \end{bmatrix}$$

with

$$A = \begin{bmatrix} 0 & 1 \\ 0 & 0 \end{bmatrix}, \qquad \mathbf{b} = \begin{bmatrix} 0 \\ 1 \end{bmatrix}, \qquad R = \rho.$$

It can be verified that $(A, Q^{1/2})$ is observable. We write the matrix Riccati equation, Equation 7.28:

$$\begin{bmatrix} 0 & 0 \\ 1 & 0 \end{bmatrix} \begin{bmatrix} P_{11} & P_{12} \\ P_{12} & P_{22} \end{bmatrix} + \begin{bmatrix} P_{11} & P_{12} \\ P_{12} & P_{22} \end{bmatrix} \begin{bmatrix} 0 & 1 \\ 0 & 0 \end{bmatrix}$$

$$- \begin{bmatrix} P_{11} & P_{12} \\ P_{12} & P_{22} \end{bmatrix} \begin{bmatrix} 0 \\ 1 \end{bmatrix} \rho^{-1} \begin{bmatrix} 0 & 1 \end{bmatrix} \begin{bmatrix} P_{11} & P_{12} \\ P_{12} & P_{22} \end{bmatrix} + \begin{bmatrix} 1 & 0 \\ 0 & 0 \end{bmatrix} = \begin{bmatrix} 0 & 0 \\ 0 & 0 \end{bmatrix}$$

or

$$\begin{bmatrix} 0 & 0 \\ P_{11} & P_{12} \end{bmatrix} + \begin{bmatrix} 0 & P_{11} \\ 0 & P_{12} \end{bmatrix} \frac{1}{\rho} \begin{bmatrix} P_{12} \\ P_{22} \end{bmatrix} \begin{bmatrix} P_{12} & P_{22} \end{bmatrix} + \begin{bmatrix} 1 & 0 \\ 0 & 0 \end{bmatrix} = \begin{bmatrix} 0 & 0 \\ 0 & 0 \end{bmatrix}.$$

Equating element by element, we have the following. For the 11 element,

$$-\frac{1}{\rho} P_{12}{}^2 + 1 = 0.$$

For the 12 (or 21) element,

$$P_{11} - \frac{1}{\rho} P_{12} P_{22} = 0.$$

For the 22 element,

$$2 P_{12} - \frac{1}{\rho} P_{22}{}^2 = 0.$$

From the "11" equation,

$$P_{12} = \pm \rho^{1/2}.$$

The "22" equation yields

$$P_{22} = \pm[2\rho P_{12}]^{1/2}.$$

Since P_{22} must be real, P_{12} must be positive. Furthermore, P_{22} must be positive in order that $P > 0$. Thus,

$$P_{12} = \rho^{1/2} \quad \text{and} \quad P_{22} = \sqrt{2}\rho^{3/4}.$$

Finally,

$$P_{11} = \frac{1}{\rho} P_{12} P_{22} = \sqrt{2}\rho^{1/4}$$

so that

$$P = \begin{bmatrix} \sqrt{2}\rho^{1/4} & \rho^{1/2} \\ \rho^{1/2} & \sqrt{2}\rho^{3/4} \end{bmatrix}.$$

We verify the positivity of P by ascertaining that all principal minors are positive. Here,

$$P_{11} = \sqrt{2}\rho^{1/4} > 0$$

and

$$\det P = \rho > 0.$$

From Equation 7.27, the optimum gain is

$$K = R^{-1}B^T P$$

$$= (1/\rho)[0 \quad 1]\begin{bmatrix} \sqrt{2}\rho^{1/4} & \rho^{1/2} \\ \rho^{1/2} & \sqrt{2}\rho^{3/4} \end{bmatrix}$$

$$= [\rho^{-1/2} \quad \sqrt{2}\rho^{-1/4}].$$

It is useful to examine the closed-loop eigenvalues. The closed-loop A matrix is

$$A - BK = \begin{bmatrix} 0 & 1 \\ 0 & 0 \end{bmatrix} - \begin{bmatrix} 0 \\ 1 \end{bmatrix}[\rho^{-1/2} \quad \sqrt{2}\rho^{-1/4}]$$

$$= \begin{bmatrix} 0 & 1 \\ -\rho^{-1/2} & -\sqrt{2}\rho^{-1/4} \end{bmatrix}.$$

This matrix is in companion form, so the characteristic polynomial is written by inspection. It is

$$s^2 + \sqrt{2}\rho^{-1/4}s + \rho^{-1/2}.$$

This is of standard second-order form, i.e., of the form $s^2 + 2\zeta\omega_0 s + \omega_0^2$. We see that $\omega_0 = \rho^{-1/4}$ and $\zeta = \sqrt{2}/2$. The optimal control yields a damping factor $\zeta = 0.707$, which, as we have seen, represents a good compromise between speed and overshoot. The bandwidth ω_0 depends on ρ. The larger ρ, the smaller ω_0 and, hence, the slower the response.

Example 7.6   (dc Servo)

The dc servo of Example 2.1 (Chapter 2), with $T_L = 0$, is to be regulated to the dc steady state $\theta = \theta_d$, $\omega = i = v = 0$. We wish to design the optimal LQ control for the performance index of the form

$$J = \int_0^\infty [Q_{11}(\theta - \theta_d)^2 + Q_{22}\omega^2 + v^2] dt.$$

(a) With $Q_{22} = 0$, choose Q_{11} so that, with the initial state $\theta = \omega = i = 0$ and $\theta_d = 1$ rad, $v(t)$ does not exceed 3 V during the transient response.

(b) Experiment with different values of Q_{22} to assess its effect on the response.

Solution   The state part of the integrand of J is

$$Q_{11}\Delta\theta^2 + Q_{22}\omega^2 = [\Delta\theta \quad \omega \quad i] \begin{bmatrix} Q_{11} & 0 & 0 \\ 0 & Q_{22} & 0 \\ 0 & 0 & 0 \end{bmatrix} \begin{bmatrix} \Delta\theta \\ \omega \\ i \end{bmatrix}.$$

The A and B matrices are as in Equation 2.19 (with $T_L = 0$). The Riccati equation and the gain are solved using the computer (MATLAB lqr). The gain is $\mathbf{k}^T$, a 1×3 matrix. The control v is generated from

$$v = v^* - \mathbf{k}^T(\mathbf{x} - \mathbf{x}^*) = 0 - [k_1 \quad k_2 \quad k_3]\left(\begin{bmatrix} \theta \\ \omega \\ i \end{bmatrix} - \begin{bmatrix} \theta_d \\ 0 \\ 0 \end{bmatrix}\right)$$

$$= k_1(\theta_d - \theta) - k_2\omega - k_3 i.$$

Transients are obtained, applying Equation 7.8 with $\Delta\theta(0) = 0 - 1 = -1$ and $\Delta\omega(0) = \Delta i(0) = 0$. Figure 7.8 shows results for $Q_{22} = 0$ and $Q_{11} = 4, 9, 20$; $Q_{11} = 4$ yields the responses with the least overshoot and smallest input, and $Q_{11} = 20$ gives the most overshoot and the largest input (because it weighs state errors correspondingly more than input effort). Curves with $Q_{11} = 4$ and 9 both meet the input requirement $v \leq 3$ V, but $Q_{11} = 9$ is selected because it gives a faster angular response. For that value of Q_{11}, the gain vector is $\mathbf{k}^T = [3 \quad 0.8796 \quad 0.1529]$.

Figure 7.9 shows a few transients for $\Delta\theta(t)$ and $\omega(t)$, with $Q_{11} = 9$ and with two values of Q_{22}. As Q_{22} is increased, the angular velocity ω is seen to decrease,

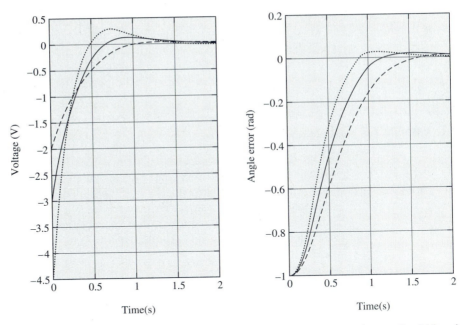

Figure 7.8 Results of LQ design for the dc servo: input voltage and angle error for $Q11 = 4$ (dashed), 9 (solid), and 20 (dotted)

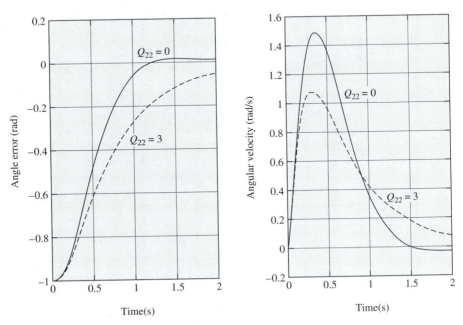

Figure 7.9 Results of LQ design for the dc servo: effect of weighting the angular velocity

which makes the error response more damped. The optimal gain for $Q_{22} = 3$ is $\mathbf{k}^T = [3 \quad 1.7165 \quad 0.2838]$; clearly, the velocity gain is greater than in the preceding case.

Example 7.7   (Pendulum on a Cart)

For the linearized model of the pendulum and cart of Example 2.9 using the numerical values of Example 2.5 (Chapter 2), design an optimal LQ feedback to regulate to the vertical equilibrium with $x = 0$. For an initial state $\theta = 15°, \omega = x = v = 0$, the transient should be such that $|x(t)| \leq 0.5$ m and $|\theta(t)| \leq 20° = 0.35$ rad. Simulate the transient response for this control system used on the *nonlinear* model of Example 2.9, with different initial state values, to explore the applicability of the linear design.

Solution

The A and B matrices are as in Example 7.3. We must carry out a trial-and-error process to choose the proper weights. (Note that there is no guarantee that the specifications can be met, even by the optimal system; they may need to be relaxed.)

One way to begin the process is to use weights that normalize the variables with respect to their largest permissible values. Thus, we try

$$J = \int_0^\infty \left[\left(\frac{x}{x_{\max}} \right)^2 + \left(\frac{\theta}{\theta_{\max}} \right)^2 + \left(\frac{F}{F_{\max}} \right)^2 \right] dt$$

where $x_{\max} = 0.5$ m and $\theta_{\max} = 0.35$ rad. The value of $F_{\max}$ is not specified. A reasonable order of magnitude is the force required to hold the cart and pendulum against gravity—say, $F_{\max} = (M + m)g = 20$ N.

From the integral, we see that $Q_{11} = 1/x_{\max}^2$, $Q_{33} = 1/\theta_{\max}^2$, and $R = 1/F_{\max}^2$. (This assumes that the states are ordered within the state vector as in Example 7.3.) All other elements of Q are zero.

Figure 7.10 shows the transient responses for $x(t)$ and $\theta(t)$ (MATLAB lqr, initial); it turns out that the initial guesses for Q and R are satisfactory. The responses should be compared to those obtained by pole placement in Example 7.3. The LQ responses are somewhat faster, and the maximum distance excursion of the cart is lessened.

In Figure 7.11, we show the angle responses using the full nonlinear model (MATLAB ode23) for three initial angles (the other three state components are 0 at $t = 0$). For $\theta(0) = 0.26$ rad, the response is close to that of the linearized system. As $\theta(0)$ increases, the response departs increasingly from that of the linearized system; for $\theta(0) \sim 0.7$ rad or greater, the angle does not go to zero. This illustrates the extent to which a linear control law can be used to control this nonlinear system.

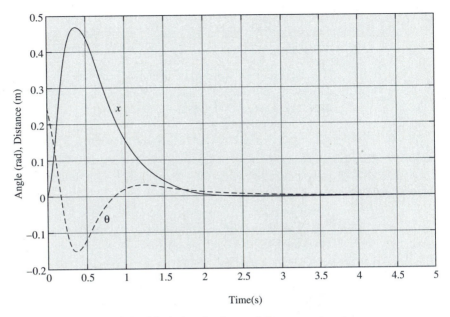

Figure 7.10   Results of the LQ design for the pendulum-on-cart system

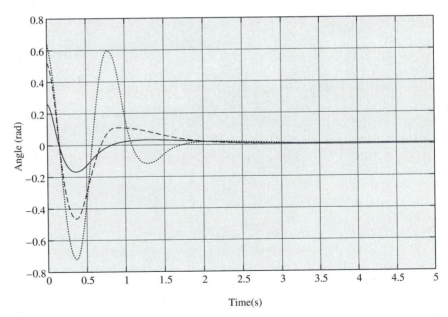

Figure 7.11   Angle time responses for different initial conditions showing the nonlinear behavior

7.5 DISTURBANCE AND REFERENCE INPUT COMPENSATION

7.5.1 Feedforward Compensation

The foregoing sections have dealt with the basic problem of taking a system from a nonzero initial state to the zero state. We now consider modifications to the basic scheme, to handle disturbances and set-point changes.

To fix ideas, consider the model of Figure 7.12, described by the equations

$$\dot{\mathbf{x}} = A\mathbf{x} + B\mathbf{u} + \Gamma\mathbf{w} \tag{7.34}$$

$$\Delta\mathbf{y} = C\mathbf{x} - \mathbf{y}_d. \tag{7.35}$$

Here, $\mathbf{w}$ is a disturbance input and $\mathbf{y}_d$ is the set-point, or reference, input.

Let us begin with the simple case $\mathbf{w} = \mathbf{w}^*$ and $\mathbf{y}_d = \mathbf{y}_d^*$, where $\mathbf{w}^*$ and $\mathbf{y}_d^*$ are constant. The dc steady state (if it exists) defined by $\Delta\mathbf{y} = \mathbf{0}$ satisfies the algebraic linear equation

$$\begin{bmatrix} A & B \\ C & 0 \end{bmatrix} \begin{bmatrix} \mathbf{x}^* \\ \mathbf{u}^* \end{bmatrix} = \begin{bmatrix} -\Gamma\mathbf{w}^* \\ \mathbf{y}_d^* \end{bmatrix} \tag{7.36}$$

If a solution exists and is unique, linearity dictates that $\mathbf{x}^*$ and $\mathbf{u}^*$ are linear functions of $\mathbf{w}^*$ and $\mathbf{y}_d^*$, i.e., of the form

$$\mathbf{u}^* = K_w^*\mathbf{w}^* + K_y^*\mathbf{y}_d^* \tag{7.37}$$

where K_w^* and K_y^* are gain matrices.

Figure 7.13 shows that Equation 7.37 represents a *feedforward* system. An input $\mathbf{u}^*$ is calculated that, in the dc steady state, will exactly cancel out the effect of the constant disturbance $\mathbf{w}^*$ and make the output equal to the set point.

Another way to view this is provided by the transfer function expression of Equation 4.1 (Chapter 4):

$$\mathbf{y}(s) = P(s)\mathbf{u}(s) + P_w(s)\mathbf{w}(s).$$

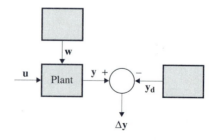

Figure 7.12 Showing the reference and disturbance signals as zero-input outputs of LTI systems

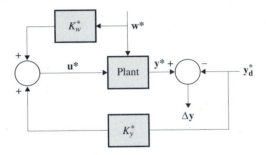

Figure 7.13 Control systems with feedforward from the steady-state and disturbance signals

If the dc steady state exists, it satisfies

$$\mathbf{y}^* = P(0)\mathbf{u}^* + P_w(0)\mathbf{w}^*$$

and thus, for a system with equal numbers of inputs and outputs,

$$\mathbf{u}^* = P^{-1}(0)[\mathbf{y}_d^* - P_w(0)\mathbf{w}^*]. \tag{7.38}$$

Equation 7.38 is clearly of the same form as Equation 7.37, and makes sense if $P^{-1}(0)$ exists. $P^{-1}(0)$ does exist if $\left[\begin{smallmatrix} A & B \\ C & 0 \end{smallmatrix}\right]$ is nonsingular, since that is also the condition for having no transmission zero at $s = 0$.

   It is possible, by feedforward, to asymptotically track any reference input of the form $\mathbf{y}_d^* e^{s^* t}$ and/or any disturbance of the same form. We need only replace $P(0)$ and $P_w(0)$ with $P(s^*)$ and $P_w(s^*)$, and write $\mathbf{u}(t) = \mathbf{u}^* e^{s^* t}$. In particular, it is possible in this manner to generate a sinusoidal input in order to track a sinusoid of the same frequency.

   Regulation to a given dc steady state was covered earlier in this chapter; the addition of a constant disturbance affects only the calculation of the steady-state input. Asymptotic convergence to a more general exponential input is carried out in precisely the same way. Essentially, then, we have a two-step procedure:

1. Calculate the steady-state (or general exponential) input so that the reference output is asymptotically tracked.

2. Using Equation 7.6, design a feedback law for $\Delta\mathbf{u}$ in terms of $\Delta\mathbf{x}$.

Example 7.8 **(dc Servo)**

The dc servo of Example 2.1 (Chapter 2) is subjected to a step load torque $T_L = T_{LS}u_{-1}(t)$.

(a) Calculate, as a function of T_{LS}, the constant input v_s required for a dc steady state $\theta = \theta_d$.

(b) Using the results of Example 7.6, obtain the control law that minimizes $J = \int_0^\infty [9(\theta - \theta_d)^2 + (v - v_s)^2]dt$.

Solution   From Equation 2.19, the dc steady state requires

$$\omega = 0$$
$$4.438i - 7.396T_L = 0$$
$$-24i + 20v = 0.$$

With $T_L = T_{LS}$, it follows that

$$i = i_s = 1.667T_{LS}$$
$$v = v_s = 1.2is = 2.0T_{LS}.$$

The dc steady-state input and states are

$$v_s = 2.0T_{LS}, \qquad \theta = \theta_d, \qquad \omega = 0, \qquad i = 1.667T_{LS}.$$

To solve part (b), recall that the incremental system of Equation 7.6 is the same as in Example 7.6. Therefore, the problem of taking the state to the reference state has already been solved. Using the gain vector derived in Example 7.6,

$$v - v_s = -3.0(\theta - \theta_d) - 0.8796\omega - 0.1529(i - i_s).$$

The complete control law, illustrated in Figure 7.14, is

$$v = 2.0T_{LS} - 3.0(\theta - \theta_d) - 0.8796\omega - 0.1529(i - 1.667T_{LS})$$
$$= 2.255T_{LS} - 3.0(\theta - \theta_d) - 0.8796\omega - 0.1529i.$$

Figure 7.15 shows the angle error response $\theta(t) - \theta_d$ to a load-torque step $T_L = 0.01u_{-1}(t)$, starting from $\theta = \theta_d$, $\omega = i = 0$.

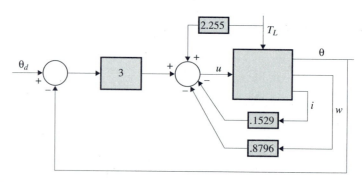

Figure 7.14   Control system with feedforward, dc servo

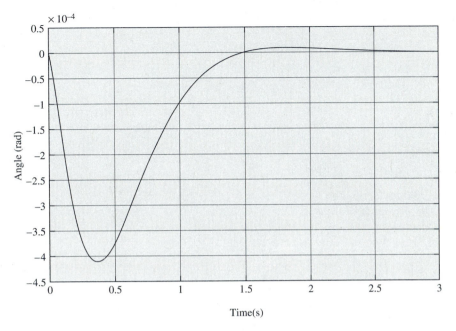

Figure 7.15 Angle error response to a step in load torque, dc servo

The procedure just followed is necessary to obtain convergence of $\Delta\mathbf{y}(t)$ to zero, in cases where the set-point and/or disturbance signals are not decreasing exponentials (i.e., they are steps, sinusoids, ramps, increasing exponentials, and so forth). If they are decreasing exponentials, they tend to zero, and so does $\mathbf{u}^*$. Tracking may be desirable (especially for slowly decreasing signals), but the LQ approach may be preferable, to reach the best compromise between control effectiveness and input effort.

To set this up, suppose that $\mathbf{w}$ and $\mathbf{y}_d$ have constant and time-varying components; i.e.,

$$\mathbf{w}(t) = \mathbf{w}^* + \Delta\mathbf{w}(t)$$

$$\mathbf{y}_d(t) = \mathbf{y}_d^* + \Delta\mathbf{y}_d(t).$$

From Equations 7.34 and 7.35,

$$\Delta\dot{\mathbf{x}} = \dot{\mathbf{x}} - \dot{\mathbf{x}}^* = \dot{\mathbf{x}} = A\mathbf{x}^* + A\Delta\mathbf{x} + B\mathbf{u}^* + B\Delta\mathbf{u} + \Gamma\mathbf{w}^* + \Gamma\Delta\mathbf{w}$$

or, since $A\mathbf{x}^* + B\mathbf{u}^* + \Gamma\mathbf{w}^* = 0$,

$$\Delta\dot{\mathbf{x}} = A\Delta\mathbf{x} + B\Delta\mathbf{u} + \Gamma\Delta\mathbf{w}. \qquad (7.39)$$

Also,

$$\Delta \mathbf{y} = C\mathbf{x}^* + C\,\Delta\mathbf{x} - \mathbf{y}_d^* - \Delta\mathbf{y}_d$$

$$\Delta \mathbf{y} = C\,\Delta\mathbf{x} - \Delta\mathbf{y}_d \tag{7.40}$$

since $C\mathbf{x}^* - \mathbf{y}_d^* = 0$.

Next, we model $\Delta\mathbf{w}(t)$ and $\Delta\mathbf{y}_d(t)$ as the zero-input responses of *stable* LTI systems:

$$\dot{\mathbf{z}}_w = A_w \mathbf{z}_w$$

$$\Delta \mathbf{w} = C_w \mathbf{z}_w \tag{7.41}$$

and

$$\dot{\mathbf{z}}_y = A_y \mathbf{z}_y$$

$$\Delta \mathbf{y}_d = C_y \mathbf{z}_y. \tag{7.42}$$

The vectors $\mathbf{z}_w$ and $\mathbf{z}_y$ may be called the disturbance and reference state vectors, respectively. They may be combined with the plant state $\Delta\mathbf{x}$ to form one complete description:

$$\frac{d}{dt}\begin{bmatrix} \Delta\mathbf{x} \\ \mathbf{z}_w \\ \mathbf{z}_y \end{bmatrix} = \begin{bmatrix} A & \Gamma C_w & 0 \\ 0 & A_w & 0 \\ 0 & 0 & A_y \end{bmatrix}\begin{bmatrix} \Delta\mathbf{x} \\ \mathbf{z}_w \\ \mathbf{z}_y \end{bmatrix} + \begin{bmatrix} B \\ 0 \\ 0 \end{bmatrix}\Delta\mathbf{u} \tag{7.43}$$

$$\Delta\mathbf{y} = \begin{bmatrix} C & 0 & -C_y \end{bmatrix}\begin{bmatrix} \Delta\mathbf{x} \\ \mathbf{z}_w \\ \mathbf{z}_y \end{bmatrix}. \tag{7.44}$$

This system is clearly not controllable, as neither $\mathbf{z}_w$ nor $\mathbf{z}_y$ is affected by the control input $\Delta\mathbf{u}$. The state feedback law has the form

$$\Delta\mathbf{u} = -\begin{bmatrix} K_x & K_w & K_y \end{bmatrix}\begin{bmatrix} \Delta\mathbf{x} \\ \mathbf{z}_w \\ \mathbf{z}_y \end{bmatrix}$$

$$= -K_x\,\Delta\mathbf{x} - K_w\mathbf{z}_w - K_y\mathbf{z}_y \tag{7.45}$$

As Figure 7.16 shows, this has both feedback and feedforward components. With this control law, the closed-loop state equations are

$$\frac{d}{dt}\begin{bmatrix} \Delta\mathbf{x} \\ \mathbf{z}_w \\ \mathbf{z}_y \end{bmatrix} = \begin{bmatrix} A - BK_x & \Gamma C_w - BK_w & -BK_y \\ 0 & A_w & 0 \\ 0 & 0 & A_y \end{bmatrix}\begin{bmatrix} \Delta\mathbf{x} \\ \mathbf{z}_w \\ \mathbf{z}_y \end{bmatrix} \tag{7.46}$$

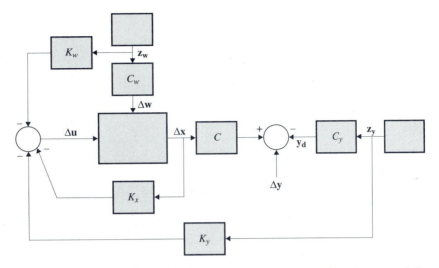

Figure 7.16 Control system with feedforward from the states of the reference and disturbance subsystems

From the block-triangular nature of the matrix, we infer that the eigenvalues are those of $A - BK_x$, A_w, and A_y. As expected, the modes associated with the disturbance and the reference input are unchanged.

If (i) (A, B) is stabilizable and (ii) A_w and A_y have only stable eigenvalues, then the system of Equation 7.46 is stabilizable. To see this, it is sufficient to notice that $A - BK_x$ can be made stable by proper choice of K_x. If that is the case, then the design of the gain matrix $\begin{bmatrix} K_x & K_w & K_y \end{bmatrix}$ can be made by the LQ technique. To summarize:

1. Calculate the dc steady-state input to yield the desired constant output. (More generally, do this for nondecreasing complex exponential signals.)

2. Append the state equations that describe those parts of the disturbance of set-point inputs corresponding to decreasing exponentials.

3. Solve the LQ control problem for the gain matrix.

Example 7.9 **(dc Servo)**

The dc servo of Example 2.1 (Chapter 2) is subjected to a load-torque disturbance $T_L(t) = 0.01e^{-t}u_{-1}(t)$. (This could model, for instance, a gust of wind acting on an antenna positioned by the motor.)

As in Example 7.6, the system is linearized about the steady state $\theta = \theta_d$, $\omega = i = v = 0$. Compute the optimal LQ control for the performance index:

$$J = \int_0^\infty [9(\theta - \theta_d)^2 + v^2]dt.$$

Solution   The equations are Equation 2.19 (Chapter 2) for the system states, and additional equations

$$\dot{z} = -z$$

$$T_L = z$$

for the disturbance state. Combining the two sets yields

$$\frac{d}{dt}\begin{bmatrix} e \\ \omega \\ i \\ z \end{bmatrix} = \begin{bmatrix} 0 & 1 & 0 & 0 \\ 0 & 0 & 4.438 & -7.396 \\ 0 & -12 & -24 & 0 \\ 0 & 0 & 0 & -1 \end{bmatrix}\begin{bmatrix} e \\ \omega \\ i \\ z \end{bmatrix} + \begin{bmatrix} 0 \\ 0 \\ 20 \\ 0 \end{bmatrix}v$$

where, as in Example 7.6, $e = \theta - \theta_d$.

The Q and R matrices are

$$Q = \text{diag}\begin{bmatrix} 9 & 0 & 0 & 0 \end{bmatrix}; \qquad R = 1.$$

The optimal gain $K = [3 \quad 0.8796 \quad 0.1529 \quad -1.8190]$ (MATLAB lqr). Figure 7.17 shows the response to the torque disturbance $T_L = 0.01e^{-t}u_1(t)$, from the initial state $e = \omega = i = 0$. [To obtain $T_L(t)$ as specified, $z(0) = 0.01$]. The response to this disturbance without feedforward—i.e., with only plant state

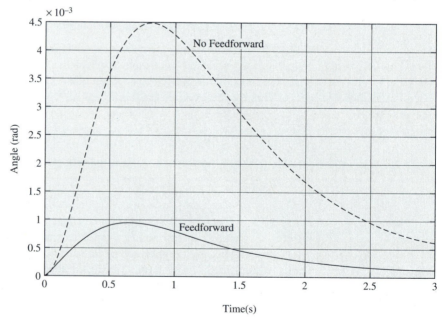

Figure 7.17   Response to a load torque input with and without feedforward, dc servo

feedback—is also shown. Note that the gains assigned to the first three states are identical to those of Example 7.6. As Problem 7.27 will explore, this is not a coincidence.

Example 7.10   (Train)

The train described in Example 2.3 (Chapter 2) is to be accelerated from rest to 25 m/s over level terrain. The head locomotive is to track the desired velocity $v_d(t) = 25(1 - e^{-t/40})$, within ± 2 m/s if possible; the traction force is not to exceed 120 kN, and the spring and damper couple forces are limited to 75,000 N and 50,000 N, respectively. Set up and solve an appropriate LQ problem for a locomotive and four cars.

Solution   First the system is linearized about the steady state. In the absence of friction and on horizontal terrain, Newton's second law implies a steady state where all velocities are 25 m/s and no forces act on the cars or the locomotive. The steady-state values $x_2, x_3, \ldots, x_N$ of the coupler lengths are all equal to the unstressed length, x_0. The state variable x_1 is not used in this problem; it is the distance traveled by the locomotive, and it does not enter into the determination of forces and velocities.

Using the desired steady state as a reference, we define

$$\Delta x_i = x_i - x_0, \qquad i = 2, 3, 4, 5$$

$$\Delta v_i = v_i - 25, \qquad i = 1, 2, 3, 4, 5.$$

From Equations 2.26 and 2.27,

$$\Delta \dot{x}_i = \Delta v_{i-1} - \Delta v_i, \qquad i = 2, 3, 4, 5$$

$$\Delta \dot{v}_1 = -12.5\Delta x_2 - .75\Delta v_1 + .75\Delta v_2 + .005 F_1$$

$$\Delta \dot{v}_i = 62.5\Delta x_i - 62.5\Delta x_{i+1} + 3.75\Delta v_{i-1}, \ 7.5\Delta v_i + 3.75\Delta v_{i+1}, \qquad i = 2, 3, 4$$

$$\Delta \dot{v}_5 = 62.5\Delta x_5 + 3.75\Delta v_4 - 3.75\Delta v_5.$$

Here, F_1 is the locomotive traction force.

The velocity error of the locomotive is described by

$$v_1 - v_d = 25 + \Delta v_1 - (25 - 25e^{-t/40})$$

$$= \Delta v_1 + z$$

where $z = 25e^{-t/40}$ satisfies

$$\dot{z} = -\frac{1}{40}z, \qquad z(0) = 25.$$

To choose a performance index, we recall that

$$\text{coupler spring force} = K(x_i - x_0) = K\,\Delta x_i$$

$$\text{damper spring force} = D(v_{i-1} - v_i) = D(\Delta v_{i-1} - \Delta v_i)$$

with $K = 2.5 \times 10^6$ N/m and $D = 1.5 \times 10^5$ N/m/s. We try an index where the forces and velocities are normalized to their desired maxima:

$$J = \int_0^\infty \left\{ \sum_{i=1}^5 \left(\frac{K\,\Delta x_i}{75,000} \right)^2 + \sum_{i=1}^5 \left(\frac{D(\Delta v_{i-1} - \Delta v_i)}{50,000} \right)^2 + \left(\frac{\Delta v_1 + z}{2} \right)^2 + \left(\frac{F_1}{120} \right)^2 \right\} dt$$

$$= \int_0^\infty \left[\sum_{i=2}^5 [3.34\Delta x_i]^2 + \sum_{i=2}^5 [3(\Delta v_{i-1} - \Delta v_i)]^2 + [.5(\Delta v_1 + z)]^2 + \left(\frac{F_1}{120} \right)^2 \right] dt.$$

To generate the Q matrix, we note that the state-dependent part of the integrand is the Euclidean length of a vector $\mathbf{v}$, where

$$\mathbf{v}^T = [3.34\,\Delta x_2 \ldots 3.34\,\Delta x_5 \quad 3(\Delta v_1 - \Delta v_2) \quad \ldots \quad 3(\Delta v_4 - \Delta v_5) \quad .5(\Delta v_1 + z)].$$

This vector is expressed as

$$
\mathbf{v} = V
\begin{bmatrix}
\Delta x_2 \\
\vdots \\
\Delta x_5 \\
--- \\
\Delta v_1 \\
\vdots \\
\Delta v_5 \\
--- \\
z
\end{bmatrix}
=
\begin{bmatrix}
3.34I & & \\
\hline
 & 3-3 & \\
 & 3-3 & \\
 & \cdots & \\
 & 3-3 & \\
\hline
.50 & & .5
\end{bmatrix}
\begin{bmatrix}
\Delta x_2 \\
\vdots \\
\Delta x_5 \\
--- \\
\Delta v_1 \\
\vdots \\
\Delta v_5 \\
--- \\
z
\end{bmatrix}
$$

Since $\mathbf{v}^T\mathbf{v} = \mathbf{x}^T V^T V \mathbf{x}$, where $\mathbf{x}$ is the state vector, we have $Q = V^T V$. This index yields the optimal gain (MATLAB lqr)

$$K = [54.53 \quad 17.28 \quad -1.303 \quad -4.361 \quad 191.7$$

$$-40.48 \quad -34.21 \quad -29.70 \quad -27.34 \quad -27.34 \quad 52.09]$$

Figure 7.18 shows the results. The velocity error is well within the desired 2-m/s magnitude, and the coupler forces are well below the limits; unfortunately, the traction force goes well above the 120-kN limit.

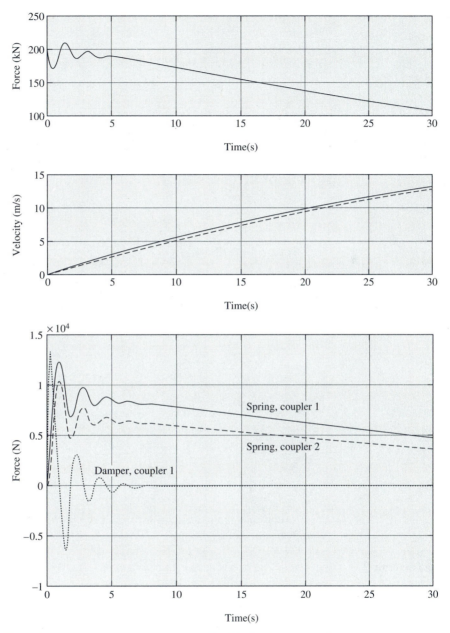

Figure 7.18 LQ control of train motion, Trial 1: Traction force, desired and actual (dotted) velocities, coupler forces

To lower $F_1(t)$, we increase the value of R, nominally $(1/120)^2$, by a factor that, after a few trials, turns out to be 35. The new gain is

$$K = [0.4559 \quad 0.3331 \quad 0.2170 \quad 0.1069 \quad 11.54$$
$$-0.2622 \quad -0.3371 \quad -0.3865 \quad -0.4110 \quad 5.373]$$

Figure 7.19 shows the responses. The coupler forces are even lower than before, and the traction does just meet the specification. Unfortunately, the velocity error is greater than 2 m/s in magnitude.

It is unlikely that the velocity specification can be met in this case. We need to either (i) relax the velocity error limit or (ii) follow a slower trajectory—for example, $25(1 - e^{-t/\tau})$, $\tau > 40s$.

7.5.2 Integral Feedback

It is not always easy to measure the dc component of a disturbance, and model inaccuracies will always lead to some error in the feedforward gains. Fortunately, there is a way around these difficulties, at the (slight) cost of introducing a few more state variables. We write

$$\dot{\mathbf{x}} = A\mathbf{x} + B\mathbf{u} + \Gamma\mathbf{w} \tag{7.47}$$

and append m integrators,

$$\dot{\mathbf{z}} = C\mathbf{x} - \mathbf{y}_d. \tag{7.48}$$

Equations 7.47 and 7.48 are combined as

$$\begin{bmatrix} \dot{\mathbf{x}} \\ \dot{\mathbf{z}} \end{bmatrix} = \begin{bmatrix} A & 0 \\ C & 0 \end{bmatrix} \begin{bmatrix} \mathbf{x} \\ \mathbf{z} \end{bmatrix} + \begin{bmatrix} B \\ 0 \end{bmatrix} \mathbf{u} + \begin{bmatrix} \Gamma \\ 0 \end{bmatrix} \mathbf{w} + \begin{bmatrix} 0 \\ I \end{bmatrix} \mathbf{y}_d. \tag{7.49}$$

We proceed to establish conditions under which this composite system is controllable from the input $\mathbf{u}$.

■ **Theorem 7.7**   Let (A, B) be controllable. The system $([\begin{smallmatrix} A & 0 \\ C & 0 \end{smallmatrix}], [\begin{smallmatrix} B \\ 0 \end{smallmatrix}])$ is controllable if, and only if, the original system (A, B, C) has no transmission zeros at the origin.

Proof: The composite A matrix is block triangular, so its eigenvalues are the eigenvalues of A, denoted by $s_1, s_2, \ldots$, plus m eigenvalues at $s = 0$. Let $\mathbf{w}_1^T, \mathbf{w}_2^T, \ldots$ be the left eigenvectors of A. Then

$$[\mathbf{w}_i^T \quad 0] \begin{bmatrix} A & 0 \\ C & 0 \end{bmatrix} = [\mathbf{w}_i^T A \quad 0] = s_i[\mathbf{w}_i^T \quad 0]$$

and $[\mathbf{w}_i^T \quad 0]$ is an eigenvector of $[\begin{smallmatrix} A & 0 \\ C & 0 \end{smallmatrix}]$ corresponding to the mode s_i.

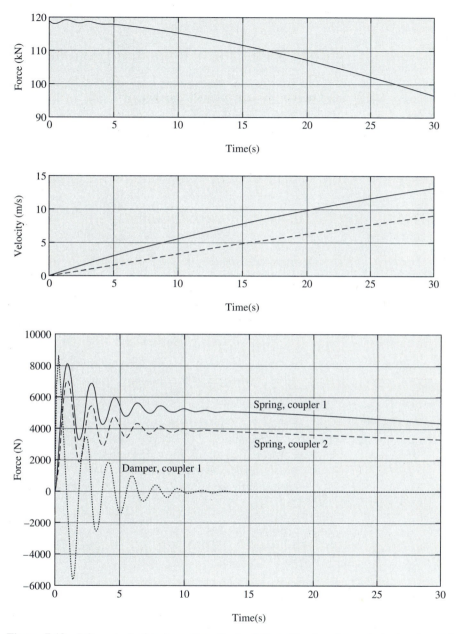

Figure 7.19 LQ control of train motion, final solution: Traction force, desired and actual (dotted) velocity coupler forces

To ascertain the controllability of that mode, we use the eigenvector test,

$$[\mathbf{w}_i^T \quad 0]\begin{bmatrix} B \\ 0 \end{bmatrix} = \mathbf{w}_i^T B \neq 0$$

because of the controllability assumption. The eigenvectors corresponding to the m eigenvalues at $s = 0$ satisfy

$$[\mathbf{w}_{i1}^T \quad \mathbf{w}_{i2}^T]\begin{bmatrix} A & 0 \\ C & 0 \end{bmatrix} = 0$$

or

$$[\mathbf{w}_{i1}^T \quad \mathbf{w}_{i2}^T]\begin{bmatrix} A \\ C \end{bmatrix} = 0. \tag{7.50}$$

Suppose a mode $s = 0$ is not controllable. Then, by the eigenvector test,

$$[\mathbf{w}_{i1}^T \quad \mathbf{w}_{i2}^T]\begin{bmatrix} B \\ 0 \end{bmatrix} = 0. \tag{7.51}$$

Putting together Equations 7.50 and 7.51,

$$[\mathbf{w}_{i1}^T \quad \mathbf{w}_{i2}^T]\begin{bmatrix} A & B \\ C & 0 \end{bmatrix} = 0. \tag{7.52}$$

This is equivalent to $\begin{bmatrix} A & B \\ C & 0 \end{bmatrix}$ having rank deficiency, which, from Chapter 3, is equivalent to the condition for a transmission zero at $s = 0$. Conversely, if Equation 7.52 is satisfied, so are Equations 7.50 and 7.51; this shows that the existence of a transmission zero at $s = 0$ implies the existence of an uncontrollable mode. ∎

With this controllability result in hand, we can now assert that the system of Equation 7.49 can be stabilized by state feedback, given the assumptions of Theorem 7.7. Let

$$\mathbf{u} = \mathbf{u}_f - [K_x \quad K_z]\begin{bmatrix} \mathbf{x} \\ \mathbf{z} \end{bmatrix}$$

where $\mathbf{u}_f$ is a feedforward term linearly dependent on $\mathbf{y}_d$ and possibly on $\mathbf{w}$. We assume that the gain $[K_x \quad K_z]$ is stabilizing. In the dc steady state,

$$\left\{\begin{bmatrix} A & 0 \\ C & 0 \end{bmatrix} - \begin{bmatrix} B \\ 0 \end{bmatrix}[K_x \quad K_z]\right\}\begin{bmatrix} \mathbf{x}^* \\ \mathbf{z}^* \end{bmatrix} = -\begin{bmatrix} B \\ 0 \end{bmatrix}\mathbf{u}_f^* - \begin{bmatrix} \Gamma \\ 0 \end{bmatrix}\mathbf{w}^* + \begin{bmatrix} 0 \\ I \end{bmatrix}\mathbf{y}_d^*. \tag{7.53}$$

The matrix on the LHS has only stable eigenvalues and is therefore nonsingular (no eigenvalues at $s = 0$). Thus, there is a unique solution for $\mathbf{x}^*, \mathbf{z}^*$. Furthermore, from Equation 7.53, that solution satisfies $C\mathbf{x}^* = \mathbf{y}_d^*$; i.e., $\mathbf{y}^* = \mathbf{y}_d^*$. The practical meaning of this fact is that, with any feedforward control $\mathbf{u}_f$ (including $\mathbf{u}_f = 0$), the system will find an equilibrium such that $\mathbf{y} = \mathbf{y}_d$. The use of a feedforward term becomes optional—it may help speed up the transient to a new steady state, but it is not necessary. Any inaccuracy in the value of $\mathbf{u}_f$ will be compensated by the integral control. This, in fact, is a generalization of the Type 1 system to the multivariable case, presented in a state-space context.

The design procedure can be summarized as follows:

1. Append m integrators, driven by the error signals $y - y_d$, to the state equations.

2. Design a stabilizing state feedback controller, using any method. If the LQ regulator method is used, all integrator outputs $\mathbf{z}$ must figure in the performance index. It is necessary that these zero-frequency modes be observable from an output $Q^{1/2}\mathbf{x}$; otherwise, the detectability condition fails.

3. If desired, design a feedforward control.

◆ ◆ ◆ REMARK

This procedure can be extended to cases where $\mathbf{y}_d$ and $\mathbf{w}$ are representable by nondecreasing complex exponentials. Integrators work with dc signals because, if a dc steady state exists, all integrator inputs must be zero (any other constant value would give rise to a ramp, and a dc steady state would not exist). To apply to other signals, we replace the integrators by all-pole blocks with poles corresponding to the exponents of the complex exponentials. For instance, to obtain zero sinusoidal steady-state error at some frequency ω_0, the integrator should be replaced by $1/(s^2 + \omega_0^2)$. In essence, the model of the disturbance is included in the system. This arrangement is known as the *internal model principle* [5].   ◆

Example 7.11   **(Active Suspension)**

In the active suspension of Example 2.2 (Chapter 2), let the roadway deviation be represented by the equation $\dot{y}_R = w(t)$.[1]

(a) Write a new set of state equations, with the state variables $\ell_1 = x_1 - x_2$, $\ell_2 =$ $x_1 - y_R$, v_1, and v_2.

(b) Design a state feedback law that will ensure $\ell_1 = \ell_{1d}$ in the steady state, with closed-loop eigenvalues at $-3 \pm j3$, $-25 \pm j25$, and -5.

(c) Simulate the response, from equilibrium, for $w(t) = 0.1u_0(t)$ and $\ell_{1d} = 0.1$.

---

[1]The reader familiar with stochastic processes will recognize that, if $w(t)$ is white noise, $y_R(t)$ is a Wiener process.

Solution   The new state equations, with an integrator appended, are

$$\dot{\ell}_1 = v_1 - v_2$$

$$\dot{\ell}_2 = v_1 - w$$

$$\dot{v}_1 = -10\ell_1 - 2(v_1 - v_2) + 0.00334u$$

$$\dot{v}_2 = 60\ell_1 + 12(v_1 - v_2) - 660(-\ell_1 + \ell_2) - 0.02u$$

$$\dot{x} = \ell_1 - \ell_{1d}.$$

In matrix form,

$$\frac{d}{dt}\begin{bmatrix} \ell_1 \\ \ell_2 \\ v_1 \\ v_2 \\ x \end{bmatrix} = \begin{bmatrix} 0 & 0 & 1 & -1 & 0 \\ 0 & 0 & 1 & 0 & 0 \\ -10 & 0 & -2 & 2 & 0 \\ 720 & -660 & 12 & -12 & 0 \\ 1 & 0 & 0 & 0 & 0 \end{bmatrix} \begin{bmatrix} \ell_1 \\ \ell_2 \\ v_1 \\ v_2 \\ x \end{bmatrix} + \begin{bmatrix} 0 \\ 0 \\ .00334 \\ -.02 \\ 0 \end{bmatrix} u + \begin{bmatrix} 0 \\ -1 \\ 0 \\ 0 \\ 0 \end{bmatrix} w - \begin{bmatrix} 0 \\ 0 \\ 0 \\ 0 \\ 1 \end{bmatrix} \ell_{1d}.$$

The new state variables have an advantage over the previous ones in that the absolute heights x_1 and x_2 are no longer needed—a definite plus if the vehicle is to travel over mountainous terrain, for example. That is made possible by the fact that y_d is described by pure integration.

By pole placement (MATLAB place), we obtain the gain

$$\mathbf{k}^T = [5.1493e4 - 2.5223e4 \quad 6.2280e3 - 1.3099e3 \quad 5.1034e4]$$

It is not difficult to show that, at equilibrium, $\ell_1 = \ell_2 = \ell_{1d}$, $v_1 = v_2 = 0$, and $u = 3000\ell_{1d}$; the state variable x may have any value at all.

The control law is

$$u = -k_1\ell_1 - k_2\ell_2 - k_3v_1 - k_4v_2 - k_5x.$$

At equilibrium, we must have

$$3000\ell_{1d} = -k_1\ell_{1d} - k_2\ell_{1d} - k_5x^*$$

or

$$x^* = -0.5735\ell_{1d}.$$

At $t = 0_-$, the state and control are at the equilibrium values. For $w(t) = 0.1u_0(t)$, $\ell_2(0_+) = \ell_2(0_-) - 0.1$; all other state variables remain the same. Figure 7.20 shows simulation results. Note that both ℓ_1 and x come back to their starting values; that is expected, since the equilibrium values are the same after the step in y_R.

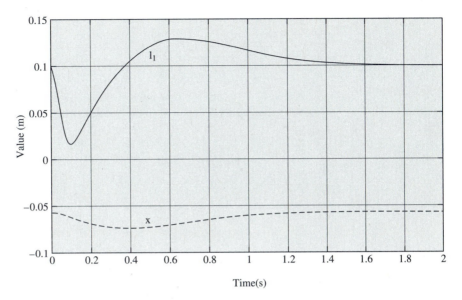

Figure 7.20 Time responses for the active suspension system pole-placement with integrator

---

7.6 STATE OBSERVERS

7.6.1 Introduction

Up to now in this chapter, we have assumed that the state vector (system or disturbance) is available for use in a state feedback law. In practice, more often than not, the complete state vector is not measured; sensors are not always available and, furthermore, they add to the system's cost and complexity.

It might appear that state feedback methods have very limited applicability, being restricted to those few cases where all components of the state are available. Fortunately, it is possible to *estimate* the state of a system, provided the system is observable. This is done by designing a dynamical system called a *state observer*, whose inputs are the system inputs and outputs and that produces as outputs estimates of the unmeasured states of the system. It is possible to recover much of the behavior of a state feedback law by using state estimates instead of states in the feedback law.

7.6.2 A Naive Observer

We begin with the LTI system:

$$\dot{\mathbf{x}} = A\mathbf{x} + B\mathbf{u}. \tag{7.54}$$

We might think of estimating the state $\mathbf{x}(t)$ by running a model in parallel with the system, driven by the system input. Let this be represented by

$$\dot{\widehat{\mathbf{x}}} = A\widehat{\mathbf{x}} + B\mathbf{u}. \tag{7.55}$$

Here, $\widehat{\mathbf{x}}(t)$ (read "x hat") is presumed to track $\mathbf{x}(t)$.

Define the estimation error $\widetilde{\mathbf{x}}$ (read "x tilde") to be

$$\widetilde{\mathbf{x}}(t) = \mathbf{x}(t) - \widehat{\mathbf{x}}(t). \tag{7.56}$$

Differentiation yields

$$\begin{aligned}
\dot{\widetilde{\mathbf{x}}} &= \dot{\mathbf{x}} - \dot{\widehat{\mathbf{x}}} \\
&= A\mathbf{x} + B\mathbf{u} - A\widehat{\mathbf{x}} - B\mathbf{u} \\
&= A(\mathbf{x} - \widehat{\mathbf{x}})
\end{aligned}$$

or

$$\dot{\widetilde{\mathbf{x}}} = A\widetilde{\mathbf{x}}. \tag{7.57}$$

The solution to Equation 7.57 is

$$\widetilde{\mathbf{x}}(t) = e^{At}\widetilde{\mathbf{x}}(0). \tag{7.58}$$

If the system is stable, $\widetilde{\mathbf{x}}(t) \to 0$ and the estimate $\widehat{\mathbf{x}}(t)$ will asymptotically track the state $\mathbf{x}(t)$. If the system is unstable, the error $\widetilde{\mathbf{x}}(t)$ will not go to zero—unless we are so fortunate as to have an initial error $\widetilde{\mathbf{x}}(0)$ that excites only stable modes.

Even if the system is stable, this observer leaves much to be desired; for one thing, we have no control of the error dynamics, because we do not choose A. Furthermore, if A and B are not precisely known, the system state $\mathbf{x}$ and the estimate $\widetilde{\mathbf{x}}$ will not track each other.

7.6.3 The Full-Order Observer

We add a measurement to the system of Equation 7.54, i.e.,

$$\begin{aligned}
\dot{\mathbf{x}} &= A\mathbf{x} + B\mathbf{u} \\
\mathbf{y} &= C\mathbf{x}
\end{aligned} \tag{7.59}$$

with $\mathbf{x}$ of dimension n and $\mathbf{y}$ of dimension m. We try the observer structure

$$\dot{\widehat{\mathbf{x}}} = A\widehat{\mathbf{x}} + B\mathbf{u} + G(\mathbf{y} - C\widehat{\mathbf{x}}). \tag{7.60}$$

Here, $\dim(\widehat{\mathbf{x}}) = n$ [same as $\dim(\mathbf{x})$] and G is a constant $n \times m$ matrix, the *observer gain*. This structure adds a correction term, $G(\mathbf{y} - C\widehat{\mathbf{x}})$, to the previous observer.

If $\widehat{\mathbf{x}} = \mathbf{x}$, then $\mathbf{y} - C\widehat{\mathbf{x}} = \mathbf{y} - C\mathbf{x} = \mathbf{0}$, and there is no correction. On the other hand, if $C\widehat{\mathbf{x}} \neq \mathbf{y}$, a correction will be applied. We must show that this correction can be made so as to drive the error to zero.

■ Theorem 7.8   Let (A, C) be observable. Then there exists an observer gain matrix G such that $\widehat{\mathbf{x}}(t) \rightarrow \mathbf{x}(t)$, asymptotically.

Proof: With $\widetilde{\mathbf{x}} = \mathbf{x} - \widehat{\mathbf{x}}$ as before, we write

$$\dot{\widetilde{\mathbf{x}}} = \dot{\mathbf{x}} - \dot{\widehat{\mathbf{x}}}$$
$$= A\mathbf{x} + B\mathbf{u} - A\widehat{\mathbf{x}} - B\mathbf{u} - G(\mathbf{y} - C\widehat{\mathbf{x}})$$
$$= A(\mathbf{x} - \widehat{\mathbf{x}}) - G(C\mathbf{x} - C\widehat{\mathbf{x}})$$
$$= (A - GC)(\mathbf{x} - \widehat{\mathbf{x}})$$

or

$$\dot{\widetilde{\mathbf{x}}} = (A - GC)\widetilde{\mathbf{x}}. \tag{7.61}$$

It remains only to show that $A - GC$ can be made stable by a proper choice of G; if that is so, then $\widetilde{\mathbf{x}}(t) \rightarrow 0$ and $\widehat{\mathbf{x}}(t) \rightarrow \mathbf{x}(t)$.

From Chapter 3, recall that the system of Equation 7.59 is observable if, and only if, the dual system

$$\dot{\mathbf{x}} = A^T\mathbf{x} + C^T\mathbf{u} \tag{7.62}$$

is controllable. If the system of Equation 7.59 is observable, then the system of Equation 7.62 is controllable and a state gain matrix G^T (an $m \times n$ matrix) can be chosen for arbitrary placement of the eigenvalues of the matrix $A^T - C^T G^T$. Since a matrix and its transpose have identical eigenvalues, and because $(A^T - C^T G^T)^T = A - GC$, it follows that $A - GC$ can always be made to have only stable eigenvalues.   ■

The proof has disclosed one way of choosing the observer gain G, as the solution to a pole-placement problem for a controllable system. That system is given by Equation 7.62, and the state feedback gain to be calculated is G^T. To summarize:

1. Solve the pole-placement problem for the dual system $\dot{\mathbf{x}} = A^T\mathbf{x} + C^T\mathbf{u}$, with the desired poles being those of the *error system*, given by Equation 7.61.

2. Write the observer as $\dot{\widehat{\mathbf{x}}} = A\widehat{\mathbf{x}} + B\mathbf{u} + G(\mathbf{y} - C\widehat{\mathbf{x}})$.

Example 7.12   (dc Servo)

The dc servo of Example 2.1 (Chapter 2) is provided with an angle sensor (θ) only—for example, a shaft encoder. We wish to estimate the states ω and i and, in addition, the value of a load torque known to be constant. The eigenvalues of the error system are to be located at $-5 \pm j5$, and $-7 \pm j7$ (somewhat faster than the system eigenvalues). Design an observer, and compute the state variables and their estimates for $v(t) = \sin t$, $T_L = 1$ N, $\theta(0) = 1$ rad, $\omega(0) = 1$ rad/s, $i(0) = 0$, and zero initial state for the observer.

Solution   Including the (constant) disturbance state equation $\dot{z} = 0$, $T_L = z$, the system is described by

$$\frac{d}{dt}\begin{bmatrix} \theta \\ \omega \\ i \\ z \end{bmatrix} = \begin{bmatrix} 0 & 1 & 0 & 0 \\ 0 & 0 & 4.438 & -7.396 \\ 0 & -12 & -24 & 0 \\ 0 & 0 & 0 & 0 \end{bmatrix}\begin{bmatrix} \theta \\ \omega \\ i \\ z \end{bmatrix} + \begin{bmatrix} 0 \\ 0 \\ 20 \\ 0 \end{bmatrix} v$$

$$y = \theta = \begin{bmatrix} 1 & 0 & 0 & 0 \end{bmatrix}\begin{bmatrix} \theta \\ \omega \\ i \\ z \end{bmatrix}.$$

It can be verified that this system is observable.

The pole placement problem to be solved refers to the controllable system of Equation 7.62, resulting here in

$$\dot{\mathbf{x}} = \begin{bmatrix} 0 & 0 & 0 & 0 \\ 1 & 0 & -12 & 0 \\ 0 & 4.438 & -24 & 0 \\ 0 & -7.396 & 0 & 0 \end{bmatrix}\mathbf{x} + \begin{bmatrix} 1 \\ 0 \\ 0 \\ 0 \end{bmatrix} u.$$

This example is simple enough to yield to solution by hand. We have

$$sI - A^T = \begin{bmatrix} s & 0 & 0 & 0 \\ -1 & s & 12 & 0 \\ 0 & -4.438 & s+24 & 0 \\ 0 & 7.396 & 0 & s \end{bmatrix}$$

$$\det(sI - A^T) = s^2(s^2 + 24s + 53.256)$$

and

$$Adj(sI - A^T) = \begin{bmatrix} s(s^2 + 24s + 53.256) & x & x & x \\ s(s+24) & x & x & x \\ 4.438s & x & x & x \\ -7.396(s+24) & x & x & x \end{bmatrix}.$$

Note that only the first column is needed, because the **b** vector's only nonzero entry is the first one.

Using Equation 7.18,

$$\det(sI - A^T) + k^T Adj(sI - A^T)C = s^4 + 24s^3 + 53.256s^2$$
$$+ k_1(s^3 + 24s^2 + 53.256s) + k_2(s^2 + 24s) + k_3(4.438s) + k_4(-7.396s - 177.5).$$

The desired closed-loop polynomial is

$$p(s) = (s + 5 + j5)(s + 5 - j5)(s + 7 + j7)(s + 7 - j7)$$
$$= s^4 + 24s^3 + 288s^2 + 1680s + 4900.$$

Matching of coefficients according to Equation 7.18 yields

$$24 + k_1 = 24$$
$$53.256 + 24k_1 + k_2 = 288$$
$$53.256k_1 + 24k_2 + 4.438k_3 - 7.396k_4 = 1680$$
$$-177.5k_4 = 4900.$$

The solution is

$$k_1 = 0; \qquad k_2 = 234.7; \qquad k_3 = -937.6; \qquad k_4 = -27.60.$$

The observer is

$$\frac{d}{dt}\begin{bmatrix} \widehat{\theta} \\ \widehat{\omega} \\ \widehat{\imath} \\ \widehat{z} \end{bmatrix} = \begin{bmatrix} 0 & 1 & 0 & 0 \\ 0 & 0 & 4.438 & -7.396 \\ 0 & -12 & -24 & 0 \\ 0 & 0 & 0 & 0 \end{bmatrix} \begin{bmatrix} \widehat{\theta} \\ \widehat{\omega} \\ \widehat{\imath} \\ \widehat{z} \end{bmatrix} + \begin{bmatrix} 0 \\ 0 \\ 20 \\ 0 \end{bmatrix} v + \begin{bmatrix} 0 \\ 234.7 \\ -937.6 \\ -27.60 \end{bmatrix} (\theta - \widehat{\theta}).$$

Figure 7.21 shows the simulation results. The estimates (dotted curves) converge to the states, albeit with some oscillations.

7.6.4 An Optimal Observer

The pole-placement observer design implies that convergence of the estimation error to zero can be made as fast as desired by moving the eigenvalues of $A - GC$ arbitrarily far into the LHP. That is indeed the case, if the model is exact. Let us now examine the case where

$$\dot{\mathbf{x}} = A\mathbf{x} + B\mathbf{u} + \mathbf{w} \tag{7.63}$$

$$\mathbf{y} = C\mathbf{x} + \mathbf{v}. \tag{7.64}$$

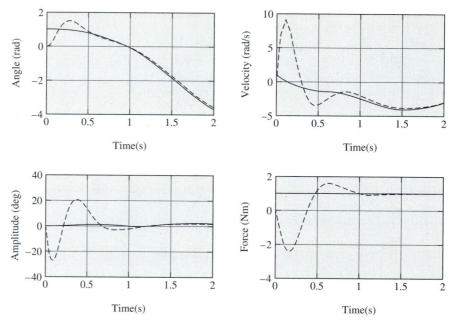

Figure 7.21 Actual and observed (dashed) variables, dc servo

This model differs from the preceding one in two respects: it allows for (i) an (unknown) disturbance input $\mathbf{w}(t)$ (the *plant noise*) and (ii) a measurement error $\mathbf{v}(t)$ (the *measurement noise*). The observer is as before; i.e.,

$$\dot{\widehat{\mathbf{x}}} = A\widehat{\mathbf{x}} + B\mathbf{u} + G(\mathbf{y} - C\widehat{\mathbf{x}}). \tag{7.65}$$

We compute the equation satisfied by $\widetilde{\mathbf{x}}$, as

$$\dot{\widetilde{\mathbf{x}}} = \dot{\mathbf{x}} - \dot{\widehat{\mathbf{x}}} = A\mathbf{x} + B\mathbf{u} + \mathbf{w} - A\widehat{\mathbf{x}} - B\mathbf{u} - G(C\mathbf{x} + \mathbf{v} - C\widehat{\mathbf{x}})$$
$$= A(\mathbf{x} - \widehat{\mathbf{x}}) - GC(\mathbf{x} - \widehat{\mathbf{x}}) + \mathbf{w} - G\mathbf{v}$$

or

$$\dot{\widetilde{\mathbf{x}}} = (A - GC)\widetilde{\mathbf{x}} + \mathbf{w} - G\mathbf{v}. \tag{7.66}$$

We consider the estimate of a scalar function

$$z = \mathbf{k}^T\mathbf{x} \tag{7.67}$$

where $\mathbf{k}^T$ is a row vector. For example, Equation 7.67 could represent a state feedback law for a system with one input. The estimation error is

$$\widetilde{z} = z - \widehat{z} = \mathbf{k}^T\mathbf{x} - \mathbf{k}^T\widehat{\mathbf{x}} = \mathbf{k}^T\widetilde{\mathbf{x}}. \tag{7.68}$$

The zero-state response $\widetilde{z}_w(t)$ to an input $\mathbf{w}(t) = \mathbf{w}_0 u_0(t)$ is, from Equations 7.66 and 7.67,

$$\widetilde{z}_w(t) = \mathbf{k}^T e^{(A-GC)t} \mathbf{w}_0, \qquad t > 0. \tag{7.69}$$

Similarly, the zero-state response $\widetilde{z}_v(t)$ to $\mathbf{v}(t) = \mathbf{v}_0 u_0(t)$ is

$$\widetilde{z}_v(t) = -\mathbf{k}^T e^{(A-GC)t} G \mathbf{v}_0, \qquad t > 0. \tag{7.70}$$

The use of an impulse to model $\mathbf{w}(t)$ is not as restrictive as it may appear. Any signal that can be modeled as the impulse response of a lumped-parameter LTI system can be included in this formulation; all we need do is derive the state-space model that generates the signal, and append it to the plant description. The impulse model used to describe the measurement noise represents a spike error in measurement. The impulses are actually the deterministic counterparts of the so-called *white-noise* processes in the stochastic framework that underlies this theory.

Examination of Equation 7.69 reveals that $\widetilde{z}_w(t)$ will be driven rapidly to zero if the eigenvalues of $A - GC$ are large and stable; this calls for a G with large elements. On the other hand, we conclude from Equation 7.70 that a large G will give a large value of $\widetilde{z}_v(t)$ for small t, which is undesirable. A compromise between speed of response and initial value is sought, by minimization of

$$J = \int_0^\infty [\widetilde{z}_w{}^2(t) + \widetilde{z}_v{}^2(t)] dt. \tag{7.71}$$

This performance index penalizes both slow convergence and large initial values. The solution is drawn directly from LQ theory. For technical reasons, we prove the result for scalar output.

■ **Theorem 7.9** Suppose (C, A) is detectable, $(A, \mathbf{w}_0)$ is stabilizable, and, $v_0 > 0$. The gain G that minimizes the performance index J of Equation 7.69 is

$$G = PC^T V^{-1} \tag{7.72}$$

where P is the positive definite solution of the Riccati equation,

$$AP + PA^T - PC^T V^{-1} CP + W = 0. \tag{7.73}$$

Here, $V = v_0{}^2$ and $W = \mathbf{w}_0 \mathbf{w}_0{}^T$.

Proof: From Equations 7.69, 7.70, and 7.71,

$$J = \int_0^\infty [\mathbf{k}^T e^{(A-GC)t} \mathbf{w}_0 \mathbf{w}_0{}^T e^{(A-GC)^T t} \mathbf{k} + \mathbf{k}^T e^{(A-GC)t} G v_0{}^2 G^T e^{(A-GC)^T t} \mathbf{k}] dt$$

where we have used the fact that $\widetilde{z}_w{}^T = \widetilde{z}_w$, since $\widetilde{z}_w$ is a scalar. We write this as

$$J = \mathbf{k}^T \left[\int_0^\infty e^{(A^T - C^T G^T)^T t} (W + GVG^T) e^{(A^T - C^T G^T)t} dt \right] \mathbf{k}. \tag{7.74}$$

We can relate this to the LQ problem as follows. Define the system

$$\dot{\theta} = A^T \theta + C^T \psi. \tag{7.75}$$

A control law $\psi = -G^T \theta$ applied to this system results in

$$\dot{\theta} = (A^T - C^T G^T)\theta.$$

The response for an initial state $\theta(0) = \mathbf{k}$ is $\theta(t) = e^{(A^T - C^T G^T)t}\mathbf{k}$. Thus, J in Equation 7.74 is written

$$J = \int_0^\infty \theta^T(t)[W + (G^T)^T V(G^T)]\theta(t)dt. \tag{7.76}$$

The minimization of J, with $\theta(t)$ described by Equation 7.75, is an LQ problem. The matrix W replaces Q, V replaces R, and G^T stands for K. The detectability of $(W^{1/2}, A^T)$ is ensured by the stabilizability of $(A, \mathbf{w}_0)$ because of duality; the stabilizability of (A^T, C^T) follows from the assumption that (C, A) is detectable.

The solution of the LQ problem is

$$G^T = V^{-1}(C^T)^T P \tag{7.77}$$

where P satisfies the Riccati equation,

$$(A^T)^T P + P(A^T) - P(C^T)V^{-1}(C^T)^T P + W = 0. \tag{7.78}$$

Transposition of Equation 7.74 yields

$$G = PC^T V^{-1}$$

where the fact that P and V are symmetric is used. Equation 7.78 is written as

$$AP + PA^T - PC^T V^{-1}CP + W = 0.$$

This completes the proof. ■

◆ ◆ ◆ R E M A R K S

1. The result is independent of $\mathbf{k}$, so the observer is optimal for *all* linear functions of the state, including individual state variables.

2. Since $\mathbf{w}_0$ is not usually given, we normally use expected averages of the terms w_{0i}^2 and $w_{0i}w_{0j}$ in W.

3. Similarly, we use an average v_0^2 in V. We can also include a vector measurement noise $\mathbf{v}$, provided that matrix V is positive definite. The dyad $\mathbf{v}_0\mathbf{v}_0^T$ is not, but its average often is. (For example, the average of the product of two independent signals v_1 and v_2 is zero, so the off-diagonal terms of V can justifiably be made zero.)

4. The properties of the dual LQ solution apply to the observer. For example, $A - GC$ is stable. ◆

A relatively large V tends to make the observer gain small, exactly as a large R makes the state feedback gain small. This happens when the observation noise $\mathbf{v}(t)$ is expected to be large; since the correction term $\mathbf{y} - C\hat{\mathbf{x}} = C\tilde{\mathbf{x}} + \mathbf{v}$ is heavily noise-contaminated, it is weighted less heavily by the gain G than when $\mathbf{v}$ is expected to be generally small. As the observation noise increases, the gain decreases, and the observer "shuts down."

Conversely, if the plant perturbations are expected to be heavy, a large W is used, making V small by comparison. The observer gain increases to reflect the fact that more updating must be done. As the plant noise increases, the observer "opens up."

This observer actually has its origin in the stochastic setting of the problem and is a special case of the celebrated *Kalman filter* [6, 7]. In the problem setup for the Kalman filter, the signals $\mathbf{w}(t)$ and $\mathbf{v}(t)$ are white-noise processes, and W and V are their respective covariances. The filter is an observer that minimizes the variance of all components of the estimation error. The general Kalman filter holds for time-varying systems with time-varying noise properties. The Kalman filter is a vital tool not only in control, but also in a great many estimation problems in such diverse fields as navigation, seismic signal processing, and speech processing.

Example 7.13   (Train)

For the train described in Examples 2.3 (Chapter 2) and 7.10, we wish to study the possibility of estimating the state vector with a rather small number of sensors. Specifically, two sensors are contemplated, to measure the velocity of the locomotive, v_1, and the distance (or coupler extension) between the locomotive and the first car, Δx_2. As in Example 7.10, the state is to be derived with respect to a 25-m/s steady speed over level terrain.

(a) Verify the observability of the system (with x_1, the distance state component, deleted).

(b) The velocity measurement is good to within ± 1 m/s, and the coupler extension to within $\pm .02$ m. This suggests using $V = \text{diag}[4 \times 10^{-4} \quad 1]$. The plant noise comes in the form of random forces (e.g., wind or sway-induced friction) added to the driving force, F. This disturbance input, of the order of 3 kN, directly affects only $\Delta \dot{v}_1$. We use

$$W = \text{diag}[0 \quad 0 \quad 0 \quad 0 \quad 9 \quad 0 \quad 0 \quad 0 \quad 0].$$

Design an optimal observer.

(c) Simulate the train behavior for an initial state $\Delta x_i = \Delta v_i = 0$, all i; a force pulse input $F(t)$ of 10 kN, of duration 10 s; and an additive, *unmeasured* force of 2 kN, of duration 4 s. Calculate the estimates, using zero initial conditions for the observer and using only the known 10-kN pulse as the $F(t)$ fed to the observer.

(d) Simulate the train as in part (c), but with a coupler stiffness $K = 3.5 \times 10^5$ N/m instead of the nominal value $K = 2.5 \times 10^5$ N/m. Estimate a few state variables, using the observer designed for the nominal value of K. (This is to assess robustness to parameter changes.)

Solution   The dynamics are as in Example 7.10, but without the disturbance state z. The observation matrix is

$$C = \begin{bmatrix} 1 & 0 & 0 & 0 & 0 & 0 & 0 & 0 & 0 \\ 0 & 0 & 0 & 0 & 1 & 0 & 0 & 0 & 0 \end{bmatrix}$$

since states $1(\Delta x_2)$ and $5(\Delta v_1)$ are measured.

The observer gain is a 9×2 matrix whose transpose is computed to be

$$G^T = \begin{bmatrix} 21.00 & 4.138 & 0.4889 & 0.0587 & 352.3 & 107.88 \\ 0.1409 & 0.0086 & -0.0031 & -0.0014 & 4.703 & 0.9010 \end{bmatrix}$$

$$\begin{bmatrix} 12.15 & -5.633 & -7.966 \\ 0.2583 & 0.2040 & 0.2030 \end{bmatrix}.$$

Figure 7.22 shows values and estimates of the last coupler extension Δx_5 and the velocity deviation Δv_5 of the last car from the steady-state 25 m/s. Since the last car is physically the farthest from the points of measurement, estimates related to it are expected to be the worst. In fact, as Figure 7.22 shows, the estimates and actual are almost identical.

Figure 7.23 shows the same quantities as Figure 7.22, but the estimates are generated by a model with an erroneous stiffness constant. The estimates are quite acceptable.

7.6.5 The Reduced-Order Observer

Occasionally, some of the state variables are measured directly with good precision. In such cases, it makes little sense to estimate them, since they are directly available. The reduced-order observer, or Luenberger observer [8], uses the available information to reduce the order of the observer by the number of directly measured state variables.

Let us assume that the first m components of the state, $m < n$, are measured. (We can always rearrange the state equations so that the measured states are numbered 1 to m.) We write

$$\mathbf{x} = \begin{bmatrix} \mathbf{x}_m \\ -- \\ \mathbf{x}_u \end{bmatrix} \begin{matrix} \}m \\ \\ \}n - m \end{matrix}$$

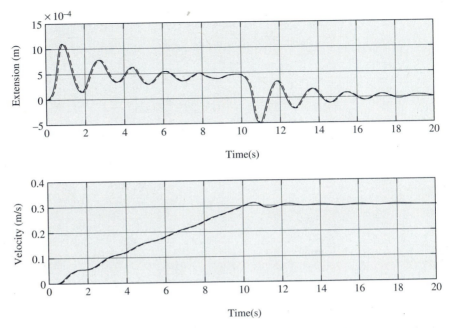

Figure 7.22   Estimates (dotted) and actual variables for the train problem: coupler 4 extension, car 5 velocity. This is the nominal case.

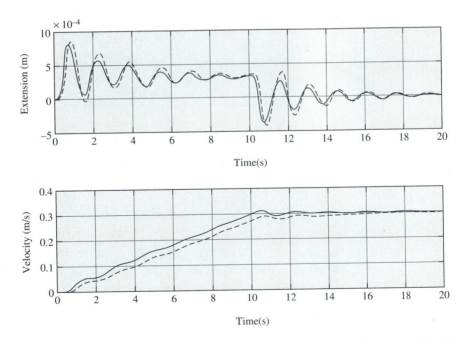

Figure 7.23   Estimates (dotted) and actual variables for the train problem: coupler 4 extension, car 5 velocity. This is the case with modeling error.

where $\mathbf{x}_m$ and $\mathbf{x}_u$ are, respectively, the measured and unmeasured portions of the state vector.

Next, we partition the system equations:

$$\dot{\mathbf{x}} = \begin{bmatrix} \dot{\mathbf{x}}_m \\ -- \\ \dot{\mathbf{x}}_u \end{bmatrix} = \begin{bmatrix} \overbrace{A_{mm}}^{m} & \vdots & \overbrace{A_{mu}}^{n-m} \\ ---- & \vdots & ---- \\ A_{um} & \vdots & A_{uu} \end{bmatrix} \begin{bmatrix} \mathbf{x}_m \\ -- \\ \mathbf{x}_u \end{bmatrix} + \begin{bmatrix} B_m \\ -- \\ B_u \end{bmatrix} \mathbf{u} \tag{7.79}$$

$$\mathbf{y} = \begin{bmatrix} \overbrace{I}^{m} & \vdots & \overbrace{0}^{n-m} \end{bmatrix} \begin{bmatrix} \mathbf{x}_m \\ \mathbf{x}_u \end{bmatrix}. \tag{7.80}$$

Note that Equation 7.80 expresses the fact that $\mathbf{y} = \mathbf{x}_m$; i.e., the vector $\mathbf{x}_m$ is the output.

We rewrite Equation 7.79 as

$$\dot{\mathbf{x}}_u = A_{uu}\mathbf{x}_u + A_{um}\mathbf{x}_m + B_u\mathbf{u} \tag{7.81}$$

$$\mathbf{z} = \dot{\mathbf{x}}_m - A_{mm}\mathbf{x}_m - B_m\mathbf{u} = A_{mu}\mathbf{x}_u. \tag{7.82}$$

Equation 7.81 may be viewed as the state equation for $\mathbf{x}_u$, of order $n - m$. Here, $\mathbf{u}$ and $\mathbf{x}_m$ serve as inputs. The vector $\mathbf{z}$ is computable since $\mathbf{x}_m$ and $\mathbf{u}$ are given; $\mathbf{z}$ is the effective output.

An observer is designed for $\mathbf{x}_u$, of the form

$$\dot{\widehat{\mathbf{x}}}_u = A_{uu}\widehat{\mathbf{x}}_u + A_{um}\mathbf{x}_m + B_u\mathbf{u} + G(\mathbf{z} - A_{mu}\widehat{\mathbf{x}}_u). \tag{7.83}$$

The observer gain G, of dimensions $(n - m) \times m$, can be chosen by pole placement of the matrix $A_{uu} - GA_{mu}$ as before, or by optimization.

Insertion of Equation 7.82 into Equation 7.83 yields

$$\dot{\widehat{\mathbf{x}}}_u = A_{uu}\widehat{\mathbf{x}}_u + A_{um}\mathbf{x}_m + B_u\mathbf{u} + G(\dot{\mathbf{x}}_m - A_{mm}\mathbf{x}_m - B_m\mathbf{u} - A_{mu}\widehat{\mathbf{x}}_u). \tag{7.84}$$

Equation 7.84 is somewhat inconvenient, because it requires differentiation of $\mathbf{x}_m$. In practice, that is to be avoided if, for example, $\mathbf{x}_m$ is actually somewhat noisy. The problem is circumvented by defining $\psi = \widehat{\mathbf{x}}_u - G\mathbf{x}_m$. We use $\widehat{\mathbf{x}}_u = \psi + G\mathbf{x}_m$ in Equation 7.84, and obtain

$$\dot{\psi} + G\dot{\mathbf{x}}_m = A_{uu}\psi + A_{uu}G\mathbf{x}_m + A_{um}\mathbf{x}_m + B_u\mathbf{u} + G\dot{\mathbf{x}}_m$$
$$-GA_{mm}\mathbf{x}_m - GB_m\mathbf{u} - GA_{mu}\psi - GA_{mu}G\mathbf{x}_m$$

or, collecting terms,

$$\dot{\psi} = (A_{uu} - GA_{mu})\psi + (A_{uu}G + A_{um} - GA_{mm} - GA_{mu}G)\mathbf{x}_m$$
$$+ (B_u - GB_m)\mathbf{u}. \tag{7.85}$$

The end product $\widehat{\mathbf{x}}_u$ is obtained from

$$\widehat{\mathbf{x}}_u = \psi + G\mathbf{x}_m. \tag{7.86}$$

Example 7.14 **(Pendulum on a Cart)**

In the pendulum-and-cart problem of Examples 2.10 (Chapter 2) and 7.7, three of the states—namely, the distance x, cart velocity v, and angle θ—are relatively easy to measure. The angular velocity ω, in contrast, is difficult to obtain; it is too small for most tachometers unless friction-inducing gears are used to "gear up" the angular velocity. Devise an estimator of order 1 with an error-system eigenvalue of -4.

Solution Example 7.3 gives the state equations, already in a form where the first three state variables are the measured ones. We write

$$\frac{d}{dt}\begin{bmatrix} x \\ v \\ \theta \\ -- \\ \omega \end{bmatrix} = \begin{bmatrix} 0 & 1 & 0 & \vdots & 0 \\ 0 & 0 & -9.8 & \vdots & 0 \\ 0 & 0 & 0 & \vdots & 1 \\ ---------&&& \vdots & -- \\ 0 & 0 & 19.6 & \vdots & 0 \end{bmatrix}\begin{bmatrix} x \\ v \\ \theta \\ -- \\ \omega \end{bmatrix} + \begin{bmatrix} 0 \\ 1 \\ 0 \\ -- \\ -1 \end{bmatrix}u.$$

Thus,

$$A_{mm} = \begin{bmatrix} 0 & 1 & 0 \\ 0 & 0 & -9.8 \\ 0 & 0 & 0 \end{bmatrix}; \quad A_{mu} = \begin{bmatrix} 0 \\ 0 \\ 1 \end{bmatrix}; \quad B_m = \begin{bmatrix} 0 \\ 1 \\ 0 \end{bmatrix}$$

$$A_{um} = \begin{bmatrix} 0 & 0 & 19.6 \end{bmatrix}; \quad A_{uu} = 0; \quad B_u = -1.$$

The observer gain is a 1×3 matrix. The error system matrix is

$$A_{uu} - GA_{mu} = 0 - \begin{bmatrix} g_1 & g_2 & g_3 \end{bmatrix}\begin{bmatrix} 0 \\ 0 \\ 1 \end{bmatrix} = -g_3.$$

For an error-system eigenvalue of -4, we use $g_3 = 4$. Since g_1 and g_2 are irrelevant, we may as well set them to zero. Thus,

$$G = \begin{bmatrix} 0 & 0 & 4 \end{bmatrix}.$$

Using the form of Equation 7.84,

$$\dot{\hat{\omega}} = \begin{bmatrix} 0 & 0 & 19.6 \end{bmatrix} \begin{bmatrix} x \\ v \\ \theta \end{bmatrix} - u$$

$$+ \begin{bmatrix} 0 & 0 & 4 \end{bmatrix} \left(\begin{bmatrix} \dot{x} \\ \dot{v} \\ \dot{\theta} \end{bmatrix} - \begin{bmatrix} v \\ -9.8\theta \\ 0 \end{bmatrix} - \begin{bmatrix} 0 \\ 1 \\ 0 \end{bmatrix} u - \begin{bmatrix} 0 \\ 0 \\ 1 \end{bmatrix} \hat{\omega} \right)$$

$$= 19.6\theta - u + 4\dot{\theta} - \hat{\omega}.$$

If we use the differentiation-free Equation 7.85,

$$\dot{\psi} = (0 - 4)\psi + \begin{bmatrix} 0 & 0 & 19.6 \end{bmatrix} - \begin{bmatrix} 0 & 0 & 4 \end{bmatrix} \begin{bmatrix} 0 \\ 0 \\ 1 \end{bmatrix} \begin{bmatrix} 0 & 0 & 4 \end{bmatrix} \begin{bmatrix} x \\ v \\ \theta \end{bmatrix} + (-1)u$$

or

$$\dot{\psi} = -4\psi + 3.6\theta - u$$

and, from Equation 7.86,

$$\hat{\omega} = \psi + 4\theta.$$

The reduced-order observer is also used when the outputs are not state variables, by transforming them into states. To do this, we start from

$$\mathbf{y} = C\mathbf{x}$$

and find an $(n - m) \times n$ matrix M such that

$$T = \begin{bmatrix} C \\ -- \\ M \end{bmatrix}$$

is nonsingular. If C has its maximal rank m (m independent rows), that is always possible. We then use the linear transformation

$$\mathbf{z} = \begin{bmatrix} C \\ -- \\ M \end{bmatrix} \mathbf{x} = T\mathbf{x}$$

and calculate the state equations for the transformed state, $\mathbf{z}$ (see Chapter 3). The first m components of $\mathbf{z}$ are $C\mathbf{x} = \mathbf{y}$, as desired, and we go on as before.

7.7 OBSERVER-BASED FEEDBACK

Since a state estimator generated by an observer converges to the actual state, it seems appropriate to close the loop by using the state estimate, rather than the (unavailable) state, in a state feedback law.

To study this situation, we recall the plant equations,

$$\dot{\mathbf{x}} = A\mathbf{x} + B\mathbf{u}.$$

Use of the control $\mathbf{u} = -K\,\widehat{\mathbf{x}}$ yields

$$\dot{\mathbf{x}} = A\mathbf{x} - BK\,\widehat{\mathbf{x}}.$$

We now use the fact that $\widehat{\mathbf{x}} = \mathbf{x} - \widetilde{\mathbf{x}}$, to obtain

$$\dot{\mathbf{x}} = A\mathbf{x} - BK\mathbf{x} + BK\widetilde{\mathbf{x}}. \tag{7.87}$$

We recall Equation 7.61, for the observer error

$$\dot{\widetilde{\mathbf{x}}} = (A - GC)\widetilde{\mathbf{x}}. \tag{7.88}$$

Equations 7.87 and 7.88 together form a description of the closed-loop system. (The estimate $\widehat{\mathbf{x}}$ is not given explicitly but can, if desired, be calculated as $\mathbf{x} - \widetilde{\mathbf{x}}$.) Taken together, Equations 7.87 and 7.88 are

$$\begin{bmatrix} \dot{\mathbf{x}} \\ \dot{\widetilde{\mathbf{x}}} \end{bmatrix} = \begin{bmatrix} A - BK & BK \\ 0 & A - GC \end{bmatrix} \begin{bmatrix} \mathbf{x} \\ \widetilde{\mathbf{x}} \end{bmatrix}. \tag{7.89}$$

The matrix in Equation 7.89 is *block triangular*. For any block-triangular matrix, the eigenvalues are those of the diagonal blocks; here, the closed-loop eigenvalues are those of (i) $A - BK$ and (ii) $A - GC$. In other words, the closed-loop eigenvalues are those of (i) the closed-loop state feedback design and (ii) the observer error system. It follows that the eigenvalues of independently designed state feedback control and observer are retained when the two are combined.

Figure 7.24 illustrates the structure of an observer-based control system for the regulator case. This figure takes into account the fact that the designs of both the state feedback controller and the observer are for the plant linearized about some constant steady state; hence the appearance of $\Delta\mathbf{u}$, $\Delta\widehat{\mathbf{x}}$, and $\Delta\mathbf{y}$. Note that $\Delta\mathbf{y} = \mathbf{y} - \mathbf{y}_d = -\mathbf{e}$. The controller is the subsystem within the dotted line. It is described by the equations

$$\Delta\dot{\widehat{\mathbf{x}}} = A\,\Delta\widehat{\mathbf{x}} + B\,\Delta\mathbf{u} + G(\Delta\mathbf{y} - C\,\Delta\widehat{\mathbf{x}})$$

$$= A\,\Delta\widehat{\mathbf{x}} - BK\,\Delta\widehat{\mathbf{x}} + G(\Delta\mathbf{y} - C\,\Delta\widehat{\mathbf{x}})$$

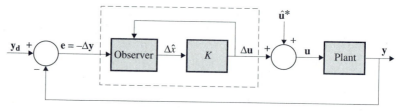

Figure 7.24 Block diagram of an observer-based control system

or

$$\Delta \dot{\hat{\mathbf{x}}} = (A - BK - GC)\Delta\hat{\mathbf{x}} + G\,\Delta\mathbf{y} \tag{7.90}$$

$$\Delta\mathbf{u} = -K\,\Delta\hat{\mathbf{x}}. \tag{7.91}$$

Equations 7.90 and 7.91 are state equations. The corresponding transfer function equation is

$$\Delta\mathbf{u}(s) = -K(sI - A + BK + GC)^{-1}G\,\Delta\mathbf{y}(s) \tag{7.92}$$

or, in terms of **e** rather than $\Delta\mathbf{y}$,

$$\Delta\mathbf{u}(s) = K(sI - A + BK + GC)^{-1}G\mathbf{e}(s). \tag{7.93}$$

If some of the states are directly measured, a reduced-order observer may be used. The measured states appear alongside $\Delta\mathbf{y}$, as additional inputs.

Example 7.15   (dc Servo)

In Example 7.9, a state feedback law was worked out for the dc servo subjected to an exponential load-torque disturbance. An observer-based regulator is to be designed for the system, with the angle being the only measurement.

(a) Design a full-order observer with error-system eigenvalues at $-5 \pm j5$ and $-7 \pm j7$.

(b) The load torque $0.01e^{-t}$ is applied at $t = 0$ when $\theta = \theta_d$ and $\omega = i = 0$. Compute and compare the responses of (i) the state feedback control system and (ii) the response of the observer-based control system, with all observer states initially at zero.

(c) Compute the controller transfer function.

Solution   From Example 7.9, the state feedback gain is

$$K = \begin{bmatrix} 3 & 0.8796 & 0.1529 & -1.8190 \end{bmatrix}.$$

(a) In Example 7.12, an observer design was worked out for this system, but for a constant load-torque input. It is reworked for this exponential torque (e^{-t}). The observer gain is

$$G = \begin{bmatrix} -1 \\ 265.7 \\ -112.7 \\ -20.66 \end{bmatrix}.$$

(b) Simulations for the load torque $-0.01e^{-t}$ are shown in Figure 7.25. Not surprisingly, the performance is better when the full state is directly available.

(c) The controller transfer function is computed from Equations 7.90 and 7.91:

$$\frac{u}{e} = \frac{92.8(s + 21.6)(s + 2.438 \pm j1.1703)}{(s + 1.332)(s + 11.47)(s + 7.127 \pm j12.14)}.$$

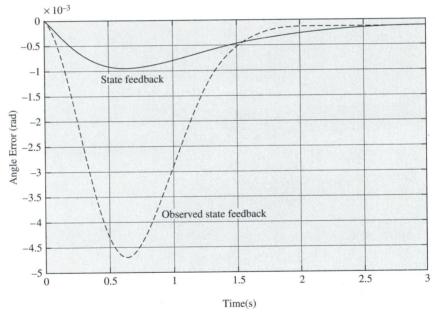

Figure 7.25   Responses for state feedback and observer-based feedback, dc servo

---

Example 7.16   (Pendulum on a Cart)

In Example 7.7, an optimal LQ state feedback law was computed for the pendulum-and-cart problem. A reduced-order observer was derived in Example 7.14. Combine the two in an observer-based control system, and calculate the angle and distance

(θ and x) responses, with an initial state $\theta = 0.26$ rad ($= 15°$), $\omega = x = v = 0$, for both the state and the observer-based control. Calculate the observer transfer function.

Solution   The optimal state feedback gain is $K = [-40.0 \quad -37.37 \quad -190.6 \quad -54.73]$. Figure 7.26 shows the angle responses for both the state feedback law and the observer-based control. The observer is implemented as

$$\dot{\psi} = -4\psi + 3.6\theta - u$$
$$= -4\psi + 3.6\theta - [+40.0x + 37.37v + 190.6\theta + 54.73\widehat{\omega}].$$

Since $\widehat{\omega} = \psi + 4\theta$, we get

$$\dot{\psi} = -40.0x - 37.37v - 405.9\theta - 58.73\psi$$

and

$$u = 40.0x + 37.37v + 409.5\theta + 54.73\psi.$$

These last two equations are the state representation of the controller.

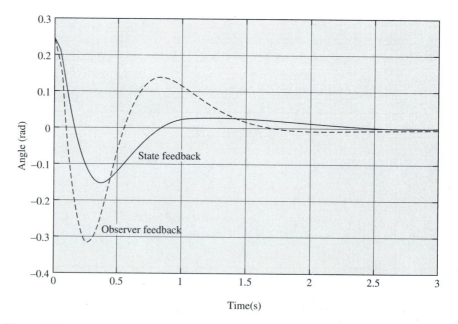

Figure 7.26 Angle responses for state feedback and observer-based feedback, pendulum-on-cart system

The transform equation for $\psi(s)$ is

$$\psi(s) = \frac{-1}{s + 58.73}(40.0x + 37.37v + 405.9\theta).$$

Hence,

$$u(s) = 40.0x + 37.37v + 409.5\theta - \frac{1}{s + 58.73}(40.0x + 37.37v + 405.9\theta).$$

7.8 DESIGN WITH UNMEASURED CONSTANT DISTURBANCES

In this section, the problem includes an unmeasured constant disturbance, so the dc steady-state input to the system is not given. It is worthwhile to study this in general, to show that the observer-based controller automatically contains integration.

We begin with the system description obtained by linearizing about the dc steady state defined by the output $\mathbf{y} = \mathbf{y}_d = $ constant. We assume that a constant input does exist that results in $\mathbf{y} = \mathbf{y}_d$ in the steady state. The linearized equations are

$$\Delta\dot{\mathbf{x}} = A\,\Delta\mathbf{x} + B\,\Delta\mathbf{u} \tag{7.94}$$

$$\Delta\mathbf{y} = C\,\Delta\mathbf{x}. \tag{7.95}$$

We express $\Delta\mathbf{u}$ as $\mathbf{u} - \mathbf{u}^*$, where $\mathbf{u}^*$ is the (unknown) constant steady-state input. Since $\mathbf{u}^*$ is constant, we write

$$\Delta\dot{\mathbf{x}} = A\,\Delta\mathbf{x} - B\mathbf{u}^* + B\mathbf{u}$$

$$\dot{\mathbf{u}}^* = 0$$

$$\Delta\mathbf{y} = C\,\Delta\mathbf{x}$$

and consider $\mathbf{u}^*$ as part of the state to be estimated, along with $\Delta\mathbf{x}$. This is rewritten as the composite system

$$\frac{d}{dt}\begin{bmatrix} \Delta\mathbf{x} \\ -- \\ \mathbf{u}^* \end{bmatrix} = \begin{bmatrix} A & -B \\ 0 & 0 \end{bmatrix}\begin{bmatrix} \Delta\mathbf{x} \\ -- \\ \mathbf{u}^* \end{bmatrix} + \begin{bmatrix} B \\ -- \\ 0 \end{bmatrix}\mathbf{u} \tag{7.96}$$

$$\Delta\mathbf{y} = \begin{bmatrix} C & 0 \end{bmatrix}\begin{bmatrix} \Delta\mathbf{x} \\ \mathbf{u}^* \end{bmatrix}. \tag{7.97}$$

The next step will be to construct an observer for this system. Before we do, however, we must verify that it is observable.

■ **Theorem 7.10**   The system of Equations 7.96 and 7.97 is observable if, and only if, (i) the original system of Equations 7.94 and 7.95 is observable and (ii) this original system has no transmission zeros at $s = 0$.

Proof: This theorem is just the dual of Theorem 7.7 and is proved in a similar manner. ■

Next, we examine the observer-based controller

$$\frac{d}{dt}\begin{bmatrix} \Delta\widehat{\mathbf{x}} \\ -- \\ \widehat{\mathbf{u}}^* \end{bmatrix} = \begin{bmatrix} A & -B \\ --- & --- \\ 0 & 0 \end{bmatrix}\begin{bmatrix} \Delta\widehat{\mathbf{x}} \\ -- \\ \widehat{\mathbf{u}}^* \end{bmatrix} + \begin{bmatrix} B \\ -- \\ 0 \end{bmatrix}\mathbf{u} + \begin{bmatrix} G_1 \\ -- \\ G_2 \end{bmatrix}(\Delta\mathbf{y} - C\,\Delta\widehat{\mathbf{x}})$$

$$\mathbf{u} = \widehat{\mathbf{u}}^* + \Delta\mathbf{u} = \widehat{\mathbf{u}}^* - K\,\Delta\widehat{\mathbf{x}}$$

where K is a state feedback gain calculated for the linearized system. Insertion of the control law for $\mathbf{u}$ yields

$$\frac{d}{dt}\Delta\widehat{\mathbf{x}} = A\,\Delta\widehat{\mathbf{x}} - B\widehat{\mathbf{u}}^* + B(\widehat{\mathbf{u}}^* - K\,\Delta\widehat{\mathbf{x}}) + G_1(\Delta\mathbf{y} - C\,\Delta\widehat{\mathbf{x}})$$

or

$$\frac{d}{dt}\Delta\widehat{\mathbf{x}} = (A - BK - G_1C)\,\Delta\widehat{\mathbf{x}} + G_1\,\Delta\mathbf{y} \tag{7.98}$$

and

$$\frac{d}{dt}\widehat{\mathbf{u}}^* = -G_2C\,\Delta\widehat{\mathbf{x}} + G_2\,\Delta\mathbf{y} \tag{7.99}$$

with the control given by

$$\mathbf{u} = \widehat{\mathbf{u}}^* - K\,\Delta\widehat{\mathbf{x}}. \tag{7.100}$$

In matrix form, the controller equations are

$$\frac{d}{dt}\begin{bmatrix} \Delta\widehat{\mathbf{x}} \\ -- \\ \widehat{\mathbf{u}}^* \end{bmatrix} = \begin{bmatrix} A - BK - G_1C & 0 \\ ----------- & --- \\ -G_2C & 0 \end{bmatrix}\begin{bmatrix} \Delta\widehat{\mathbf{x}} \\ -- \\ \widehat{\mathbf{u}}^* \end{bmatrix} + \begin{bmatrix} G_1 \\ -- \\ G_2 \end{bmatrix}\Delta\mathbf{y}$$

$$\mathbf{u} = \begin{bmatrix} -K & I \end{bmatrix}\begin{bmatrix} \Delta\widehat{\mathbf{x}} \\ -- \\ \widehat{\mathbf{u}}^* \end{bmatrix}. \tag{7.101}$$

The controller "A matrix" is block triangular, so its eigenvalues are those of the diagonal blocks, the $n \times n$ matrix $A - BK - G_1C$ and the $r \times r$ zero matrix. Because of the latter, the controller has r integrations, so the system is of Type 1.

The procedure is summarized as follows:

1. Design a state feedback control law for the incremental linear system.

2. Design an observer (full- or reduced-order) for the system, augmented by the equation for the unknown steady-state input, as in Equations 7.96 and 7.97.

Example 7.17   **(dc Servo)**

The dc servo of Example 7.9 is subject to a constant (but unmeasured) load torque, so as to cause the value of the dc steady-state input voltage to vary within $\pm.5$ V. The only measurement is that of the shaft angle θ, with a precision of $\pm.03$ rad.

(a) Set up the system equations, including u^* as a state, as in Equations 7.96 and 7.97.

(b) Design an optimal observer, using the weights

$$W = \text{diag} \begin{bmatrix} 0 & 0 & 0 & (.5)^2 \end{bmatrix}, \qquad V = (0.03)^2.$$

(c) Using the optimal state feedback gain $K = \begin{bmatrix} 3 & 0.8796 & 0.1529 \end{bmatrix}$ derived in Example 7.6, derive the observer-based controller and calculate its transfer function.

Solution   From Equations 7.96 and 7.97, we have

$$\frac{d}{dt} \begin{bmatrix} \Delta\theta \\ \Delta\omega \\ \Delta i \\ u^* \end{bmatrix} = \begin{bmatrix} 0 & 1 & 0 & \vdots & 0 \\ 0 & 0 & 4.438 & \vdots & 0 \\ 0 & -12 & -24 & \vdots & -20 \\ \cdots & \cdots & \cdots & & \cdots \\ 0 & 0 & 0 & \vdots & 0 \end{bmatrix} \begin{bmatrix} \Delta\theta \\ \Delta\omega \\ \Delta i \\ u^* \end{bmatrix} + \begin{bmatrix} 0 \\ 0 \\ 20 \\ -- \\ 0 \end{bmatrix} v.$$

(Note the difference from Example 7.12, where the problem was to estimate the load torque; here, we estimate the constant input required to counter that torque.) Equations 7.72 and 7.73 are solved, and the observer gain is

$$G = \begin{bmatrix} \mathbf{G}_1 \\ -- \\ \mathbf{G}_2 \end{bmatrix} = \begin{bmatrix} 7.174 \\ 19.06 \\ 4.823 \\ ------ \\ -17.67 \end{bmatrix}.$$

To apply Equation 7.101, we need

$$
A - BK - G_1C = \begin{bmatrix} 0 & 1 & 0 \\ 0 & 0 & 4.438 \\ 0 & -12 & -24 \end{bmatrix} - \begin{bmatrix} 0 \\ 0 \\ 20 \end{bmatrix} \begin{bmatrix} 3 & 0.8796 & 0.1529 \end{bmatrix}
$$

$$
- \begin{bmatrix} 7.174 \\ 19.06 \\ 4.823 \end{bmatrix} \begin{bmatrix} 1 & 0 & 0 \end{bmatrix}
$$

$$
= \begin{bmatrix} -7.174 & 1 & 0 \\ -19.06 & 0 & 4.438 \\ -64.82 & -29.59 & -27.06 \end{bmatrix}.
$$

From Equation 7.101, the controller, in state form, is

$$
\frac{d}{dt} \begin{bmatrix} \Delta\widehat{\theta} \\ \Delta\widehat{\omega} \\ \Delta\widehat{\imath} \\ \widehat{u}^* \end{bmatrix} = \begin{bmatrix} -7.174 & 1 & 0 & 0 \\ -19.06 & 0 & 4.438 & 0 \\ -64.82 & -29.59 & -27.06 & 0 \\ -17.67 & 0 & 0 & 0 \end{bmatrix} \begin{bmatrix} \Delta\widehat{\theta} \\ \Delta\widehat{\omega} \\ \Delta\widehat{\imath} \\ \widehat{u}^* \end{bmatrix} + \begin{bmatrix} 7.174 \\ 19.06 \\ 4.823 \\ -17.67 \end{bmatrix} \Delta y
$$

$$
u = \begin{bmatrix} -3 & -0.8796 & -0.1529 & 1 \end{bmatrix} \begin{bmatrix} \Delta\widehat{\theta} \\ \Delta\widehat{\omega} \\ \Delta\widehat{\imath} \\ \widehat{u}^* \end{bmatrix}.
$$

The transfer function of this controller is computed to be

$$
\frac{u}{\Delta y} = \frac{-52.69(s + 3.958 \pm j5.494)(s + 21.92)}{s(s + 5.971 \pm j6.531)(s + 22.29)}.
$$

Since $\Delta y = \theta - \theta_d = -e$, we have

$$
\frac{u}{e} = \frac{52.69(s + 3.958 \pm j5.494)(s + 21.92)}{s(s + 5.971 \pm j6.531)(s + 22.29)}.
$$

Problems

7.1   For the system

$$
\dot{\mathbf{x}} = \begin{bmatrix} 0 & 1 \\ -1 & -1 \end{bmatrix} \mathbf{x} + \begin{bmatrix} 0 \\ 1 \end{bmatrix} u
$$

$$
y = \begin{bmatrix} 1 & 1 \end{bmatrix} x
$$

a. Calculate the transfer function y/u.

b. Write the state equations, with u_{ex} as the input, if $u = -[k_1 \quad k_2]\mathbf{x} + u_{ex}$.

c. Calculate the transfer function y/u_{ex} and verify that the zero is as in part (a).

7.2 Repeat Problem 7.1 for the system

$$\dot{\mathbf{x}} = \begin{bmatrix} 0 & 1 \\ 0 & 0 \end{bmatrix} \mathbf{x} + \begin{bmatrix} 1 \\ -1 \end{bmatrix} u$$

$$y = \begin{bmatrix} 1 & 0 \end{bmatrix} \mathbf{x}.$$

7.3 Verify that the system of Problem 7.1 is controllable, and that it remains controllable under the control law of part (b), for any k_1 and k_2.

7.4 Repeat Problem 7.3 for the system of Problem 7.2.

7.5 A system is given as

$$\dot{\mathbf{x}} = \begin{bmatrix} 0 & 1 \\ 0 & -1 \end{bmatrix} \mathbf{x} + \begin{bmatrix} -1 \\ 1 \end{bmatrix} u.$$

a. Show that this system is uncontrollable, and find the uncontrollable mode.

b. Write the state equations, with u_{ex} as input, if $u = -\begin{bmatrix} k_1 & k_2 \end{bmatrix}\mathbf{x} + u_{ex}$.

c. Show that the system of part (b) is uncontrollable for any k_1 and k_2, and that the uncontrollable mode is the same as that of the original system.

d. For $y = [c_1 \quad c_2]\mathbf{x}$, calculate the transfer functions y/u_{ex} and show that there is always a pole-zero cancellation for any c_1, c_2, k_1, and k_2.

7.6 For the system of Problem 7.1, place the closed-loop poles at $s = -1, -1$ by:

a. Using the transfer function of part (c) of Problem 7.1 and adjusting k_1 and k_2 to obtain the correct closed-loop characteristic polynomial.

b. Applying Equation 7.18.

7.7 Repeat Problem 7.6 for the system of Problem 7.2.

7.8 For the system

$$\dot{\mathbf{x}} = \begin{bmatrix} 0 & 1 & 0 \\ 0 & -1 & 1 \\ 0 & -1 & 2 \end{bmatrix} \mathbf{x} + \begin{bmatrix} 0 \\ 0 \\ 1 \end{bmatrix} u$$

a. Verify controllability.

b. Calculate the state feedback law to place poles at -2 and $-1 \pm j$.

M **7.9** The intent of this problem is to study the interplay between robustness and the closed-loop pole locations.

a. For the system of Example 7.2, calculate the state feedback gain to place poles at $-6 \pm j6$ and -48.

b. With reference to Figure 7.26, compute the loop gain $L(j\omega) = K(j\omega I - A)^{-1}B$ and the corresponding complementary sensitivity $T(j\omega) = L(j\omega)/[1 + L(j\omega)]$. Plot $|T(j\omega)|$ versus ω.

c. Repeat parts (a) and (b) to place poles at $-12 \pm j12$ and -96 and at $-24 \pm j24$ and -192. In view of the results of Chapter 5, Section 5.6.5, what, in each case, is the frequency range over which the multiplicative modeling uncertainty $|\Delta(j\omega)| = \ell(j\omega)$ must be less than 1 to guarantee stability, according to the results of Section 5.6.5?

7.10   *Servo, simplified model*   For the dc servo of Problem 2.4 (Chapter 2):

a. Design a state feedback gain such that the closed-loop poles are complex, with damping factor $\zeta = \sqrt{2}/2$ and arbitrary distance ω_0 from the origin.

b. Design the control law for regulation to a set point $\theta = \theta_d$, and calculate the transfer function θ/θ_d.

c. Obtain the transfer function v/θ_d, and discuss its variation as a function of ω_0.

(M)   **7.11**   *Servo with flexible shaft*   For the servo with flexible shaft of Problem 2.5 (Chapter 2):

a. Design a state feedback such that the closed-loop poles are located at -4, -5, -21.52, and $-150 \pm j150$.

b. The pole placement called for in part (a) represents modest displacement of the open-loop real poles at 0, -2.47, and -21.52 and considerable movement of the open-loop imaginary poles at $\pm j171.3$. [See Problem 3.14 (Chapter 3) for the open-loop transfer functions.] To ascertain how much control effort is required to displace the high-frequency imaginary poles, repeat part (a) for the closed-loop location 0, -2.47, -21.52, $-150 \pm j150$. Compare the two gains.

c. Using the gain of part (a), design the control law for regulation to a set point $\theta_2 = \theta_d$, and calculate the transfer function θ_2/θ_d.

d. Compute the unit-step response for the transfer functions of part (c).

(M)   **7.12**   *Heat exchanger*   The heat exchanger model of Problem 2.8 (Chapter 2), linearized as in Problem 2.14, has open-loop poles at -0.1914, $-0.5252 \pm j0.1278$, -1.2872, and $-0.9531 \pm j0.1278$. [See Problem 3.18 (Chapter 3) for the transfer function.] The poles have sufficient damping, but we wish to speed up the response, i.e., move some poles farther into the LHP. The system input is ΔF_H, the change in hot liquid flow from its steady-state value.

a. Obtain the state feedback gain matrix to place the closed-loop poles at -0.70, $-0.5252 \pm j0.5252$, -1.2872, and $-0.9531 \pm j0.1278$.

b. Using the linearized model of Problem 2.14, design a control law, based on the gain of part (a), for regulation of ΔT_{c3} to a set point ΔT_d.

c. Calculate the transfer function $\Delta T_{c3}/\Delta T_d$, and compute the unit-step response.

M   **7.13**   *Chemical reactor*   The chemical reactor model of Problem 2.9 (Chapter 2) was linearized in Problem 2.15. It was shown in Problem 3.48 (Chapter 3) that the differential equations for Δc_A and ΔT constitute a minimal realization for the input ΔQ and the output ΔT.

a. Use this minimal realization to design a state feedback control law to regulate ΔT to a set point ΔT_d, with closed-loop poles $-0.5 \pm j0.5$.

b. The closed-loop control system of part (a) is described by a new set of state equations, with ΔT_d as the input. Write this new set, starting from the fourth-order model of Problem 2.15. Identify the controllable and uncontrollable modes.

c. Apply this linear control law to the nonlinear model of Problem 2.9, and compute the responses to steps in ΔT_d of $10°$K, $20°$K, and $40°$K. Comment on the performance of the linear controller for this highly nonlinear system.

M   **7.14**   *High-wire artist*   The high-wire artist model of Problem 2.12 was linearized in Problem 2.18.

a. Using the linearized model, design a pole-placement control law for closed-loop poles at -11, -12, $-0.5 \pm j0.5$.

b. Using the nonlinear model of Problem 2.12, simulate the closed-loop response for the conditions $\theta(0) = \theta_o$, $\omega_\theta(0) = \phi(0) = \omega_\phi(0) = 0$, for $\theta_o = 0.5$, 1 and 1.5 rad. Discuss the results.

M   **7.15**   *Crane*   The crane model of Problem 2.11 (Chapter 2) was linearized in Problem 2.17.

a. Using the linearized model of Problem 2.17, design a pole-placement-based control law to regulate x to a set point x_d, and to place the closed-loop poles at -1, -2, and $-2 \pm j2$. Compute the response x to a unit step in x_d.

   i. Repeat part (a), but "speed up" the response by placing poles at -4, -8, and $-8 \pm j8$.

   ii. Modify the nonlinear model of Problem 2.11 by including the control law of part (a). Simulate the response for steps in x_d of 1 m and 10 m.

   iii. Repeat part (c) for the control law of part (b). Discuss differences between the results of parts (c) and (d).

M **7.16** *Maglev* In Problem 3.22 (Chapter 3), the Maglev equations were rewritten in terms of three redefined inputs. It makes physical sense to say that heave (i.e., Δz) should be controlled through Δu_{LC}, roll (i.e., θ) should be controlled through Δu_{LD}, and sway (i.e., Δy) should be controlled through Δu_{GD}.

    a. Compute the eigenvectors, and identify each with heave, roll, or sway.

    b. Using pole placement, use u_{LC} to shift the heave modes to $-x \pm jx$ for $x = 100, 150,$ and 200 rad/s.

    c. Use u_{LD} to shift the roll modes to $-1.5x \pm j1.5x$ for x as in part (b).

    d. Use u_{GD} to shift the sway modes to $-0.5x \pm j0.5x$ for x as in part (b).

    e. Compute the responses Δz, $\Delta\theta$, and Δy and the inputs Δu_{L1}, Δu_{L2}, Δu_{G1}, and Δu_{G2} for the three initial conditions $\Delta z(0) = 1$, $\Delta\theta(0) = 1$, $\Delta y(0) = 1$, and all other state variables equal to zero (i.e., three separate simulations, each with one nonzero state variable). If it is required that $|\Delta u_{L1}|, |\Delta u_{L2}| \leq 700$ and $|\Delta u_{G1}|, |\Delta u_{G2}| \leq 170$ (because the inputs are the squares of currents, they may not be negative and cannot be less than minus the nominal value), what are the maximum allowable values of $\Delta z(0)$, $\Delta\theta(0)$, and $\Delta y(0)$?

M **7.17** For the system of Problem 7.1, with $u = 0$, set up and solve the matrix Lyapunov equation to compute $J = \int_0^\infty y^2(t)dt$, given $x(0) = x_0$. From the solution of the Lyapunov equation, verify that the system is stable.

M **7.18** A control law $u = -ky$ is used for the system of Problem 7.2.

    a. Write the closed-loop state equation.

    b. Set up and solve the matrix Lyapunov equation to obtain $J = \int_0^\infty y^2(t)dt$ as a function of k.

    c. Plot $J(k)$ versus k and find the minima for $x(0) = \begin{bmatrix} 1 \\ 0 \end{bmatrix}$ and $\begin{bmatrix} 0 \\ 1 \end{bmatrix}$. Ascertain from the solution of the Lyapunov equation that the minimizing value of k leads, in each case, to a stable closed-loop system.

7.19 For $A = \begin{bmatrix} 0 & 1 \\ -1 & -2 \end{bmatrix}$, solve the matrix Lyapunov equation with $Q = I$ to show that A is unstable.

7.20 For the system

$$\dot{\mathbf{x}} = \begin{bmatrix} 0 & 1 \\ -1 & 0 \end{bmatrix} \mathbf{x} + \begin{bmatrix} 0 \\ 1 \end{bmatrix} u$$

$$y = \begin{bmatrix} 1 & 0 \end{bmatrix} \mathbf{x}$$

Calculate (analytically) the state feedback control matrix that minimizes $J = \int_0^\infty [y^2(t) + \rho u^2(t)]dt$, $\rho > 0$.

7.21 For the system

$$\dot{\mathbf{x}} = \begin{bmatrix} 0 & 1 & 0 \\ 0 & 0 & 1 \\ 0 & 0 & 0 \end{bmatrix} \mathbf{x} + \begin{bmatrix} 0 \\ 0 \\ 1 \end{bmatrix} u$$

$$y = \begin{bmatrix} 1 & 0 & 0 \end{bmatrix} \mathbf{x}$$

verify stabilizability and detectability.

H i n t   Controllability guarantees stabilizability, and observability ensures detectability.

Calculate (analytically) the state feedback gain matrix that minimizes $J = \int_0^\infty [y^2(t) + \rho u^2(t)] dt$, $\rho > 0$.

M **7.22** *Servo with flexible shaft* We wish to design an LQ regulator for the dc servo of Problem 2.5 (Chapter 2), for the set point $\theta_2 = \theta_d$. The control objective is to achieve the fastest possible response consistent with constraints on the motor voltage and on the torsion force on the shaft. Specifically, the control should be such that the zero-state response to a step θ_d of 1 rad should not call for a motor voltage magnitude of more than 5 V, and $|\theta_1 - \theta_2| = |\Delta|$ should be held below 0.03 radians.

   a. Calculate the dc steady state corresponding to $\theta_2 = \theta_d$, and write the state equations for $\Delta \mathbf{x} = \mathbf{x} - \mathbf{x}^*$, the incremental state variables.

   b. Define $J = \int_0^\infty [Q_{11} \Delta^2 + Q_{33}(\theta_2 - \theta_d)^2 + Rv^2] dt$. Use as starting values $Q_{11} = 1/(.03)^2$, $Q_{33} = 1$, and $R = 1/(5)^2$, and calculate the optimal LQ gain.

   c. Compute the closed-loop zero-state response to a unit step $\theta_d = 1$ rad.

   d. Repeat parts (b) and (c), increasing weights on those terms of J that correspond to constraints that are violated, and decreasing weights if the constraints are satisfied.

M **7.23** *Drum speed control* An LQ control system is to be designed to regulate the drum speed ω_0 in Problem 2.6 (Chapter 2). To prevent undue torsion stress, the shaft torsion angles should be less than 0.02 rad in magnitude. The voltage inputs u_1 and u_2 should be held to magnitudes less than 180 V. The load torque T_0 is assumed to be zero.

   The control system should maximize the speed of response of ω_0, from the zero state to a constant steady-state $\omega_d = 2.0$ rad/s, while satisfying constraints. Use

$$J = \int_0^\infty [Q_{33} \Delta\omega_0{}^2 + Q_{44}\Delta_1{}^2 + Q_{55}\Delta_2{}^2 + R_{11}u_1{}^2 + R_{22}u_2{}^2] dt$$

as the performance index.

a. Start with $Q_{33} = 1/(2)^2$, $Q_{44} = Q_{55} = 1/(0.02)^2$, and $R_{11} = R_{22} = 1/(180 - u_i^*)^2$, and calculate the LQ gain. Here, u_i^*, $i = 1, 2$, are the steady-state values of u_i.

b. Compute the zero-state response for $\omega_d = 2.0$ rad/s, $u_1^* = u_2^*$. Increase weights where constraints are violated; decrease if some "play" is still possible. Repeat parts (a) and (b) until the constraints are reasonably close to being just satisfied.

c. Repeat the process with only one input, u_1. (You will need to reduce the model.) To get a fair comparison, allow u_1 to have a magnitude of 360 V (to get the same total torque). Compare the responses ω_0 for the cases of one and two control variables.

M **7.24** **Blending tank**   A linearized model was found in Problem 2.13 (Chapter 2) for the blending tank of Problem 2.7.

a. Using the linearized model, calculate the steady-state values of the control and state variables for $\Delta C_A^* = 0.1$, $\Delta \ell^* = 0$, $\Delta F_B^* = 1 \times 10^{-5}$ m^3/s.

b. We wish to move from the steady state of Problem 2.13 to the new steady state of part (a) without undue liquid flow variations or changes in liquid level. Specifically, we require $|\Delta \ell(t)| \leq 0.02$ m and $|\Delta F_A|$, $|\Delta F_0| \leq 0.5 \times 10^{-4}$ m^3/s. Use

$$J = \int_0^\infty [Q_{11} \, \Delta \ell^2 + R_{11} \, \Delta F_A^{\,2} + R_{22} \, \Delta F_0^{\,2}] dt$$

with $Q_{11} = 1/(0.02)^2$ and $R_{11} = R_{22} = 1/(0.5 \times 10^{-4})^2$ as initial guesses. Compute the optimal gain, and express the control law for ΔF_A and ΔF_0, referred to the *original* steady state. *Note*: This control law should have a linear feedforward term in F_B, also referred to the original values.

c. Simulate the transient to the new steady state and adjust R_{11} and R_{22} until (i) the specifications are met and (ii) the speed of response of $\Delta C(t)$ is maximized.

M **7.25** **High-wire artist**   A pole-placement controller was designed in Problem 7.14. We wish to design a LQ controller to satisfy the following specifications: for $\theta(0) = 0.5$ rad, the response must satisfy $|\theta(t)| \leq 0.5$ rad, $|\phi(t)| \leq 0.5$ rad, $|\omega_\phi(t)| \leq 0.5$ rad/s, $|\tau(t)| \leq 75$ Nm.

a. Use the linearized model of Problem 2.18 with a LQ law that minimizes

$$J \int_0^\infty [Q_{11}\theta^2 + Q_{33}\phi^2 + Q_{44}\omega_\psi^{\,2} + R_{11}\tau^2] dt$$

To start, use $Q_{11} = Q_{33} = Q_{44} = 1/0.25$, $R_{11} = 1/75^2$, and compute the gain matrix.

b. Adjust the weights until the specifications are met.

c. Simulate your control law with the nonlinear model of Problem 2.12, for the initial states of Problem 7.14.

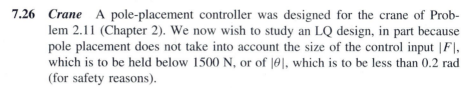

7.26 **_Crane_** A pole-placement controller was designed for the crane of Problem 2.11 (Chapter 2). We now wish to study an LQ design, in part because pole placement does not take into account the size of the control input $|F|$, which is to be held below 1500 N, or of $|\theta|$, which is to be less than 0.2 rad (for safety reasons).

a. Use the linearized model of Problem 2.17 to regulate to a set point x_d, with an LQ control law that minimizes

$$ J = \int_0^\infty [Q_{11}(x - x_d)^2 + Q_{33}\theta^2 + R_{11}F^2]dt. $$

To start, use $Q_{11} = 1/10^2$, $Q_{33} = 1/0.2^2$, and $R_{11} = 1/1500^2$, and calculate the gain matrix.

b. Adjust Q_{33} and R_{11} so that, for $x_d = 10$ m, $\theta(0) = 0$, and zero linear and angular velocities, the constraints on θ and F are met and the speed of response of x is approximately maximized.

c. Try the linear control law on the nonlinear model of Problem 2.11, for the same conditions.

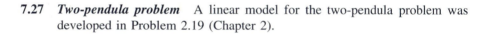

7.27 **_Two-pendula problem_** A linear model for the two-pendula problem was developed in Problem 2.19 (Chapter 2).

a. This system is unstable and requires a minimum of control effort in order to become stable. For $\ell_1 = 1$ m and $\ell_2 = 0.5$ m, solve the LQ gain for

$$ J = \int_0^\infty F^2 dt. $$

b. Solve for the responses $\theta_1(t)$, $\theta_2(t)$, and $F(t)$ with the initial state $\theta_1(0) = 0.1$ rad, $\theta_2(0) = -0.1$ rad, $\omega_1(0) = \omega_2(0) = x(0) = v(0) = 0$. Find the maximum value of $|F(t)|$ and the maximum power $F(t)v(t)$ supplied to the system.

c. It was found in Problem 3.36 (Chapter 3) that this system is uncontrollable if $\ell_2 = \ell_1$. It should be expected that control will be difficult if $\ell_2 \approx \ell_1$. Repeat parts (a) and (b) for $\ell_1 = 1$ m, $\ell_2 = 0.9$ m.

d. It is desired to limit the travel of the cart to ± 4 m for the initial conditions of part (b). Use $\ell_1 = 1$ m, $\ell_2 = 0.5$ m, and

$$ J = \int_0^\infty [Q_{11}x^2 + F^2]dt $$

and adjust Q_{11} until $|x(t)| \le 4$ m. What is the maximum power required in this case?

M   **7.28**   *Maglev*   A linearized model for a magnetically levitated vehicle was derived in Problem 2.20 (Chapter 2). A control system is required to stabilize the vehicle and to maintain the gaps between the magnets and the guideway. (The nominal gap width is only 14 mm.) We wish to design an LQ control system such that gap changes are held within ± 6 mm with $|\Delta u_{L1}|$, $|\Delta u_{L2}| \leq 700$, and $|\Delta u_{G1}|$, $|\Delta u_{G2}| \leq 170$.

a. Compute the optimal LQ gain, with

$$J = \int_0^\infty [\alpha(\Delta S_{L1}^2 + \Delta S_{L2}^2 + \Delta S_{G1}^2 + \Delta S_{G2}^2)$$
$$+ r(\Delta u_{L1}^2 + \Delta u_{L2}^2 + \Delta u_{G1} + \Delta u_{G2}^2)]dt.$$

(To determine the Q matrix, you will need to use the expressions for the gap changes in terms of the changes in the state variables.) Use several increasing values of r. You will note that the gain tends to a limit as you increase r.

b. Repeat the work of Problem 7.16(e) for a value of r sufficient to result in a gain near the asymptotic value.

7.29   Suppose the system in Equation 7.43 is stabilizable, and consider the LQ

$$J = \int_0^\infty \left\{ [\Delta \mathbf{x}^T \mathbf{z}_w^T \mathbf{z}_y^T] \begin{bmatrix} Q_{xx} & Q_{xw} & Q_{xy} \\ Q_{xw}^T & Q_{ww} & Q_{wy} \\ Q_{xy}^T & Q_{wy}^T & Q_{yy} \end{bmatrix} \begin{bmatrix} \Delta \mathbf{x} \\ \mathbf{z}_w \\ \mathbf{z}_y \end{bmatrix} + \Delta \mathbf{u}^T R \, \Delta \mathbf{u} \right\} dt$$

solution with the performance index where the partitioning of the Q matrix is compatible with that of the state vector.

Show that K_x in Equation 7.45 is the gain matrix obtained by solving the LQ problem without disturbance or reference inputs, with $J = \int_0^\infty [\Delta \mathbf{x}^T Q_{xx} \Delta \mathbf{x} + \Delta \mathbf{u}^T R \Delta \mathbf{u}]dt$.

Hint   Partition the P matrix in the same way as Q, and write the Riccati equation in terms of the matrix blocks.

7.30   For the system

$$\dot{x}_1 = x_1 + u$$
$$y = x_1$$

a. Calculate the constant input u^* corresponding to a dc steady-state output y_d^*.

b. For $y_d = y_d{}^* + Ae^{-t}$, derive the control law that minimizes

$$J = \int_0^\infty [(y - y_d)^2 + (u - u^*)^2]dt.$$

7.31 Repeat Problem 7.30, but with the constant $y_d{}^*$ replaced by the sinusoid $y_d{}^*(t) = B \sin t$.

(M) **7.32** *Servo, simplified model*   For the linear, second-order model of Problem 2.4 (Chapter 2):

    a. Calculate the steady-state v^* required to yield a constant angle θ_d in the face of a constant disturbance $T_L{}^*$. Express v^* as a linear combination of θ_d and $T_L{}^*$.

    b. Use the control law derived in Problem 7.10 plus the feedforward term for $T_L{}^*$. Show that this law cancels out T_L; i.e., show that the state does not change in response to changes in T_L.

    c. Remove the feedforward term in $T_L{}^*$, and compute the response $\Delta\theta/T_L$.

(M) **7.33** *Drum speed control*   In Problem 7.23, an LQ control system was designed for the system of Problem 2.6 (Chapter 2). The design was tested for a step response in the desired speed ω_0. We wish to investigate the response for a larger, but more gradual, speed increase. Accordingly, let $\omega_d(t) = 25 - 25e^{-t/\tau}$, where τ is a time constant to be determined.

    a. Compute the LQ gain matrix for the same performance index as in Problem 7.22, but with $Q_{33}\Delta\omega_0{}^2$ replaced by $Q_{33}[\Delta\omega_0(t) - z(t)]^2$, where $z(t) = -25e^{-t/\tau}$ and incremental quantities are defined with respect to the steady state $\omega_0 = 25$ rad/s. Use your design of Problem 7.23 for the Q's and R's, and $\tau = 10$ s.

    b. Compute the response for zero initial state. Check the constrained variables u_1, u_2, Δ_1, and Δ_2 against their limits. Go back to part (a) and repeat for other values of τ until a value is found where the constraints are just met. Give τ and the control law.

(M) **7.34** *Drum speed control*   We wish to design a disturbance feedforward control for the drum speed control.

    a. Calculate the dc steady-state input and state variables for $\omega_0 = \omega_d$, $T_0 = T_0{}^*$, and $\Delta_1{}^* = \Delta_2{}^*$. Express $u_1{}^*$ and $u_2{}^*$ as functions of ω_d and $T_0{}^*$.

    b. Compute the steady state for $\omega_0 = 25$ rad/s and $T_0 = 0$. Apply a step torque $T_0 = 10^4$ Nm, and calculate the response for the LQ control of Problem 7.23. Leave $u_1{}^*$ and $u_2{}^*$ unchanged; i.e., do not use disturbance feedforward. Note the peak values of Δ_1 and Δ_2 and the response $\omega_0(t)$.

    c. Repeat part (b), but with disturbance feedforward; i.e., apply step changes to $u_1{}^*$ and $u_2{}^*$ corresponding to the expressions derived in part (a). Compare the responses with those of part (b).

M   **7.35**   *Blending tank*   The blending tank model of Problem 2.7 (Chapter 2) was linearized about a given dc steady state in Problem 2.13.

a. Using the linearized model, calculate the changes in ΔF_0^* and ΔF_A^* in the dc steady-state inputs that cancel out steady-state disturbance change ΔF_B^*. Express ΔF_0^* and ΔF_A^* as functions of ΔF_B^*, and show that the application of this linear feedforward law will cancel out the effect of the disturbance, not only as the steady state but in the transient state as well.

b. Let $\Delta C_{Ad}(t) = \Delta C_{Ad}^*(1 - e^{-t/50})$, with $\Delta F_B = 0$. Solve for the optimal LQ gain, with

$$ J = \int_0^\infty [Q(\Delta C_A(t) - \Delta C_{Ad}(t))]^2 + R\,\Delta F_0^2 + R\,\Delta F_A^2] dt $$

and give the control law. Start with $Q = 1/(.8)^2$ and $R = 1/(10^{-4})^2$. Adjust R so that, for $\Delta \ell(0) = \Delta V_A(0) = 0$ and $\Delta C_{Ad}^* = 0.1$, the error $\Delta C_A(t) - \Delta C_{Ad}$ is minimized, while $|\Delta F_A(t)|$ and $|\Delta F_0(t)|$ are held below 0.5×10^{-4} m^3/s.

M   **7.36**   *Heat exchanger*   A heat exchanger is often a good candidate for the use of feedforward control. A pole-placement design was carried out in Problem 7.12, based on the linearized model of Problem 2.14 (Chapter 2). The major disturbance input is ΔT_{C0}, the cold liquid inlet temperature.

a. Use the linearized model of Problem 2.14 with the control law of Problem 7.12. For the system initially in the nominal dc steady state, compute $\Delta T_{C3}(t)$ for a step increase of 2°C in T_{C0}.

b. We now add feedforward. Use the linearized model to calculate the steady-state incremental input ΔF_H^* that corresponds to the same steady-state outlet temperature as before ($\Delta T_{C3}^* = 0$), but with $\Delta T_{C0}^* \neq 0$. Use the linear expression for ΔF_H^* as a function of ΔT_{C0}^* as a feedforward term in the control law, and repeat the simulation of part (a).

7.37   For the system of Problem 7.1, design a second-order observer such that the poles of the error system are at $-3 \pm j\sqrt{3}$. Calculate the state $\mathbf{x}(t)$ and the estimate $\hat{\mathbf{x}}(t)$, given that $u(t)$ is a unit step, $\mathbf{x}(0) = [\begin{smallmatrix}1\\1\end{smallmatrix}]$, and $\hat{\mathbf{x}}(0)$ is the zero state.

7.38   Repeat Problem 7.37 for the system of Problem 7.2.

M   **7.39**   *Heat exchanger*   The heat exchanger model of Problem 2.8 (Chapter 2) was linearized about a particular dc steady state in Problem 2.14. The outlet temperature ΔT_{C3} is measured, and we wish to design an observer to estimate the other state variables.

a. With $\Delta T_{C0} = \Delta T_{H0} = 0$ and with ΔF_H and ΔF_C available for measurement, design an observer with error system poles at $-1 \pm j$, $-.8 \pm j.8$, -1, and -1.1.

b. Using the linearized model, compute the state variables, given:

$$\Delta T_{C0}(t) = \Delta T_{H0}(t) = 0, \Delta F_H(t) = 0.05 \text{ m}^3/\text{s},$$
$$\Delta F_C(t) = -0.01 \text{ m}^3/\text{s},$$
$$\Delta T_{C1}(0) = \Delta T_{C2}(0) = \Delta T_{C3}(0) = 7°\text{C},$$
$$\Delta T_{H1}(0) = \Delta T_{H2}(0) = \Delta T_{H3}(0) = -10°\text{C}.$$

Simulate the observer of part (a), driven with $\Delta T_{C3}(t)$, $\Delta F_C(t)$, and $\Delta F_H(t)$ from the zero state, and compare the estimated and true state variables.

(M) **7.40** *Heat exchanger*   Repeat Problem 7.39 for the case where the disturbance flow ΔF_C is not directly available. Assume that $\Delta F_C(t)$ is of the form $A + Be^{-t}$, with A and B constants. Thus, $\Delta F_C(t)$ is modeled by the second-order system

$$\dot{\mathbf{z}} = \begin{bmatrix} 0 & 0 \\ 0 & -1 \end{bmatrix} \mathbf{z}$$
$$\Delta F_C(t) = z_1 + z_2.$$

Add these to the system description, and add the poles $-.8$ and -0.9 to the observer (now of order 8). In part (b), use $z_1 = -0.01$ and $z_2 = 0.01$, and observe $\Delta F_c(t)$ in addition to the system state variables.

(M) **7.41** *Chemical reactor*   A model for a chemical reactor was derived in Problem 2.9 (Chapter 2) and linearized in Problem 2.15. Often the difficulty of measuring concentration "on line" makes it necessary to remove samples from time to time for laboratory analysis. Between laboratory measurements, it is useful to estimate concentration from on-line measurements, such as temperature.

a. Suppose ΔT, ΔQ, and Δc_{A0} are available, with $\Delta F = \Delta T_0 = 0$. Use the linearized model to show that Δc_B and Δc_c cannot be estimated by an observer.

b. It was shown in Problem 3.48 (Chapter 3) that the first two equations (for $\Delta \dot{c}_A$ and $\Delta \dot{T}$) constitute a minimal realization for the input ΔQ and the output ΔT. Show that this is also the case if the input Δc_{A0} is included.

c. Design an observer to estimate the state variables Δc_A and ΔT, with the error-system poles at $-0.8 \pm j0.8$.

d. Simulate the linear system response with $\Delta Q(t) = 10,000$ J/min, $\Delta c_{A0}(t) = 0.01$ kg moles/m^3, $\Delta c_A(0) = 0.5$ kg moles/m^3, and $\Delta T(0) = -10°$C. Drive the observer of part (c) with $\Delta T(t)$, $\Delta Q(t)$, and $\Delta c_{A0}(t)$, with the observer initially in the zero state. Compare the estimates with the true values of the state variables.

(M) **7.42** **Chemical reactor**   With reference to Problem 7.41, it is noted that the inlet concentration $\Delta c_{A0}(t)$ may often be unavailable. We wish to modify the observer to observe Δc_{A0} of the form $Ae^{-0.3t}$. Design a modified (third-order) observer, adding a pole to the error system at -1. Repeat the simulation of Problem 7.41(d), but with $\Delta c_{A0}(t) = 0.01e^{-0.3t}$, and estimate Δc_{A0} as well as the system state.

7.43 Design an optimal observer for the system

$$\dot{x} = ax + u + w$$

$$y = x + v.$$

Express the observer gain in terms of a and the design parameters W and V, and discuss its variation as a function of those parameters. (Note that $W \geq 0, V > 0$, and a may be positive or negative.) Calculate the transfer function $\widehat{x}/y$.

7.44 The design of a differentiator may be approached from the observer point of view. A signal $x(t)$ is modeled as

$$\dot{x} = z$$

$$\dot{z} = w$$

with a noisy observation

$$y = x + v.$$

Here, $z(t)$ is the derivative of $x(t)$.

Use the design parameters $W = \begin{bmatrix} 0 & 0 \\ 0 & w_0^2 \end{bmatrix}$ and $V > 0$ to obtain an optimal observer. Calculate the transfer function $\widehat{z}/y$ of the differentiator, and discuss its variations with w_0^2 and V.

7.45 An observer can be designed to recover a sinusoid of known frequency from a noisy observation.

a. Show that the model

$$\dot{\mathbf{x}} = \begin{bmatrix} 0 & \omega \\ -\omega & 0 \end{bmatrix} \mathbf{x}$$

$$y = x_1$$

is such that $y(t)$ is a sinusoid of frequency ω whose phase depends on $\mathbf{x}(0)$.

b. With $W = \left[\begin{smallmatrix} 0 & 0 \\ 0 & 1 \end{smallmatrix}\right]$ and $V > 0$ as design parameters, obtain an optimal observer. Calculate the transfer function $\widehat{x}_1/y$, and discuss its variation with ω and V.

M **7.46** *dc servo with flexible shaft* For the dc servo with flexible shaft of Problem 2.5 (Chapter 2), we wish to study the effectiveness of state estimation, given different combinations of measurements. Suppose that θ_2, ω_2, and i are available, with respective precisions of ± 0.001 rad, ± 0.05 rad/s, and $\pm .005$ A. We model the disturbance input by assuming there is an external torque acting directly on the output shaft; this amounts to adding the vector

$$\begin{bmatrix} 0 \\ 0 \\ 0 \\ w \\ 0 \end{bmatrix}$$

to the right-hand side of the state equations.

a. Design an optimal observer, using θ_2, ω_2, and i as measurements. Use the design parameters

$$V = \operatorname{diag}\begin{bmatrix} 0.001^2 & 0.05^2 & 0.005^2 \end{bmatrix}$$

and

$$W = \begin{bmatrix} 0 & 0 & 0 & w_0 & 0 \end{bmatrix}^T \begin{bmatrix} 0 & 0 & 0 & w_0 & 0 \end{bmatrix}.$$

with $w_0 = 5$.

b. Simulate the system and the observer with $w(t) = 5u_0(t)$, with both the system and the observer in the zero state at $t = 0_-$.

c. Repeat parts (a) and (b), but without the ω_2 measurement. Discuss the usefulness of the angular velocity sensor in this case.

M **7.47** *Drum speed control* The system of Problem 2.6 (Chapter 2) is to be provided with angular velocity sensors accurate to $\pm v_0$ rad/s. External disturbances are taken into account by modeling the load torque T_0 as an impulse $10^5 u_0(t)$.

a. Using the design parameters

$$V = \operatorname{diag}\begin{bmatrix} v_0^2 & v_0^2 & v_0^2 \end{bmatrix}$$

$$W = \begin{bmatrix} 0 & 0 & 10^{-4} & 0 & 0 \end{bmatrix}^T 10^{10} \begin{bmatrix} 0 & 0 & 10^{-4} & 0 & 0 \end{bmatrix}$$

with $v_0 = 1.0$ rad/s, design an optimal observer, using ω_0, ω_1, and ω_2 as outputs.

b. Simulate the system for $T_0(t) = 10^5 u_0(t)$ and $\omega_1(0_-) = \omega_2(0_-) = 66.68$ rad/s, $\omega_0(0_-) = 4.67$ rad/s, $\Delta_1(0_-) = \Delta_2(0_-) = 0$, $u_1(t) = u_2(t) = 400$ V. Simulate the observer, using the same initial conditions as for the plant. Compare the estimator and actual state variable values.

c. Repeat parts (a) and (b) for $v_0 = 0.1$ rad/s and 0.01 rad/s; this represents improved sensor accuracy. Comment on the degree of improvement in the estimator.

7.48 **M** *Two-pendula problem*   In Problem 2.19 (Chapter 2), a linearized model was derived for a system with two pendula. In this instance, let $\ell_1 = 1$ m and $\ell_2 = 0.5$ m. We wish to design an observer, assuming that θ_1, θ_2, and x are available for measurement. The assumption of errors of ± 0.001 m on x and ± 0.001 rad on θ_1 and θ_2 suggests using

$$V = (0.001)^2 I.$$

To model plant disturbances, assume external torque impulses of magnitude w_0 acting directly on the rods. This suggests using a W that is diagonal, with elements w_0^2 in the positions corresponding to ω_1 and ω_2.

Design optimal observers for several values of w_0, and calculate the eigenvalues of the error system (i.e., of $A-GC$). Select a value of w_0 just sufficient to ensure that every eigenvalue has a real part of -2 or less (i.e., $0.5a$ time constant of 0.5 s or less).

7.49 For the system of Problem 7.2, design a reduced-order observer of order 1 with the error-system eigenvalue at -2.

7.50 The system of Problem 7.1 has an output that is not a single state variable; i.e., $y = x_1 + x_2$.

a. Choose the second row of $T^{-1} = \begin{bmatrix} 1 & 1 \\ x & x \end{bmatrix}$ so that T^{-1} is nonsingular. Use T to carry out a linear transformation to a coordinate system **z**, with $\mathbf{x} = T\mathbf{z}$. Note that $y = z_1$.

b. Design a reduced-order observer of order 1 for the transformed system, with the error-system eigenvalue at -4. Use z_1 and $\widehat{z}_2$ to generate estimates of the components of **x**.

7.51 **M** *dc servo with flexible shaft*   For the servo with flexible shaft of Problem 2.5 (Chapter 2):

a. Design a reduced-order observer (of order 4), using θ_2 as the output. The eigenvalues of the error system should be $-200 \pm j200$ and $-10 \pm j10$.

b. With θ_2, i, and ω_2 as outputs, design a reduced-order observer of order 2. The observer gain matrix should be chosen so that the error system has eigenvalues at $-10 \pm j10$. (*Note*: Infinitely many gain matrices can do this.)

c. Simulate the system and the two observers, with $v = 2$ V (a constant), $\omega_2(0) = 5$ rad/s, and all other initial state and observer variables equal to zero.

7.52 *Crane* The crane model of Problem 2.11 (Chapter 2) was linearized in Problem 2.17. Using the linearized model:

a. Design a reduced-order observer (of order 3) with x as the measured output. The eigenvalues of the error system should be -10 and $-10 \pm j10$.

b. Design a reduced-order observer (of order 2) with x and θ as measured outputs. The eigenvalues of the error system should be $-10 \pm j10$. (This is achievable with infinitely many observer gains.)

c. With $F = 0, \theta(0) = 0.5$ rad, and all other system and observer state variables equal to zero, simulate the linearized system with both observers.

7.53 *High-wire artist* For the linearized system of Problem 2.18, with θ and ϕ as outputs, design a reduced-order observer (of order 2) to estimate the two angular velocities. The error system eigenvalues should be at $-20 \pm j20$.

7.54 *Two-pendula problem* In Problem 2.19 (Chapter 2), a linearized model was derived for a system with two pendula. (Assume $\ell_1 = 1$ m and $\ell_2 = 0.5$ m.) We wish to use optimal observer theory to design a reduced-order observer with θ_1, θ_2, and x as measurements.

Assume $W = I$ and $V = v_0^2 I$. Design a few third-order observers for ω_1, ω_2, and v, and adjust v_0 so that each of the three eigenvalues has a real part less than -2. Calculate (approximately) the value of v_0 that just achieves this.

7.55 *Maglev*

a. For the Maglev model of Problem 2.20 (Chapter 2), design a reduced-order observer with the gap lengths ΔS_{L1}, ΔS_{L2}, Δ_{G1}, and ΔS_{G2} as measurements. Place the poles of the second-order error system at $-300 \pm j300$.

b. Obtain, by simulation, the time and observed state variables for zero inputs and (i) $\Delta z(0) = 1$, (ii) $\Delta \theta(0) = 1$, (iii) $\Delta y(0) = 1$, and all other state variables being zero.

7.56 *Maglev* Sensors do fail, and failure is often detectable. Suppose ΔS_{L1} becomes unavailable. We wish to reconfigure the observer to handle this contingency.

a. For the Maglev model, design a reduced-order observer using the gap lengths ΔS_{L2}, ΔS_{G1}, and ΔS_{G2} as measurements. Place the poles of the error system at $-300 \pm j300$ and -300.

b. Simulate for the conditions of Problem 7.55, and compare.

c. In the event of a failure, how would you initialize the third-order observer to ensure smooth transfer from the second-order observer?

7.57 In Problem 7.6, a pole-placement state feedback was designed for the system of Problem 7.1. A second-order observer for this plant was designed in Problem 7.37.

a. Write the state description of the controller that results from generating a state estimate and using this in the control law.

b. Compute the transfer function u/y of the control law.

7.58 Repeat Problem 7.57 for the plant of Problem 7.2, the state feedback law of Problem 7.7, and the observer of Problem 7.38.

7.59 The plant of Problem 7.1 is to be regulated to a nonzero reference value, with zero steady-state error. This requires a steady-state control component, u^*. We wish to design a controller that uses an estimate of u^* and an estimate of the incremental state $\Delta \mathbf{x}$.

a. Design a third-order observer for u^* and $\Delta \mathbf{x}$.

b. Using the state feedback gain $\mathbf{k}^T$ of Problem 7.6, design a controller where $u = \widehat{u}^* + \mathbf{k}^T \widehat{\mathbf{x}}$.

c. Calculate the transfer function u/e, where $e = y_d - y = -\Delta y$. Verify that this transfer function contains an integrator.

(M) **7.60** *Servo, simplified model* The dc servo of Problem 2.4 (Chapter 2) is subjected to an unmeasured constant torque disturbance. The output angle θ is to be regulated to a value θ_d. This requires a constant input u^*, whose value depends on the load torque.

a. Write, as in Equations 7.96 and 7.97, equations for the incremental state, and append $\dot{u}^* = 0$.

b. Assume that θ and ω are measured without noise, and design a reduced-order observer of order 1 to estimate u^*. The error system should have an eigenvalue of -8. (There are many solutions for the observer gain.)

c. With $u = \widehat{u}^* - K[\begin{smallmatrix} \Delta\theta \\ \Delta\omega \end{smallmatrix}]$, where K is the state feedback gain of Problem 7.10, write the state description of the controller.

d. Write the controller in transfer-function form $u/\Delta\theta$, using the fact that $\Delta\omega(s) = s\Delta\theta(s)$. Write u/e, where $e = \theta_d - \theta$.

M   **7.61**   ***Servo with flexible shaft***   An optimal LQ control law for the servo model of Problem 2.5 (Chapter 2) was designed in Problem 7.22. An optimal observer was designed in Problem 7.46. The observer is also valid when applied to the incremental system, defined with respect to a dc steady state $\theta_2 = \theta_d$. With θ_2, ω_2, and i as measurements, it is necessary to know the dc steady-state values of those quantities and of the input v in order to generate $\Delta\theta_2$, $\Delta\omega_2$, Δi, and Δv. Fortunately, it was shown in Problem 7.21 that, except for θ_2, these variables all have zero steady state.

    a. Write the state descriptions of the controller formed from the observer of Problem 7.46 with the LQ gain of Problem 7.22.

    b. In transfer-function form, the input is of the form $v = H_1 e + H_2\omega_2 + H_3 i$, with $e = \theta_d - \theta_2$. Compute H_1, H_2, and H_3.

    c. Compute the closed-loop zero-state responses for a unit step $\theta_d = 1$ rad with (i) state feedback and (ii) the observer-based control system.

M   **7.62**   ***Servo with flexible shaft***   For the servo of Problem 2.5 (Chapter 2), a pole-placement control law was designed in Problem 7.11, and two reduced-order observers, of orders 4 and 2, respectively, in Problem 7.51.

    a. Modify the observers for use with the incremental system referred to the dc steady state $\theta_2 = \theta_d$.

    b. Write the state descriptions for the controllers formed by using the state estimate from each observer in turn, with the pole-placement control law.

    c. Compute the closed-loop zero-state responses for the two controllers.

M   **7.63**   ***Drum speed control***   A state feedback–feedforward system was designed in Problem 7.33 for the system of Problem 2.7 (Chapter 2). An optimal observer was designed in Problem 7.47(a).

    a. Modify the observer for use with the incremental system refered to $\omega_0 = \omega_d^*$ (ω_d^* = dc steady state for the reference). Note that the ratio R/r relating ω_0 and ω_1, ω_2 is usually well known, so that the steady-state relationship $\omega_1^* = \omega_2^* = (R/r)\omega_0^*$ is reliable.

    b. Write the state description of the controller obtained by combining the observer and the LQ control. It is assumed that the set point $\omega_d(t)$ is measured.

    c. Compute the zero-state response to $\omega_d(t) = 25 - 25e^{-t/\tau}$ for the value of τ found in Problem 7.33(c). Do this for both the state feedback system and the observer-based controller.

M   **7.64**   ***Heat exchanger***   The heat exchanger model of Problem 2.8 (Chapter 2) was linearized in Problem 2.14. The system is subjected to constant, unmeasured disturbances. A pole-placement controller was designed in Problem 7.12. The observer designed in Problem 7.39 is to be modified to include observation of the dc steady-state input ΔF_H^*.

a. Append the equation $\Delta \dot{F}_H{}^* = 0$ to the linearized state description as in Equations 7.96 and 7.97, and, with ΔT_{C3} as the measured output, design an observer of order 7 to estimate the incremental states and $\Delta F_H{}^*$. The error-system eigenvalues are to be those of the observer of Problem 7.37, with the addition of -5.

b. Let the controller be $\Delta F_H = \Delta \widehat{F}_H{}^* - K \, \Delta\widehat{\mathbf{x}}$, where K is the gain obtained in Problem 7.12. Compute the controller transfer function $\Delta F_H/e$, where $e = \Delta T_{C3d} - \Delta T_{C3}$.

c. Compute the response $\Delta T_{C3}(t)$ to (i) a 5°C step change in ΔT_{C3d} and (ii) a 0.01 m³/s step change in ΔF_C.

Ⓜ   **7.65**   *Chemical reactor*   A model for a chemical reactor was derived in Problem 2.9 (Chapter 2) and linearized in Problem 2.15. Problem 3.48 (Chapter 3) showed that the equations for $\Delta \dot{c}_A$ and $\Delta \dot{T}$ constitute a minimal realization. A pole-placement controller was designed in Problem 7.13, but control against unmeasured constant disturbances must be added.

a. Append the equation $\Delta \dot{Q}^* = 0$ to the two linear state equations, as in Equations 7.96 and 7.97.

b. With ΔT as measured output, design a reduced-order observer for the variables Δc_A and ΔQ^*. The error-system eigenvalues are to be $-0.8 \pm j0.8$.

c. Express the observer in a form that does not require the evaluation of derivatives. Write the state description of the controller with $\Delta Q = \Delta \widehat{Q}^* - K[\begin{smallmatrix} \Delta T \\ \Delta \hat{c}_A \end{smallmatrix}]$, where K is the gain from Problem 7.13.

d. Calculate the controller transfer function $\Delta Q/\Delta T_d$.

e. Apply this linear control to the nonlinear model of Problem 2.9 (Chapter 2), and compute the responses to steps in ΔT_d of 10°K, 20°K, and 40°K.

Ⓜ   **7.66**   *Crane*   A model of a crane system was derived in Problem 2.11 (Chapter 2) and linearized in Problem 2.17. An LQ state feedback was obtained in Problem 7.26, and reduced-order observers in Problem 7.52.

a. Modify the two reduced-order observers for use on the incremental system referred to the dc steady state $x = x_d$.

b. Design two controllers by combining each observer in turn with the feedback gain of Problem 7.26.

c. Compute the zero-state response to a step x_d of 5 m, for (i) state feedback and (ii) the two controllers.

d. Apply the controller based on the third-order observer to the nonlinear model of Problem 2.11, for step x_d values of 1 m, 5 m, and 10 m. Discuss the results.

(M) 7.67 *High-wire artist*

  a. Combine the LQ controller of Problem 7.25 with the reduced-order observer of Problem 7.53 to form a controller.

  b. Simulate this controller with the nonlinear model of Problem 2.12, for the initial states of Problem 7.14.

(M) 7.68 *Two-pendula problem* A linear model for the two-pendula problem was derived in Problem 2.19 (Chapter 2). An LQ control gain was found in Problem 7.27, and an LQ observer in Problem 7.48.

  a. Combine the observer and control gain to generate a controller.

  b. Compute the response for the initial state $\theta_1(0) = 0.1$ rad, $\theta_2(0) = -0.1$ rad, and $\omega_1(0) = \omega_2(0) = x(0) = v(0) = 0$ for the controls with (i) state feedback and (ii) the observer-based controller.

(M) 7.69 *Maglev* Combine the state-feedback controller of Problem 7.28 with the observer of Problem 7.55 to form a controller.

  a. Simulate with the initial state variables: (i) $\Delta z(0) = 1$, (ii) $\Delta \theta(0) = 1$, (iii) $\Delta y(0) = 1$, with all other system and observer state variables equal to zero.

  b. Compute the matrix transfer function of the controller with the gap lengths as inputs and the input variables as outputs.

(M) 7.70 *Maglev* The nominal value of one of u_{G1} and u_{G2} is arbitrary (the other is nominally equal to the first, but may not be exactly so if the magnets are slightly different). We wish to use the design technique of Chapter 6, Section 6.8.

  a. Append the three equations $\Delta \dot{u}_{L1} = \Delta \dot{u}_{L2} = \Delta \dot{u}_{G2} = 0$ to the original set of state equations, and design a reduced-order observer of order 5. Place the poles of the error system at -300, $-300 \pm j300$, and $-250 \pm j250$.

  b. Proceed to design the controller as in Section 7.8, using the state feedback law of Problem 7.28.

  c. Simulate for the conditions of Problem 7.69(a).

  d. Compute the matrix transfer function of the controller, with the gap lengths as inputs and the input variables as outputs. Verify the presence of poles at $s = 0$.

References

[1] Kailath, T., *Linear Systems*, Prentice-Hall (1980).

[2] Friedland, B., *Control System Design: An Introduction to State-Space Methods*, McGraw-Hill (1986).

[3] Moore, B. C., "On the Flexibility Offered by State Feedback in Multivariable Systems Beyond Closed Loop Eigenvalue Assignment," *IEEE Trans. Aut. Control*, vol. AC-21, pp. 689–692 (1977).

[4] Wonham, W. H., *Linear Multivariable Control: A Geometric Approach*, 2nd ed., Springer-Verlag (1979).

[5] Bartels, R. H., and G. W. Stewart, "Solution of the Equation $AX + XB = C$ (Algorithm 432)," *Comm. Assoc. Computing Mach.*, vol. 15, pp. 820–826 (1972).

[6] Kalman, R. E., "A New Approach to Linear Filtering and Prediction Problems," *ASME J. of Basic Eng.*, vol. 820, pp. 35–45 (1960).

[7] Kalman, R. E., and R. S. Bucy, "New Results in Linear Filtering and Prediction Problems," *ASME J. of Basic Eng.*, vol. 830, pp. 95–108 (1961).

[8] Luenberger, D. G., "Observing the State of a Linear System," *IEEE Trans. on Military Electronics*, vol. MIL-8, pp. 74–80 (1964).

Chapter 8

Multivariable Control

8.1 INTRODUCTION

During the past 15 to 20 years, developments in linear control theory have concentrated on the control of multivariable systems. Many systems, particularly in technologically advanced areas such as aerospace systems, are represented by models with several inputs, with each input having a significant effect on several outputs. Such cross-coupling makes the use of single-input, single-output (SISO) methods difficult.

In parallel with developments in multi-input, multi-output (MIMO) systems, the past 15 years have seen a renewed emphasis on frequency response. The ability of the state framework to handle uncertainty, especially nonparametric uncertainty, proved deficient. In contrast, uncertainty fits quite naturally in an input–output setting such as frequency response. The state framework has not been cast aside; rather, connections have been made between it and the frequency-response approach.

Developments in frequency-response design have been greatly abetted by the introduction of H^∞ theory. Quadratic performance indices do lead to integrals in the frequency domain, through Parseval's integral. However, engineers are usually concerned with specifications expressed pointwise in the frequency domain, not with the averages yielded by integrals. The H^∞ theory provides a direct handle on this type of specification.

In this chapter, we introduce the basic closed-loop expressions for the multivariable case. We then discuss norms, with emphasis on the 2-norm and the ∞-norm. Stability, uncertainty, and robust stability follow. The standard design problem is defined, and H^2 and H^∞ solutions are worked out.

8.2 BASIC EXPRESSION FOR MIMO SYSTEM

The purpose of this section is to present a few basic expressions in MIMO linear systems theory.

In Figure 8.1, the signals are all vector quantities, and $F(s)$ and $P(s)$ are matrices. The dimensions are $m \times r$ for P and $r \times m$ for F, where $m = \dim(\mathbf{y})$

379

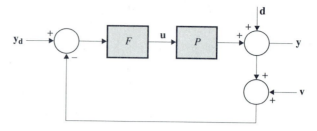

Figure 8.1 Basic 1-DOF block diagram

and $r = \dim(\mathbf{u})$. The most significant differences between this and the SISO case are that division becomes inversion, and the order of multiplication matters. From Figure 8.1,

$$\mathbf{y} = \mathbf{d} + P\mathbf{u}$$
$$= \mathbf{d} + PF(\mathbf{y}_d - \mathbf{y} - \mathbf{v})$$

or

$$(I + PF)\mathbf{y} = \mathbf{d} + PF(\mathbf{y}_d - \mathbf{v})$$
$$\mathbf{y} = (I + PF)^{-1}\mathbf{d} + (I + PF)^{-1}PF(\mathbf{y}_d - \mathbf{v}). \qquad (8.1)$$

The sensitivity and complementary sensitivity are defined as

$$S = (I + PF)^{-1} \qquad (8.2)$$
$$T = (I + PF)^{-1}PF. \qquad (8.3)$$

Both are $m \times m$ matrices. Note that

$$I - S = (I + PF)^{-1}(I + PF) - (I + PF)^{-1}$$
$$= (I + PF)^{-1}(I + PF - I)$$
$$= T$$

as in the SISO case.

Equation 8.1 is rewritten as

$$\mathbf{y} = T\mathbf{y}_d + S\mathbf{d} - T\mathbf{v}. \qquad (8.4)$$

We also write

$$\mathbf{e} = \mathbf{y}_d - \mathbf{y} = S\mathbf{y}_d - S\mathbf{d} + T\mathbf{v} \qquad (8.5)$$

and

$$\mathbf{u} = F(\mathbf{y}_d - \mathbf{y} - \mathbf{v}) = FS(\mathbf{y}_d - \mathbf{d} - \mathbf{v}). \qquad (8.6)$$

Those expressions are analogous to their SISO counterparts.

8.3 SINGULAR VALUES

The ℓ_2 norm of a complex vector $\mathbf{w}$ is defined as

$$\|\mathbf{x}\|_2 = \left(\sum_i |w_i|^2\right)^{1/2} \qquad (8.7)$$

$$= (\mathbf{w}^*\mathbf{w})^{1/2} \qquad (8.8)$$

where the superscript "*" refers to the conjugate transpose. This norm is the so-called ℓ_2 norm; it is the Euclidean length of the complex vector $\mathbf{w}$.

Next, the norm of an output signal spectrum must be related in some manner to that of the input. To fix ideas, consider $\mathbf{w} = S\mathbf{d}$, the part of the output $\mathbf{y}$ that is due to the disturbance. We write

$$\|\mathbf{w}\|^2 = (S\mathbf{d})^*(S\mathbf{d})$$

$$= \mathbf{d}^* S^* S \mathbf{d}. \qquad (8.9)$$

Suppose $\|\mathbf{d}(j\omega)\|$ is given. Geometrically, we may then think of $\mathbf{d}$ as a vector of given length, whose direction is unknown. For given $\|\mathbf{d}\|$, the right-hand side (RHS) of Equation 8.9 will change as the direction of $\mathbf{d}$ varies. It is known, however, that $\|\mathbf{w}\|$ has maximum and minimum values, given by

$$\max \|\mathbf{w}\| = \sqrt{\lambda_{\max}(S^*S)} \|\mathbf{d}\| \qquad (8.10)$$

$$\min \|\mathbf{w}\| = \sqrt{\lambda_{\min}(S^*S)} \|\mathbf{d}\|. \qquad (8.11)$$

The matrix S^*S is Hermitian (i.e., is its own conjugate transpose), and the eigenvalues of such a matrix are real and positive. The quantities $\lambda_{\max}$ and $\lambda_{\min}$ are the largest and smallest eigenvalues of S^*S.

The singular values of S are the square roots of the eigenvalues of S^*S. It can be shown [1] that any $n \times n$ matrix S can be written as

$$S = U \Sigma V^* \qquad (8.12)$$

where U is an $n \times n$ unitary matrix, V is an $n \times n$ unitary matrix, and Σ is an $n \times n$ diagonal matrix whose elements are the singular values of S. The number of nonzero singular values is equal to the rank of S. (Recall that a unitary matrix has the property that $UU^* = I$.) The largest and smallest singular values of S are denoted by $\overline{\sigma}(S)$ and $\underline{\sigma}(S)$, respectively. Thus,

$$\overline{\sigma}(S) = \sqrt{\lambda_{\max}(S^*S)} \tag{8.13}$$

$$\underline{\sigma}(S) = \sqrt{\lambda_{\min}(S^*S)} \tag{8.14}$$

and, from Equations 8.10 and 8.11,

$$\underline{\sigma}(S)\|\mathbf{d}\|_2 \le \|\mathbf{w}\|_2 \le \overline{\sigma}(S)\|\mathbf{d}\|_2. \tag{8.15}$$

We shall need a few properties of singular values.

◆ **Property I**   If S^{-1} exists,

$$\underline{\sigma}(S) = \frac{1}{\overline{\sigma}(S^{-1})}$$

and

$$\overline{\sigma}(S) = \frac{1}{\underline{\sigma}(S^{-1})}. \qquad ◆$$

◆ **Property II** $\det S = \sigma_1\sigma_2\ldots\sigma_n.$ ◆

◆ **Property III** $\overline{\sigma}(A+B) \le \overline{\sigma}(A) + \overline{\sigma}(B).$ ◆

◆ **Property IV** $\overline{\sigma}(AB) \le \overline{\sigma}(A)\overline{\sigma}(B).$ ◆

◆ **Property V** $\max[\overline{\sigma}(A), \overline{\sigma}(B)] \le \overline{\sigma}([A\ \ B]) \le \sqrt{2}\max[\overline{\sigma}(A), \overline{\sigma}(B)].$ ◆

From Equation 8.15, we see that $\overline{\sigma}(S)$ describes the worst-case situation for a given $\|\mathbf{d}\|_2$. In other words, given that the squares of the magnitudes of the components of the disturbance at a frequency ω sum to $\|\mathbf{d}\|_2^2$, the most unfavorable distribution of disturbance components will lead to $\|\mathbf{y}(j\omega)\|_2 = \overline{\sigma}(S(j\omega))\|\mathbf{d}\|_2$. It follows that $\overline{\sigma}[S(j\omega)]$ is the key scalar quantity describing the maximum amplification of $S(j\omega)$ and plays the role that was given to $|S(j\omega)|$ in the SISO case. The same is true of the other contributions in Equations 8.4, 8.5, and 8.6. For example, the magnitude of the measurement noise contribution at ω is measured by $\overline{\sigma}[T(j\omega)]\|\mathbf{v}(j\omega)\|_2$.

Singular values also describe the magnitude of the loop gain. In the SISO case, it was found that $S(j\omega)$ is small and $T(j\omega) \approx 1$ if $|L(j\omega)| \gg 1$, and that $S(j\omega) \approx 1$ and $T(j\omega) \approx L(j\omega)$ if $|L(j\omega)| \ll 1$. To arrive at a MIMO equivalent, consider

$$(I + L)\mathbf{w} = \mathbf{w} + L\mathbf{w} \tag{8.16}$$

where **w** is a complex vector of appropriate dimensions and $L = PF$ is the loop-gain matrix. The approximation $(I+L) \approx L$ is valid if, for any vector **w**, $(I+L)\mathbf{w} \approx L\mathbf{w}$. From Equation 8.16, that is true if $\|L\mathbf{w}\|_2 \gg \|\mathbf{w}\|_2$. Since $\|L\mathbf{w}\|_2$ is, at minimum, equal to $\underline{\sigma}(L)\|\mathbf{w}\|_2$, the approximation holds if $\underline{\sigma}[L(j\omega)] \gg 1$. This is precisely what is meant by "large loop gain." In such a case,

$$S(j\omega) \approx L^{-1}(j\omega)$$

$$T(j\omega) \approx L^{-1}(j\omega)L(j\omega) = I.$$

On the other hand, $(I + L) \approx I$ if $\|L\mathbf{w}\|_2$ is small compared to $\|\mathbf{w}\|_2$. That is so for all **w** if $\bar{\sigma}(L) \ll 1$; this is the precise meaning of "small loop gain." In such a case,

$$S(j\omega) \approx I$$

$$T(j\omega) \approx L(j\omega).$$

Example 8.1   The loop-gain matrix for a two-input, two-output system is

$$L(s) = \begin{bmatrix} \frac{1}{s} & -\frac{.5}{s+1} \\ 1 & \frac{1}{s(s+1)} \end{bmatrix}.$$

(a) Compute and display $\bar{\sigma}[L(j\omega)]$ and $\underline{\sigma}[L(j\omega)]$.

(b) Calculate $S(s)$ and $T(s)$.

(c) Compute and display $\bar{\sigma}[S(j\omega)]$.

Solution   (a) The maximum and minimum singular values of L are shown in Figure 8.2 (MATLAB sigma).

(b) We have

$$S(s) = [I + L(s)]^{-1} = \begin{bmatrix} \frac{s+1}{s} & -\frac{.5}{s+1} \\ 1 & \frac{s^2+s+1}{s(s+1)} \end{bmatrix}^{-1}.$$

We calculate

$$\det[I + L(s)] = \frac{(s + 1)(s^2 + s + 1)}{s^2(s + 1)} + \frac{.5}{s + 1}$$

$$= \frac{s^2 + 2.5s^2 + 2s + 1}{s^2(s + 1)}$$

and

$$S = \frac{1}{s^3 + 2.5s^2 + 2s + 1} \begin{bmatrix} s(s^2 + s + 1) & .5s^2 \\ -s^2(s + 1) & s(s + 1)^2 \end{bmatrix}.$$

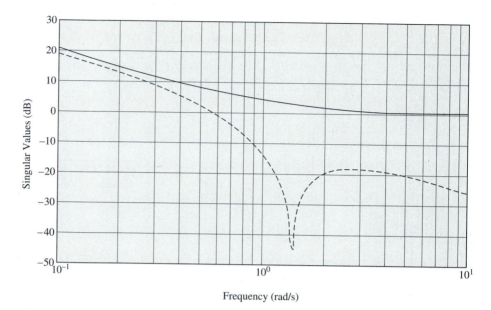

Figure 8.2 Singular values of the loop gain

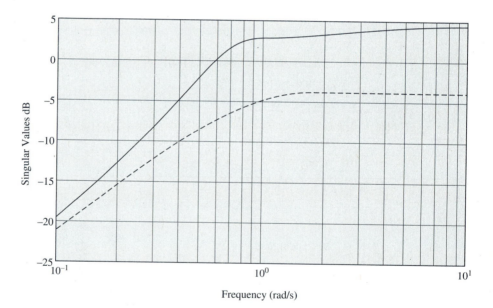

Figure 8.3 Singular values of the sensitivity

Finally,

$$T(s) = I - S(s) = \frac{1}{s^3 + 2.5s^2 + 2s + 1} \begin{bmatrix} 1.5s^2 + s + 1 & -.5s^2 \\ s^2(s+1) & .5s^2 + s + 1 \end{bmatrix}.$$

(c) The singular value $\overline{\sigma}[S(j\omega)]$ is computed numerically and is displayed in Figure 8.3. Note that $\overline{\sigma}[S(j\omega)]$ is small where $\underline{\sigma}\,[L(j\omega)] \gg 1$, and near 1 where $\overline{\sigma}[L(j\omega)] \ll 1$.

8.4 STABILITY

We write

$$P(s)F(s) = L(s) = \frac{N(s)}{D(s)}.$$

where $N(s)$ is an $m \times m$ matrix of polynomials and $D(s)$ is a polynomial. It is assumed that, if $D(s_0) = 0$, then $\det[N(s_0)] \neq 0$. This is the equivalent of the SISO condition that excludes pole-zero cancellations. Then, from Equation 8.2,

$$S(s) = \left(I + \frac{N(s)}{D(s)}\right)^{-1}$$

$$= D(s)[D(s)I + N(s)]^{-1}$$

$$= D(s)\frac{Adj[D(s)I + N(s)]}{\det[D(s)I + N(s)]}.$$

The characteristic equation is

$$\det\left(D(s)I + N(s)\right) = 0. \tag{8.17}$$

Using the fact that, for an $n \times n$ matrix A and a scalar k, $\det(kA) = k^n \det A$, we write Equation 8.17 as

$$[D(s)]^n \det\left(I + \frac{N(s)}{D(s)}\right) = 0. \tag{8.18}$$

If $D(s_0) = 0$, then s_0 is *not* a root of the characteristic equation; from Equation 8.17, that would require $\det N(s_0) = 0$, which is excluded. Therefore, in Equation 8.18, the factor $[D(s)]^n$ must be cancelled out by the determinant, so the roots of the characteristic equation satisfy

$$\det\left(I + \frac{N(s)}{D(s)}\right) = \det[I + L(s)] = 0. \tag{8.19}$$

The Routh criterion may be applied to the characteristic equation, Equation 8.19, as in the SISO case. To use the Nyquist criterion in its usual form, we write

$$1 + \{\det[I + L(s)] - 1\} = 0. \tag{8.20}$$

The Nyquist plot is that of $\det[I + L(j\omega)] - 1$; N is counted as in the SISO case, and P is the number of RHP poles of $L(s)$, i.e., the number of RHP roots of $D(s)$.

Example 8.2   A 2×2 plant has a matrix transfer function $P(s)$ equal to the loop gain $L(s)$ of Example 8.1. The controller is $F(s) = kI$. Study the stability for $k = 0.1$, 1, and 10.

Solution

$$\det[I + L(s)] = \det\begin{bmatrix} 1 + \frac{k}{s} & -\frac{.5k}{s+1} \\ k & 1 + \frac{k}{s(s+1)} \end{bmatrix}$$

$$= 1 + \frac{k}{-s} + \frac{k}{s(s+1)} + \frac{k^2}{s^2(s+1)} + \frac{0.5k^2}{s+1}$$

$$= 1 + \frac{(-.5k^2 + k)s^2 + 2ks + k^2}{s^2(s+1)}.$$

According to Equation 8.17, we should study the Nyquist locus for

$$\det[I + L(s)] - 1 = \frac{(-0.5k^2 + k)s^2 + 2ks + k^2}{s^2(s+1)}.$$

This is *not* of the form $kL'(s)$; it is not possible, as in the SISO case, to plot the locus for $L'(s)$ and count encirclements of the point $(-1/k, 0)$. It is necessary to plot different loci for different values of k, each time ascertaining the number of encirclements of the point $(-1, 0)$.

In this case, it is impossible to put the three Nyquist plots on the same graph because of the widely different scales. Figure 8.4 shows Bode plots for $k = 0.1$, 1.0, and 10. Here, $P = 0$. There are no real axis crossings for $k = 0.1$ and $k = 1$,

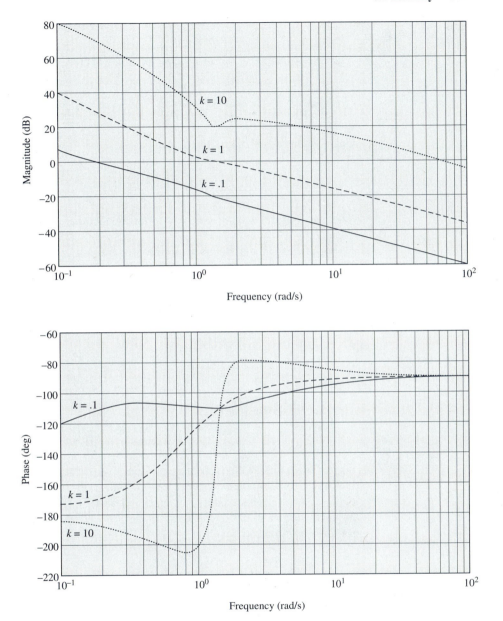

Figure 8.4 Determinant bode plots for three values of the gain

but one occurs for $k = 10$ at a frequency slightly greater than 1 rad/s and at a magnitude clearly greater than 1. Therefore, the system is stable for $k = 0.1$ and 1, and unstable for $k = 10$.

8.5   NORMS

The Euclidean, or ℓ_2, norm of a vector $\mathbf{x}$ has been defined as

$$\|\mathbf{x}\|_2 = \left(\sum_{i=1}^{n} x_i^2 \right)^{1/2} = (\mathbf{x}^T \mathbf{x})^{1/2}. \tag{8.21}$$

For a vector signal $\mathbf{x}(t)$, the ℓ_2 norm is

$$\|\mathbf{x}\|_2 = \left[\int_{-\infty}^{\infty} \mathbf{x}^T(t)\mathbf{x}(t)dt \right]^{1/2}. \tag{8.22}$$

This norm is the square root of the sum of the energy in each component of the vector.

For power signals, we may use the root-mean-square (rms) value

$$rms(\mathbf{x}) = \left(\lim_{T \to \infty} \frac{1}{2T} \int_{-T}^{T} \mathbf{x}^T(t)\mathbf{x}(t)dt \right)^{1/2}. \tag{8.23}$$

For an $m \times r$ matrix, we can define the *Frobenius* norm,

$$\|A\|_2 = \left(\sum_{i=1}^{m} \sum_{j=1}^{r} A_{ij}^2 \right)^{1/2}. \tag{8.24}$$

It can be shown that

$$\|A\|_2^2 = \text{tr}(A^T A) = \text{tr}(A A^T) \tag{8.25}$$

where "tr" refers to the trace, i.e., the sum of the diagonal elements.

Linear, time-invariant systems are generalizations of matrices; a matrix operates on a vector to produce another vector, and an LTI system operates on a signal to produce another signal. By analogy to the Frobenius norm, we define the L_2 norm for an $m \times r$ transfer function $G(s)$ as

$$\|G\|_2 = \left(\frac{1}{2\pi} \int_{-\infty}^{\infty} \text{tr}[G^T(-j\omega)G(j\omega)]d\omega \right)^{1/2}. \tag{8.26}$$

It is easy to show that $\|G\|_2$ exists [2] if, and only if, each element of $G(s)$ is strictly proper and has no poles on the imaginary axis. In that case, we write $G \in L_2$. Under those conditions, the norm can be evaluated as an integral in the complex plane:

$$\|G\|_2^2 = \frac{1}{2\pi j} \int_{-j\infty}^{j\infty} \text{tr}[G^T(-s)G(s)]ds$$

$$= \frac{1}{2\pi j} \oint \text{tr}[G^T(-s)G(s)]ds \tag{8.27}$$

where the last integral is taken over a contour that runs up the j-axis and around an infinite semicircle in either half plane. Since $G(s)$ is strictly proper, it is easily shown that the integrand vanishes over the semicircle, so that the residue theorem can be used.

If $G \in L_2$ and, in addition, G is stable, then we say that $G \in H^2$; H^2 is the so-called Hardy space defined with the 2-norm.

Example 8.3　　Calculate the L_2 norm of

$$G(s) = \frac{1}{s^2 + 3s + 2} \begin{bmatrix} s+3 & -(s+2) \\ -2 & s+2 \end{bmatrix}.$$

Solution　　We compute

$$\text{tr}G^T(-s)G(s) = \frac{(s+3)(-s+3) + 4 + (s+2)(s+2) - (-s+2)(s+2)}{(s+1)(s+2)(-s+1)(-s+2)}$$

$$= \frac{-3s^2 + 21}{(s+1)(s+2)(-s+1)(-s+2)}.$$

If we integrate about a contour enclosing the LHP (positive angle direction), the integral of Equation 8.27 is the sum of the residues at $s = -1$ and $s = -2$:

$$\|G\|_2^2 = \frac{18}{(1)(2)(3)} + \frac{9}{(-1)(3)(4)} = \frac{9}{4}.$$

Therefore, $\|G\|_2^2 = 3/2$.

A different type of norm is the *induced* norm, which applies to operators and is essentially a type of maximum gain. For a matrix, the *induced Euclidean* norm is

$$\|A\|_{2i} = \max_{\|d\|_2 = 1} \|Ad\|_2$$

$$= \bar{\sigma}(A) \tag{8.28}$$

as we have seen.

To obtain the induced norm for an LTI system, consider first a stable, strictly proper SISO system. If the input $u(\cdot) \in \ell_2$ (see Equation 8.22), the output $y(\cdot) \in \ell_2$. By Parseval's theorem,

$$\|y\|_2^2 = \frac{1}{2\pi} \int_{-\infty}^{\infty} |G(j\omega)|^2 |u(j\omega)|^2 d\omega. \tag{8.29}$$

Clearly,

$$\|y\|_2^2 \leq \sup_{\omega} |G(j\omega)|^2 \frac{1}{2\pi} \int_{-\infty}^{\infty} |u(j\omega)|^2 d\omega$$

or

$$\|y\|_2^2 \leq \sup_{\omega} |G(j\omega)|^2 \|u\|_2^2. \qquad (8.30)$$

We argue that the RHS of Equation 8.30 can be approached arbitrarily closely, for a fixed value of $\|u\|_2$ that is chosen to be 1 with no loss of generality. Suppose $|u(j\omega)|^2$ approaches an impulse of area 2π in the frequency domain at $\omega = \omega_0$. Then the integral of Equation 8.29 approaches $|G(j\omega_0)|^2$. If $|G(j\omega)|$ has a maximum at some finite value of ω, we may choose ω_0 to be that frequency. If not, then $|G(j\omega)|$ must approach a supremum as $\omega \to \infty$; we can make ω_0 as large as we like, and $|G(j\omega_0)|$ will be as close to the supremum as we wish. The RHS of Equation 8.30 can be approached arbitrarily closely, and

$$\sup_{\|u\|_2=1} \|y\|_2 = \sup_{\omega} |G(j\omega)|. \qquad (8.31)$$

This norm is also the infinity norm of G, given by

$$\|G\|_\infty = \lim_{p \to \infty} \left(\frac{1}{2\pi} \int_{-\infty}^{\infty} |G(j\omega)|^p d\omega \right)^{1/p}. \qquad (8.32)$$

The ∞-norm of $G(s)$ exists if, and only if, G is proper with no poles on the j-axis. In that case, we write $G \in L^\infty$. If, in addition, G is stable, then we say $G \in H^\infty$, the Hardy space with the infinity norm.

For multivariable systems,

$$\|\mathbf{y}\|_2^2 = \frac{1}{2\pi} \int_{-\infty}^{\infty} \|G(j\omega)\mathbf{u}(j\omega)\|_2^2 d\omega$$

$$\leq \frac{1}{2\pi} \int_{-\infty}^{\infty} \overline{\sigma}[G(j\omega)] \|\mathbf{u}(j\omega)\|^2 d\omega \qquad (8.33)$$

$$\leq \sup_{\omega} \left[\overline{\sigma}[G(j\omega)] \right]^2 \frac{1}{2\pi} \int_{-\infty}^{\infty} \|\mathbf{u}(j\omega)\|^2 d\omega$$

or

$$\|\mathbf{y}\|_2^2 \leq \left[\sup_{\omega} \overline{\sigma}[G(j\omega)] \right]^2 \|\mathbf{u}\|_2^2. \qquad (8.34)$$

Note that $\|\mathbf{u}(j\omega)\|_2$ in the integrand refers to the 2-norm of the vector $\mathbf{u}(j\omega)$, whereas in Equation 8.34 $\|\mathbf{u}\|_2$ refers to the 2-norm of the signal.

Here again, the RHS of Equation 8.34 can be approached arbitrarily closely, by proper choice of $\mathbf{u}(j\omega)$. Essentially, we pick $\mathbf{u}(j\omega)$ to be the eigenvector of $G^*(j\omega)G(j\omega)$ corresponding to the largest eigenvalue, and we concentrate the spectrum of $\mathbf{u}(j\omega)$ at the frequency where $\bar{\sigma}$ is the largest (or at some frequency that is arbitrarily large, if $\bar{\sigma}$ has no maximum but a supremum). Therefore,

$$\sup_{\|\mathbf{u}\|_2=1} \|\mathbf{y}\|_2 = \sup_\omega [G(j\omega)] \tag{8.35}$$

and we define

$$\|G\|_\infty = \sup_\omega \bar{\sigma}[G(j\omega)]. \tag{8.36}$$

Example 8.4   Compute the infinity norm for the transfer functions of Example 8.3.

Solution   Figure 8.5 shows the singular values (MATLAB sigma) as a function of ω, for $s = j\omega$. The supremum of the magnitude is approximately 2.26, so $\|G\|_\infty = 2.26$.

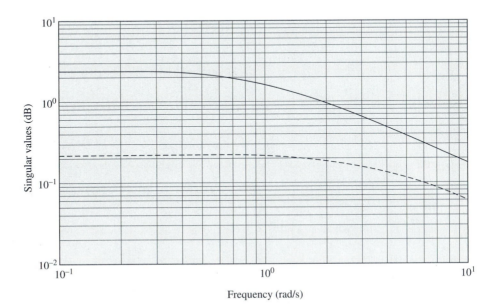

Figure 8.5   Singular values, magnitude plots

8.6 THE CALCULATION OF SYSTEM NORMS

We shall calculate system norms from state-space data, because good algorithms have been devised for this purpose. First let us compute the H^2 norm. Let $g(t)$ be the matrix impulse response corresponding to $G(s)$, i.e.,

$$g(t) = Ce^{At}B.$$

Since A is assumed stable ($G \in H_2$), we may use Parseval's theorem and write Equation 8.27 as

$$\|G\|_2^2 = \mathrm{tr} \int_0^\infty g^T(t)g(t)dt$$

$$= \mathrm{tr} \int_0^\infty B^T C^{A^T t} C^T Ce^{At} B dt$$

or

$$\|G\|_2^2 = \mathrm{tr} B^T L_c B \tag{8.37}$$

where

$$L_c = \int_0^\infty e^{A^T t} C^T Ce^{At} dt. \tag{8.38}$$

From Chapter 7, L_c satisfies the Lyapunov equation,

$$A^T L_c + L_c A = -C^T C. \tag{8.39}$$

Since $\mathrm{tr}\mathbf{xy} = \mathrm{tr}\mathbf{yx}$, we may also use

$$\|G\|_2^2 = \mathrm{tr} \int_0^\infty g(t)g^T(t)dt$$

which leads to

$$\|G\|_2^2 = \mathrm{tr} CL_o C^T \tag{8.40}$$

where

$$AL_o + L_o A^T = -BB^T. \tag{8.41}$$

Example 8.5   A minimal realization for the transfer function of Example 8.3 is

$$\dot{\mathbf{x}} = \begin{bmatrix} 0 & 1 \\ -2 & -3 \end{bmatrix} \mathbf{x} + \begin{bmatrix} 1 & -1 \\ 0 & 1 \end{bmatrix} \mathbf{u}$$

$$\mathbf{y} = \mathbf{x}.$$

Compute the H^2 norm of G.

Solution   The Lyapunov equation, Equation 8.39, is simple enough to be solved by hand. The result is

$$L_c = \begin{bmatrix} \frac{5}{4} & \frac{1}{4} \\ \frac{1}{4} & \frac{1}{4} \end{bmatrix}.$$

From Equation 8.37,

$$\|G\|_2^2 = \text{tr} \begin{bmatrix} 1 & 0 \\ -1 & 1 \end{bmatrix} \begin{bmatrix} \frac{5}{4} & \frac{1}{4} \\ \frac{1}{4} & \frac{1}{4} \end{bmatrix} \begin{bmatrix} 1 & -1 \\ 0 & 1 \end{bmatrix}$$

$$= \frac{9}{4}$$

as expected.

As a check, we may use Equations 8.40 and 8.41. We calculate

$$L_o = \begin{bmatrix} \frac{17}{12} & -1 \\ -1 & \frac{5}{6} \end{bmatrix}$$

$$\|G\|_2 = \text{tr} I L_o I = \frac{9}{4}.$$

To derive an algorithm for the calculation of the H^∞ norm, we need the following results.

Lemma 8.1   Let

$$H = \begin{bmatrix} A & BB^T \\ -C^T C & -A^T \end{bmatrix}$$

where (A, B, C) is a realization of a strictly proper $G(s)$. Then

$$[I - G^T(-s)G(s)]^{-1} = I + [0 \quad B^T](sI - H)^{-1} \begin{bmatrix} B \\ 0 \end{bmatrix}.$$

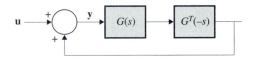

Figure 8.6 Pertaining to Lemma 8.1

Proof: In the block diagram of Figure 8.6, the sensitivity (note the positive feedback) is

$$\mathbf{y} = [I - G^T(-s)G(s)]^{-1}\mathbf{u}. \tag{8.42}$$

Now, the realization of $G(s)$ is

$$\dot{\mathbf{x}}_1 = A\mathbf{x}_1 + B\mathbf{u}$$

$$\mathbf{y} = C\mathbf{x}_1$$

and the transfer function is $C(sI - A)^{-1}B$. The transfer function $G^T(-s)$ is

$$G^T(-s) = [C(-sI - A)^{-1}B]^T$$

$$= B^T[sI - (-A^T)]^{-1}(-C^T)$$

which has the realization

$$\dot{\mathbf{x}}_2 = -A^T\mathbf{x}_2 - C^T\mathbf{u}$$

$$\mathbf{y} = B^T\mathbf{x}_2.$$

The realization for the system of Figure 8.6, concatenating the two state vectors, is

$$\dot{\mathbf{x}}_1 = A\mathbf{x}_1 + B(\mathbf{u} + B^T\mathbf{x}_2)$$

$$\dot{\mathbf{x}}_2 = -A^T\mathbf{x}_2 - C^T(C\mathbf{x}_1)$$

$$\mathbf{y} = \mathbf{u} + B^T\mathbf{x}_2$$

or

$$\begin{bmatrix} \dot{\mathbf{x}}_1 \\ \dot{\mathbf{x}}_2 \end{bmatrix} = \begin{bmatrix} A & BB^T \\ -C^TC & -A^T \end{bmatrix} \begin{bmatrix} \mathbf{x}_1 \\ \mathbf{x}_2 \end{bmatrix} + \begin{bmatrix} B \\ 0 \end{bmatrix}\mathbf{u} \tag{8.43}$$

$$\mathbf{y} = \mathbf{u} + [0 \quad B^T]\begin{bmatrix} \mathbf{x}_1 \\ \mathbf{x}_2 \end{bmatrix}.$$

The input–output relationship for this is

$$\mathbf{y} = \left(I + [0 \quad B^T](sI - H)^{-1}\begin{bmatrix} B \\ 0 \end{bmatrix}\right)\mathbf{u}. \tag{8.44}$$

Comparison of Equations 8.42 and 8.44 proves the result. ∎

Lemma 8.2   Let the system (A, B, C) be stabilizable and detectable. Then, the realization equation, Equation 8.43, may not have unobservable or uncontrollable modes on the j-axis.

Proof: Let αj be an unobservable eigenvalue of H, with $\mathbf{v}$ the corresponding eigenvector. Let $\mathbf{v}$ be partitioned into two n-vectors, $\mathbf{v}_1$ and $\mathbf{v}_2$.
   From Equation 8.43, and because this mode is unobservable,

$$[0 \quad B^T] \begin{bmatrix} \mathbf{v}_1 \\ \mathbf{v}_2 \end{bmatrix} = B^T \mathbf{v}_2 = 0. \tag{8.45}$$

Since $\mathbf{v}$ is an eigenvector,

$$H\mathbf{v} = \begin{bmatrix} A & BB^T \\ -C^T C & -A^T \end{bmatrix} \begin{bmatrix} \mathbf{v}_1 \\ \mathbf{v}_2 \end{bmatrix} = \alpha j \begin{bmatrix} \mathbf{v}_1 \\ \mathbf{v}_2 \end{bmatrix}. \tag{8.46}$$

Combining 8.45 and 8.46,

$$A\mathbf{v}_1 = \alpha j \mathbf{v}_1 \tag{8.47}$$

$$-C^T C \mathbf{v}_1 - A^T \mathbf{v}_2 = \alpha j \mathbf{v}_2. \tag{8.48}$$

We conclude from Equation 8.47 that either (i) $\mathbf{v}_1 = 0$ or (ii) A has an eigenvalue αj, with $\mathbf{v}_1$ as the corresponding eigenvector.
   If $\mathbf{v}_1 = 0$, then, from Equation 8.48,

$$A^T \mathbf{v}_2 = -\alpha j \mathbf{v}_2$$

which, combined with Equation 8.45, shows that A^T (hence A) has an eigenvalue $-\alpha j$, and that mode is uncontrollable. That is contrary to our assumption of stabilizability, so $\mathbf{v}_1 \neq 0$.
   If $\mathbf{v}_1 \neq 0$, then Equation 8.45 shows that A has an eigenvalue αj with $\mathbf{v}_1$ as the corresponding eigenvector. Premultiplying Equation 8.48 by $\mathbf{v}_1^{*T}$,

$$-\mathbf{v}_1^{*T} C^T C \mathbf{v}_1 = \mathbf{v}_1^{*T} (A^T + \alpha j I) \mathbf{v}_2$$
$$= \mathbf{v}_2^T (A + \alpha j I) \mathbf{v}_1^*$$

where the last step uses the fact that a scalar is its own transpose. Now, if αj is an eigenvalue of A, so is $-\alpha j$, and its associated eigenvector is $\mathbf{v}_1^*$. Hence, $A\mathbf{v}_1^* = -\alpha j \mathbf{v}_1^*$ and

$$\mathbf{v}_1^{*T} C^T C \mathbf{v}_1 = \mathbf{v}_2^T (-\alpha j \mathbf{v}_1^* + \alpha j \mathbf{v}_1^*) = 0$$

and

$$\mathbf{v}_1^{*T} C^T C \mathbf{v}_1 = \|C\mathbf{v}_1\|^2 = 0$$

or

$$Cv_1 = 0$$

which shows that the mode αj is not observable, contrary to our detectability assumption. We see that the assumed existence of an unobservable j-axis mode for the realization of Equation 8.43 leads to a contradiction and hence must be rejected. A similar proof goes through if we assume a j-axis uncontrollable mode. ∎

We now come to our central result.

■ **Theorem 8.1** Let $G(s)$ be strictly proper with a stabilizable and detectable realization (A, B, C). Then $\|G\|_\infty < 1$ if, and only if, H has no eigenvalues on the imaginary axis.

Proof: By Lemma 8.1, the poles of $[I - G^T(-s)G(s)]^{-1}$ must be eigenvalues of H. Conversely, because of Lemma 8.2, all j-axis eigenvalues of H will appear as poles of $[I - G^T(-s)G(s)]^{-1}$.

If: Suppose $\|G\|_\infty < 1$. For all ω and complex vectors $\mathbf{v} \neq \mathbf{0}$,

$$
\begin{aligned}
\mathbf{v}^T[I - G^T(-j\omega)G(j\omega)]\mathbf{v} &= \|\mathbf{v}\|^2 - \mathbf{v}^{*^T}G^T(-j\omega)G(j\omega)\mathbf{v} \\
&\geq \{1 - \overline{\sigma}^2[G(j\omega)]\}\|\mathbf{v}\|^2 \\
&\geq (1 - \|G\|_\infty^2)\|\mathbf{v}\|^2 \\
&> 0.
\end{aligned}
$$

This shows that $I - G^T(-j\omega)G(j\omega)$ is Hermitian for all ω; hence, its determinant is strictly positive so that $[I - G^T(-s)G(s)]^{-1}$ has no poles at $s = j\omega$, any ω.

Only if: Suppose $\|G\|_\infty \geq 1$. Since G is strictly proper, $G(j\omega) \to 0$ as $\omega \to \infty$; because of continuity, $\overline{\sigma}[G(j\omega)] = 1$ for some $\omega = \omega_0$. We may then choose a $\mathbf{v}_0$ such that $\mathbf{v}_0^{*^T}G^T(-j\omega_0)G(j\omega_0)\mathbf{v}_0 = 1$ and, following the steps in the "if" part, show that $\det[I - G^T(-j\omega_0)G(j\omega_0)] = 0$. In that case, $[I - G^T(-s)G(s)]^{-1}$ does have a pole at $s = j\omega_0$, and that pole appears as an eigenvalue of H. ∎

In actual fact, we want a test of the statement $\|G\|_\infty < \gamma$. This is achieved by testing $\|\gamma^{-1}G\|_\infty < 1$; to see this, note that

$$\|\gamma^{-1}G\|_\infty = \gamma^{-1}\|G\|_\infty$$

so that

$$\|\gamma^{-1}G\|_\infty < 1 \Rightarrow \|G\|_\infty < \gamma.$$

The realization for $\gamma^{-1}G$ is constructed simply by replacing B with $\gamma^{-1}B$ in the matrix H.

The calculation of $\|G\|_\infty$ is an iterative process, a search for that value of γ such that the matrix $H(\gamma) = \begin{bmatrix} A & \gamma^{-2}BB^T \\ -C^TC & -A^T \end{bmatrix}$ has one or more eigenvalues on the imaginary axis. The procedure can begin with an upper bound for γ, and the search can commence from there. It helps to realize that H is a *sympletic* matrix—it has the property that its eigenvalues are symmetrically located with respect to the j-axis.

Example 8.6   Calculate the ∞-norm for the system of Example 8.4, using the realization of Example 8.5.

Solution   We form the matrix

$$H(\mu) = \begin{bmatrix} A & \mu BB^T \\ -C^TC & -A^T \end{bmatrix} = \begin{bmatrix} 0 & 1 & 2\mu & -\mu \\ -2 & -3 & -\mu & \mu \\ -1 & 0 & 0 & 2 \\ 0 & -1 & -1 & 3 \end{bmatrix}$$

where $\mu = \gamma^{-2}$ has been used. The characteristic polynomial of H is

$$\det(sI - H) = s^4 + s^2(3\mu - 5) + \mu^2 - 21\mu + 4.$$

Note that only even powers of s are present. This leads to roots that are symmetrically located with respect to both the real and the imaginary axes.

For $\mu = 0 (\gamma = \infty)$, the characteristic polynomial is

$$s^4 - 5s^2 + 4$$

with roots $s^2 = +4, +1$, and thus $s = \pm 2, \pm 1$. Indeed, with $\mu = 0$, H is block triangular and its eigenvalues are those of A and $-A^T$, i.e., of A and $-A$; those are $-1, -2, +1$, and $+2$.

Now, H has imaginary-axis eigenvalues when $s^{*^2} < 0$, where s^* is a root of the characteristic polynomial. For $\mu = 0$, s^{*^2} has the values 1 and 4; as μ increases, s^{*^2} crosses zero to become negative. We have $s^{*^2} = 0$ for

$$\mu^2 - 21\mu + 4 = 0$$

or

$$\mu = \frac{21 \pm \sqrt{425}}{2}.$$

We take the smaller of the two values, corresponding to the largest value of γ that results in imaginary-axis eigenvalues. Thus, $\mu = 0.1922$ and $\|G\|_\infty = \gamma = 1/\sqrt{\mu} = 2.281$.

8.7 ROBUST STABILITY

Figure 8.7 shows a simple MIMO feedback loop consisting of a linear $m \times r$ transfer function $G(s)$ and a no-memory $r \times m$ operator Δ that, for the moment, may be nonlinear. The ∞-norm of Δ is defined as

$$\|\Delta\|_\infty = \sup_{x \neq 0} \frac{\|\Delta(x)\|_2}{\|x\|_2}. \tag{8.49}$$

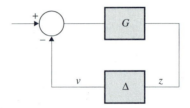

Figure 8.7   Feedback system used in the small-gain theorem

If Δ is linear, i.e., is a matrix, this definition comes down to the one already given. The *small-gain theorem* [3] asserts that sufficient conditions for stability of this loop are:

1. $G(s)$ is stable.

2. $G(s)$ is strictly proper.

3. $\|G\|_\infty \|\Delta\|_\infty < 1$.

Let us now specialize to the case in which Δ is a matrix, with $\overline{\sigma}(\Delta) \leq 1$. From the small-gain theorem, the condition $\|G\|_\infty < 1$ is clearly sufficient to guarantee stability. We now establish that this condition is also necessary, if stability is to be maintained for all Δ of unit norm or less.

Since G is stable, the number of encirclements of the origin by the locus $\det[I + G(j\omega)\Delta]$ must be zero for all admissible Δ. For $\Delta = 0$, $\det I = 1$ for all frequencies, and there are no encirclements. The number of encirclements changes if, for some admissible Δ, the locus moves across the origin for some ω. That happens if, for some $\omega = \omega_0$ and Δ, $\|\Delta\|_\infty \leq 1$, $\det[I + G(j\omega_0)\Delta] = 0$. We show that this is always the case if $\|G\|_\infty \geq 1$.

Suppose $\|G\|_\infty \geq 1$. Then, for some $\omega = \omega_0$, $\overline{\sigma}[G(j\omega_0)] = 1$ and there exists a complex vector $\mathbf{v}_1$ such that (i) $\|\mathbf{v}_1\| = 1$ and (ii) $\mathbf{v}_2 = G(j\omega_0)\mathbf{v}_1$, $\|\mathbf{v}_2\| = 1$. Pick $\Delta = \mathbf{v}_1(\mathbf{v}_2{}^*)^T$. It is not difficult to show that $\lambda_{\max}(\Delta^*\Delta) = 1$, so that $\overline{\sigma}(\Delta) = 1$. Now, form

$$[I + G(j\omega_0)\Delta]\mathbf{v}_2 = \mathbf{v}_2 - G(j\omega_0)\mathbf{v}_1(\mathbf{v}_2{}^*)^T\mathbf{v}_2$$

$$= \mathbf{v}_2 - G(j\omega_0)\mathbf{v}_1$$

$$= 0.$$

This shows that $I + G(j\omega_0)\Delta$ is singular and that its determinant is therefore zero.

To summarize, if $\|G\|_\infty \geq 1$, there always exists a Δ, $\|\Delta\|_\infty \leq 1$, that will yield an unstable closed loop; if $\|G\|_\infty < 1$, the loop is always stable if G is stable and $\|\Delta\|_\infty \leq 1$.

We may use this result to obtain sufficient conditions for robust stability in several situations. We shall represent the uncertainty as $\Delta W(s)$, where Δ is a constant matrix such that $\|\Delta\|_\infty \leq 1$; the function $W(s)$ is used to incorporate the frequency information concerning the uncertainty.

We consider the following:

1. Multiplicative uncertainty referred to the input (Figure 8.8)
2. Multiplicative uncertainty referred to the output (Figure 8.9)
3. Additive uncertainty (Figure 8.10)

In each case, we reduce the diagram to that of Figure 8.7; this requires calculation of the transfer function $G(s)$ to which Δ is connected—i.e., the transfer functions from $\mathbf{v}$ to $\mathbf{z}$, with Δ taken out.

The reader is invited to show that:

- For the input multiplicative uncertainty,

$$G(s) = -W(I + FP)^{-1}FP = -WT. \tag{8.50}$$

- For the output multiplicative uncertainty,

$$G(s) = -W(I + PF)^{-1}PF. \tag{8.51}$$

- For the additive uncertainty,

$$G(s) = -W(I + FP)^{-1}F. \tag{8.52}$$

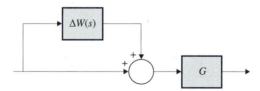

Figure 8.8 Illustration of a weighted input multiplicative uncertainty

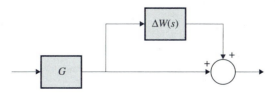

Figure 8.9 Illustration of a weighted output multiplicative uncertainty

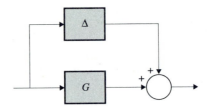

Figure 8.10 Illustration of an additive uncertainty

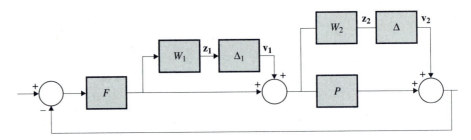

Figure 8.11 A system with both multiplicative and additive uncertainties

Simultaneous perturbations are handled by using the fact that several inputs (outputs) can be gathered into one multivariable input (outputs). Consider, for instance, Figure 8.11, featuring a multiplicative input uncertainty and an additive uncertainty. It can be shown straightforwardly that

$$\mathbf{z}_1 = -W_1(I + FD)^{-1}(FP\mathbf{v}_1 + F\mathbf{v}_2)$$

$$\mathbf{z}_2 = W_2(I + FP)^{-1}(\mathbf{v}_1 - F\mathbf{v}_2).$$

Therefore, we may view the system as in Figure 8.7, where

$$G = \begin{bmatrix} -W_1(I + FP)^{-1}FP & -W_1(I + FP)^{-1}F \\ W_2(I + FP)^{-1} & -W_2(I + FP)^{-1}F \end{bmatrix}. \tag{8.53}$$

The uncertainty block for this problem is

$$\Delta_T = \begin{bmatrix} \Delta_1 & 0 \\ 0 & \Delta_2 \end{bmatrix}$$

whose norm we need to bound.

To do this, assume $\mathbf{v}_1$ and $\mathbf{v}_2$ to be of unit Euclidean length and of dimensions equal to the numbers of columns of Δ_1 and Δ_2, respectively. Given $\alpha, 0 \leq \alpha \leq 1$, the vector $\mathbf{v}_T = [\, \frac{\sqrt{\alpha}}{\sqrt{1-\alpha}} \, \begin{smallmatrix} \mathbf{v}_1 \\ \mathbf{v}_2 \end{smallmatrix} \,]$ is of unit length. We form

$$(\mathbf{v}_T{}^*)^T \Delta_T{}^* \Delta_T{}^* \mathbf{v}_T = \alpha(\mathbf{v}_1{}^*)^T \Delta_1 \mathbf{v}_1 + (1 - \alpha)(\mathbf{v}_2{}^*)^T \Delta_2{}^* \Delta_2 \mathbf{v}_2. \tag{8.54}$$

Given $\|\Delta_1\|_\infty = \|\Delta_2\|_\infty = 1$, the RHS is maximized for any α between 0 and 1 by choosing $\mathbf{v}_1$ and $\mathbf{v}_2$ such that

$$(\mathbf{v}_1{}^*)^T \, \Delta_1{}^* \Delta_2 \mathbf{v}_1 = (\mathbf{v}_2{}^*)^T \Delta_2{}^* \, \Delta_2 \mathbf{v}_2 = 1$$

in which case the RHS of Equation 8.54 is 1. Thus, $\|\Delta_T\|_\infty \leq 1$ if $\|\Delta_1\|_\infty$, $\|\Delta_2\|_\infty \leq 1$.

8.8 DEFINITION OF THE DESIGN PROBLEM

We refer to Figure 8.12, where the inputs $\mathbf{u}$ and $\mathbf{w}$ and the outputs $\mathbf{y}$ and $\mathbf{z}$ are vectors. The vector $\mathbf{w}$ groups exogenous signals such as disturbance, set point, or test inputs. The vector $\mathbf{z}$ represents performance variables that must, in some sense, be kept small; more precisely, the design objective is to keep the norm of the transmission $T_{\mathbf{wz}}$ small. The vector $\mathbf{u}$ contains the control inputs, and $\mathbf{y}$ the measurements used for feedback purposes. It is easy to identify the vectors $\mathbf{u}$ and $\mathbf{y}$, since they are given at the outset. The problem definition consists in the identification of the inputs $\mathbf{w}$ and the outputs $\mathbf{z}$. Because specifications for both performance and robustness are given in terms of the weighted norms of certain transmissions, we must locate input–output pairs that have the required transmissions and form $\mathbf{w}$ and $\mathbf{z}$ from the unions of all required inputs and outputs, respectively.

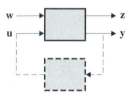

Figure 8.12 The basic design configuration

Example 8.7 The most common problem definition is probably the one illustrated in Figure 8.13. The transmissions of interest are the sensitivity $S(s)$, the set point–to–input transmission $U(s)$, and the transmission of the system connected to the uncertainty matrix Δ (from $\mathbf{r}$ to $\mathbf{q}$, with the Δ-block removed). Weight functions $W_1(s)$ and $W_2(s)$ are assigned to S and U, respectively. Define an input vector $\mathbf{w}$ and an output vector $\mathbf{z}$ such that all transmissions of interest are present in the matrix transmission from $\mathbf{w}$ to $\mathbf{z}$.

Solution Since S is the transmission from $\mathbf{y}_d$ to $\mathbf{e}$ and U is the transmission from $\mathbf{y}_d$ to $\mathbf{u}$, the outputs $\mathbf{z}_1$ and $\mathbf{z}_2$ are as shown. Clearly, $\mathbf{z}_1 = W_1 S \mathbf{y}_d$ and $\mathbf{z}_2 = W_2 U \mathbf{y}_d$. We could also include $\mathbf{r}$ among the inputs and $\mathbf{q}$ among the outputs; it was shown, however, that the transmission from one side of the input multiplicative uncertainty to the other is T, the complementary sensitivity. Since this is also the transmission from

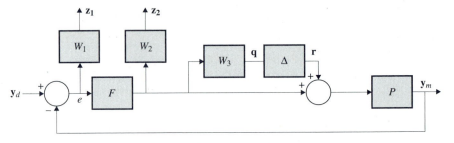

Figure 8.13   Setup for the solution of the design problem

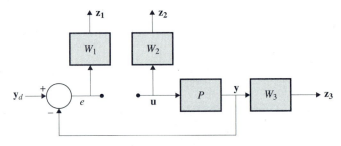

Figure 8.14   Design setup with weighting functions

$\mathbf{y}_d$ to $\mathbf{y}_m$, we need only run $\mathbf{y}_m$ through W_3 to generate $W_3 T$. Thus, we define

$$\mathbf{w} = \mathbf{y}_d, \qquad \mathbf{z} = \begin{bmatrix} \mathbf{z}_1 \\ \mathbf{z}_2 \\ \mathbf{z}_3 \end{bmatrix}$$

as shown in Figure 8.14.

The solution algorithms will minimize or bound $\|T_{\mathbf{wz}}\|_k$, with $k = 2$ or ∞. In the preceding example,

$$\|T_{\mathbf{wz}}\| = \left\| \begin{bmatrix} W_1 S \\ W_2 U \\ W_3 T \end{bmatrix} \right\|.$$

The quantities of interest are $\|W_1 S\|$, $\|W_2 U\|$, and $\|W_3 T\|$; we must therefore know how these quantities are related to $\|T_{\mathbf{wz}}\|$. To proceed, we need the following result.

■ **Theorem 8.2**   Let $T(s)$ be decomposed into an $n \times m$ array of submatrices $T_{ij}(s)$. Then

$$\max_{ij}[\|T_{ij}\|_2] \le \|T\|_2 \le \sqrt{mn} \max_{ij}[\|T_{ij}\|_2]. \tag{8.55}$$

The same result holds for the ∞-norm.

Proof: For the 2-norm, it is easy to show that

$$\mathrm{tr}T^*T = \sum_{ij} \mathrm{tr}T_{ij}{}^*T_{ij}.$$

Since all the terms in the same are positive,

$$\mathrm{tr}T^*T \geq \max_{ij}[\mathrm{tr}T_{ij}{}^*T_{ij}].$$

Also,

$$\mathrm{tr}T^*T \leq nm \max_{ij}[\mathrm{tr}T_{ij}{}^*T_{ij}]$$

so that

$$\max_{ij}[\mathrm{tr}T_{ij}{}^*T_{ij}] \leq \mathrm{tr}T^*T \leq nm \max_{ij}[\mathrm{tr}T_{ij}{}^*T_{ij}]. \qquad \textbf{(8.56)}$$

If T is a function of $j\omega$, Equation 8.56 also holds for the Parseval integral over frequency, since it holds pointwise. Taking square roots everywhere yields the desired result.

For the ∞-norm, we define a vector

$$\mathbf{v}^T = [\mathbf{v}_1{}^T\mathbf{v}_2{}^T \ldots \mathbf{v}_m{}^T]$$

where the dimension of $\mathbf{v}_j$ is the column dimension of T_{ij}. Then

$$\|T\mathbf{v}\|_2 = \left\| \begin{bmatrix} T_{11}\mathbf{v}_1 + T_{12}\mathbf{v}_2 + \cdots + T_{1m}\mathbf{v}_m \\ T_{21}\mathbf{v}_1 + \quad\quad \cdots \quad\quad + T_{2m}\mathbf{v}_m \\ \vdots \quad\quad\quad \vdots \quad\quad\quad \vdots \\ T_{m1}\mathbf{v}_1 + \quad\quad \cdots \quad\quad + T_{mm}\mathbf{v}_m \end{bmatrix} \right\|.$$

We can select the T_{ij} with the largest ∞-norm, choose the $\mathbf{v}_j$ of unit norm that maximizes $\|T_{ij}\mathbf{v}_j\|_2$, and set the remainder of the vector $\mathbf{v}$ to zero. In that case, $\|T\mathbf{v}\|_2 = \|T_{ij}\|_\infty$; this value is therefore a lower bound, which proves the left inequality (Equation 8.55).

To obtain the upper bound, we write

$$\|T\mathbf{v}\|_2{}^2 = \sum_{i=1}^{n} \|T_{i1}\mathbf{v}_1 + T_{i2}\mathbf{v}_2 + \cdots + T_{im}\mathbf{v}_m\|_2{}^2$$

and

$$\sup_{\|\mathbf{v}\|=1} \|T_{i1}\mathbf{v}_1 + \cdots + T_{im}\mathbf{v}_m\|^2 \leq m \max_{j} \overline{\sigma}^2(T_{ij}).$$

Since there are n such terms,

$$\|T\mathbf{v}\|_2^2 \leq nm \max_{ij} \bar{\sigma}^2(T_{ij}).$$

The theorem follows if we take square roots and note that the result holds at all frequencies. ∎

The preceding theorem shows that the norm of each submatrix transfer function is bounded above by $\|T_{\mathbf{wz}}\|$, so that making $\|T_{\mathbf{wz}}\|$ small forces all components to be at least as small. The lower bound applies only to the submatrix of maximum norm; it is therefore possible for some submatrices to be appreciably smaller in norm than $(1/\sqrt{mn})\|T_{\mathbf{wz}}\|$. The design can proceed by trial and error. For example, in Example 8.7, if it is required that $\|W_2 U\| \leq 1$ and the algorithm actually returns the value $\|W_2 U\| = 0.2$, another try can be made with a scaled-down W_2 (or, equivalently, an increased upper bound on $\|W_2 U\|$).

8.9 THE AUGMENTED STATE-SPACE MODEL

Current solution algorithms are based on the state-space representation

$$\dot{\mathbf{x}} = A\mathbf{x} + B_1\mathbf{w} + B_2\mathbf{u}$$

$$\mathbf{z} = C_1\mathbf{x} + D_{11}\mathbf{w} + D_{12}\mathbf{u}$$

$$\mathbf{y} = C_2\mathbf{x} + D_{21}\mathbf{w} + D_{22}\mathbf{u}. \tag{8.57}$$

This representation combines the system model and the models of the various weight functions, as shown by the following example.

Example 8.8   In Example 8.7, let the following be given:

$$P(s) = \frac{1}{s^2 + 0.2s + 2}; \qquad W_1(s) = \frac{s+1}{s(s+10)}$$

$$W_2(s) = 1; \qquad W_3(s) = 2 \times 10^{-3}(s+10)^2.$$

Write a state-space representation in the form of Equation 8.57 for the system shown in Figure 8.14 (the same as in Figure 8.13, but with the controller removed). The measured output is taken to be e in order that the result be a standard 1-DOF controller.

Solution   Because no state-space model can be written for $W_3(s)$, we look upon the combination of P and W_3 as one system, with an input u and outputs y and z_3.

Since

$$\frac{z_3}{u} = \frac{2 \times 10^{-3}(s^2 + 20s + 100)}{s^2 + .2s + 1} = 2 \times 10^{-3}\left[1 + \frac{19.8s + 99}{s^2 + .2s + 1}\right]$$

the controllable canonical form is

$$\begin{bmatrix} \dot{x}_1 \\ \dot{x}_2 \end{bmatrix} = \begin{bmatrix} 0 & 1 \\ -1 & -0.2 \end{bmatrix}\begin{bmatrix} x_1 \\ x_2 \end{bmatrix} + \begin{bmatrix} 0 \\ 1 \end{bmatrix}u$$

$$y = x_1$$

$$e = y_d - x_1$$

$$z_3 = 2 \times 10^{-3}[99x_1 + 19.8x_2] + 2 \times 10^{-3}u.$$

Now,

$$\frac{z_1}{e} = W_1(s) = \frac{s + 1}{s^2 + 10.01s + 0.1}$$

so that

$$\begin{bmatrix} \dot{x}_3 \\ \dot{x}_4 \end{bmatrix} = \begin{bmatrix} 0 & 1 \\ -.1 & -10.01 \end{bmatrix}\begin{bmatrix} x_3 \\ x_4 \end{bmatrix} + \begin{bmatrix} 0 \\ 1 \end{bmatrix}(y_d - x_1).$$

$$z_1 = x_3 + x_4.$$

Putting these expressions together, with $w = y_d$,

$$\dot{\mathbf{x}} = \begin{bmatrix} 0 & 1 & 0 & 0 \\ -1 & -0.2 & 0 & 0 \\ 0 & 0 & 0 & 1 \\ -1 & 0 & -.1 & -10.01 \end{bmatrix}\mathbf{x} + \begin{bmatrix} 0 \\ 0 \\ 0 \\ 1 \end{bmatrix}y_d + \begin{bmatrix} 0 \\ 1 \\ 0 \\ 0 \end{bmatrix}u$$

$$y = \begin{bmatrix} -1 & 0 & 0 & 0 \end{bmatrix}\mathbf{x} + y_d$$

$$\mathbf{z} = \begin{bmatrix} 0 & 0 & 1 & 1 \\ 0 & 0 & 0 & 0 \\ .198 & .0396 & 0 & 0 \end{bmatrix}\mathbf{x} + \begin{bmatrix} 0 \\ 1 \\ 2 \times 10^{-3} \end{bmatrix}u.$$

8.10 H^2 SOLUTION

For the realization of Equation 8.57, the problem is to minimize $\|T_{\mathbf{wz}}(s)\|_2$. The following assumptions are required:

1. (A, B_1) and (A, B_2) are stabilizable.
2. (C_1, A) and (C_2, A) are detectable.
3. $D_{12}{}^T D_{12}$ and $D_{21} D_{21}{}^T$ are nonsingular.
4. $D_{11} = 0$.

We write $T_{\mathbf{wz}}$ by columns, as

$$T_{\mathbf{wz}} = [(T_{\mathbf{wz}})_1 (T_{\mathbf{wz}})_2 \cdots (T_{\mathbf{wz}})_m].$$

To calculate the H^2-norm, we need to form

$$\mathrm{tr} T_{\mathbf{wz}}{}^T(-s) T_{\mathbf{wz}}(s) = \sum_{i=1}^{m} (T_{\mathbf{wz}})_i{}^T(-s)(T_{\mathbf{wz}})_i(s). \tag{8.58}$$

The square of the H^2-norm is obtained by integrating the RHS of Equation 8.58 over the j-axis. We may use Parseval's theorem to translate this into the time-domain integral

$$\|T_{\mathbf{wz}}\|_2{}^2 = \sum_{i=1}^{m} \int_0^{\infty} \|(T_{\mathbf{wz}})_i(t)\|^2 dt \tag{8.59}$$

where $(T_{\mathbf{wz}})_i(t) = \mathcal{L}^{-1}[(T_{\mathbf{wz}})_i(s)]$ is the response $\mathbf{z}(t)$ to a unit impulse in $w_i(t)$. Now, "hitting" the system of Equation 8.57 with such an impulse is the same as giving it an initial state $\mathbf{b}_i$, the ith column of the matrix B_1. Let $\mathbf{z}(\mathbf{x}_0, t)$ be the response to an initial state $\mathbf{x}_0$. We may then write Equation 8.59 as

$$\|T_{\mathbf{wz}}\|_2{}^2 = \sum_{i=1}^{m} \int_0^{\infty} \|\mathbf{z}(\mathbf{b}_i, t)\|^2 dt$$

$$= \sum_{i=1}^{m} \int_0^{\infty} \|C_1 \mathbf{x}(\mathbf{b}_i, t) + D_{12}\mathbf{u}(\mathbf{b}_i, t)\|^2 dt. \tag{8.60}$$

where the notation for $\mathbf{x}$ and $\mathbf{u}$ parallels that for $\mathbf{z}$. The problem, then, is to minimize the RHS of Equation 8.60, with

$$\dot{\mathbf{x}} = A\mathbf{x} + B_2\mathbf{u} \tag{8.61}$$

$$\mathbf{y} = C_2\mathbf{x} + D_{21}\mathbf{w} + D_{22}\mathbf{u}. \tag{8.62}$$

Let us examine the case $\mathbf{y} = \mathbf{x}$, i.e., full state feedback, and expand the term corresponding to $i = 1$ in the RHS of Equation 8.60. We write

$$J_1 = \int_0^\infty [\mathbf{x}^T(\mathbf{b}_1, t)C_1{}^TC_1\mathbf{x}(\mathbf{b}_1, t) + \mathbf{x}^T(\mathbf{b}_1, t)C_1{}^TD_{12}\mathbf{u}(\mathbf{b}_1, t)$$

$$+ \mathbf{u}^T(\mathbf{b}_1, t)D_{12}{}^TC_1\mathbf{x}(\mathbf{b}_1, t) + \mathbf{u}^T(\mathbf{b}_1, t)D_{12}{}^TD_{12}\mathbf{u}(\mathbf{b}_1, t)]dt. \quad \textbf{(8.63)}$$

Equations 8.61 and 8.63 form an LQ problem, even with the inclusion of cross terms (mixed $\mathbf{x}$ and $\mathbf{u}$) in J_1. The solution of that problem is of the form $\mathbf{u} = -K\mathbf{x}$ and is *independent of the initial state*. The *same* control law minimizes every term of the sum in Equation 8.60, and hence minimizes the sum itself.

To remove the cross terms, define a new input $\mathbf{v}$,

$$\mathbf{v} = \mathbf{u} + (D_{12}{}^TD_{12})^{-1}D_{12}{}^TC_1\mathbf{x}. \quad \textbf{(8.64)}$$

It is easy to show that

$$\dot{\mathbf{x}} = [A - B_2(D_{12}{}^TD_{12})^{-1}D_{12}{}^TC_1]\mathbf{x} + B_2\mathbf{v} \quad \textbf{(8.65)}$$

$$J_1 = \int_0^\infty \{\mathbf{x}^TC_1{}^T[I - D_{12}(D_{12}{}^TD_{12})^{-1}D_{12}{}^T]C_1\mathbf{x} + \mathbf{v}^T(D_{12}{}^TD_{12})\mathbf{v}\}dt. \quad \textbf{(8.66)}$$

Equations 8.65 and 8.66 present a standard LQ problem. Its solution is calculated by the methods of Chapter 7. With $\mathbf{v} = -\kappa\mathbf{x}$,

$$\mathbf{u} = -[\kappa + (D_{12}{}^TD_{12})^{-1}D_{12}{}^TC_1]\mathbf{x} = -K\mathbf{x}. \quad \textbf{(8.67)}$$

It is usually not necessary to work out these transformations, since most packages will solve the LQ problem for an integrand with cross terms.

If the state is not directly available, it is necessary to use a state estimator. We assume a control law $\mathbf{u} = -\kappa\hat{\mathbf{x}}$, simply replacing the state with its estimate in Equation 8.67. The problem is to choose the observer gain G so as to minimize $\|T_{\mathbf{wz}}\|_2$. We define

$$\tilde{\mathbf{x}} = \mathbf{x} - \hat{\mathbf{x}}$$

and

$$\nu = \mathbf{u} + K\mathbf{x} = K\tilde{\mathbf{x}}. \quad \textbf{(8.68)}$$

We write

$$\dot{\mathbf{x}} = A\mathbf{x} + B_1\mathbf{w} + B_2\mathbf{u}$$

$$= (A - B_2K)\mathbf{x} + B_1\mathbf{w} + B_2\nu \quad \textbf{(8.69)}$$

$$\mathbf{z} = C_1\mathbf{x} + D_{12}\mathbf{u}$$

$$= (C_1 - D_{12}K)\mathbf{x} + D_{12}\nu. \quad \textbf{(8.70)}$$

Also,

$$\dot{\mathbf{x}} = A\widehat{\mathbf{x}} + B_2\mathbf{u} + G(\mathbf{y} - D_{22}\mathbf{u} - C_2\widehat{\mathbf{x}})$$

which leads to

$$\dot{\widetilde{\mathbf{x}}} = (A - GC_2)\widetilde{\mathbf{x}} + (B_1 - GD_{21})\mathbf{w}. \tag{8.71}$$

From Equations 8.69 and 8.70, we deduce that $\mathbf{z}$ is of the form

$$\mathbf{z} = T_1\mathbf{w} + T_2\nu.$$

From Equations 8.68 and 8.71, we have

$$\nu = T_{w\nu}\mathbf{w}$$

so that

$$\mathbf{z} = (T_1 + T_2T_{\mathbf{w}\nu})\mathbf{w}. \tag{8.72}$$

Clearly,

$$T_{\mathbf{wz}}{}^T(-s)T_{\mathbf{wz}}(s) = T_1{}^T(-s)T_1(s) + T_1{}^T(-s)T_2(s)T_{\mathbf{w}\nu}(s)$$
$$+ T_{\mathbf{w}\nu}{}^T(-s)T_2{}^T(-s)T_1(s) + T_{\mathbf{w}\nu}{}^T(-s)T_2{}^T(-s)T_2(s)T_{\mathbf{w}\nu}(s). \tag{8.73}$$

Key results in Doyle et al. [4] are that (i) $T_2{}^T(-s)T_2(s) = I$ and (ii) the Parseval integrals of the two middle terms in Equation 8.73 are zero. Therefore,

$$\|T_{\mathbf{wz}}\|_2{}^2 = \|T_1\|_2{}^2 + \|T_{\mathbf{w}\nu}\|^2. \tag{8.74}$$

Note that T_1 is the $\mathbf{w}$-to-$\mathbf{z}$ transmissions of the system

$$\dot{\mathbf{x}} = (A - B_2K)\mathbf{x} + B_1\mathbf{w}$$
$$\mathbf{z} = (C_1 - D_{12}K)\mathbf{x} \tag{8.75}$$

which is precisely the $T_{\mathbf{wz}}$ for the case of full state feedback. This transmission is not a function of G, so the minimization is carried over the second term in Equation 8.74.

From Equations 8.68 and 8.71,

$$T_{\mathbf{w}\nu} = K(sI - A + GC_2)^{-1}(B_1 - GD_{21})$$

and

$$\text{tr}T_{\mathbf{w}\nu}{}^T(-s)T_{\mathbf{w}\nu}(s) =$$
$$\text{tr}(B_1 - GD_{21})^T(-sI - A + GC_2)^{-T}K^TK \cdot$$
$$(sI - A + GC_2)^{-1}(B_1 - GD_{21}).$$

The fact that $\text{tr}AB = \text{tr}BA$ allows us to write

$$\text{tr}T_{\mathbf{w}\nu}{}^T(-s)T_{\mathbf{w}\nu}(s) =$$
$$\text{tr}K(sI - A + GC_2)^{-1}(B_1 - GD_{21}) \cdot$$
$$(B_1 - GD_{21})^T(sI - A + GC_2)^{-T}K^T.$$

Because $\text{tr}A^T = \text{tr}A$,

$$\text{tr}T_{\mathbf{w}\nu}{}^T(-s)T_{\mathbf{w}\nu}(s) =$$
$$\text{tr}K(-sI - A + GC_2)^{-T}(B_1 - GD_{21}) \cdot$$
$$(B_1 - GD_{21})^T(sI - A + GC_2)^{-T}K^T. \tag{8.76}$$

For the case of full state feedback, the corresponding quantity (see Equation 8.75) was

$$\text{tr}T_{\mathbf{wz}}{}^T(-s)T_{\mathbf{wz}}(s) =$$
$$\text{tr}B_1{}^T(-sI - A + B_2K)^{-T}(C_1 - D_{12}K)^T \cdot$$
$$(C_1 - D_{12}K)(sI - A + B_2K)^{-1}B_1. \tag{8.77}$$

It is clear that Equation 8.76 is obtained from Equation 8.77 by making the following substitutions: K^T for B_1, A^T for A, G^T for K, $C_2{}^T$ for B_2, $B_1{}^T$ for C_1, and $D_{21}{}^T$ for D_{12}. We solve for G as we solved for K, as the results of an LQ problem. Just as K did not depend on B_1, so G is independent of K.

The solution process can be summarized as follows. The system is

$$\dot{\mathbf{x}} = A\mathbf{x} + B_1\mathbf{w} + B_2\mathbf{u}$$
$$\mathbf{z} = C_1\mathbf{x} + D_{12}\mathbf{u}$$
$$\mathbf{y} = C_2\mathbf{x} + D_{21}\mathbf{w} + D_{22}\mathbf{u}.$$

1. **The control gain.** Solve the LQ control problem for

$$\dot{\mathbf{x}} = A\mathbf{x} + B_2\mathbf{u}$$

$$J = \int_0^\infty [\mathbf{x}^T \mathbf{u}^T] \begin{bmatrix} C_1{}^T C_1 & C_1{}^T D_{12} \\ D_{12}{}^T C_1 & D_{12}{}^T D_{12} \end{bmatrix} \begin{bmatrix} \mathbf{x} \\ \mathbf{u} \end{bmatrix} dt$$

and obtain the gain matrix K.

2. **The filter.** Solve the LQ estimation problem (Kalman filter) for

$$\dot{\mathbf{x}} = A\mathbf{x} + \mathbf{w}$$

$$\mathbf{y} = C_2\mathbf{x} + \mathbf{v}$$

with the noise intensity matrix

$$E\left\{ \begin{bmatrix} \mathbf{w} \\ \mathbf{v} \end{bmatrix} [\mathbf{w}^T \mathbf{v}^T] \right\} = \begin{bmatrix} B_1 B_1{}^T & B_1 D_{21}{}^T \\ D_{21} B_1{}^T & D_{21} D_{21}{}^T \end{bmatrix}$$

and obtain the filter gain G.

3. **The controller.**

$$\dot{\widehat{\mathbf{x}}} = A\widehat{\mathbf{x}} + B_2\mathbf{u} + G(\mathbf{y} - D_{22}\mathbf{u} - C_2\widehat{\mathbf{x}})$$

or

$$\dot{\widehat{\mathbf{x}}} = (A - B_2 K - G C_2 + G D_{22} K)\widehat{\mathbf{x}} + G\mathbf{y}$$

$$\mathbf{u} = -K\widehat{\mathbf{x}}.$$

4. **The optimal norm.**

$$\min \|T_{\mathbf{wz}}\|_2{}^2 = \|T_c\|_2{}^2 + \|T_f\|_2{}^2$$

where T_c is the transmission of

$$\dot{\mathbf{x}} = (A - B_2 K)\mathbf{x} + B_1\mathbf{w}$$

$$\mathbf{z} = (C_1 - D_{12} K)\mathbf{x}$$

and T_f is the input–output transmission of

$$\dot{\widetilde{\mathbf{x}}} = (A - G C_2)\widetilde{\mathbf{x}} + (B_1 - G D_{21})\mathbf{w}$$

$$\nu = K\widetilde{\mathbf{x}}.$$

8.11 H^∞ SOLUTION

One important difference between the H^2 and H^∞ solutions is that, for the latter, we look not for the optimal solution but for a solution (not unique) for which $\|T_{\mathbf{wz}}\|_\infty < \gamma$. A search for the minimum is carried out simply by decreasing γ until a solution fails to exist.

The solution is taken from [4, 5]. It requires more technical machinery than we can display here, so proofs will not be given. The starting point is the Hamiltonian matrix,

$$H = \begin{bmatrix} A & -R \\ -Q & -A^T \end{bmatrix}$$

with Q and R symmetric but not necessarily positive definite. All submatrices in H are of dimensions $n \times n$. The matrix H is symplectic and has the property that its eigenvalues are located symmetrically with respect to the j-axis as well as the real axis.

We assume that H has no eigenvalues on the j-axis; this assumption is called the *stability* condition. We form the $2n \times n$ matrix X with columns from the eigenvectors and generalized eigenvectors of H that correspond to LHP eigenvalues; for complex conjugate pairs, we use the real and imaginary parts. The resulting matrix is partitioned into $n \times n$ matrices X_1 and X_2 as follows:

$$X = \begin{bmatrix} X_1 \\ X_2 \end{bmatrix}.$$

If X_1 is nonsingular, H is said to satisfy the *complementarity* condition. If H has no j-axis eigenvalues and X_1^{-1} exists, then $P = X_2 X_1^{-1}$ satisfies the Riccati equation,

$$A^T P + PA - PRP + Q = 0$$

and, in addition, $A - RP$ is a stable matrix. The converse is not true: P may satisfy the Riccati equation and $A - RP$ may be stable without the complementarity and stability conditions being satisfied.

If $R > 0$ and $Q \geq 0$, a square-root method can be used to express R as BB^T, and the Riccati equation is the familiar one. If, in addition, (A, B) is stabilizable and $(Q^{1/2}, A)$ is detectable, then the complementarity and stability conditions can be shown to hold.

Doyle et al., the major reference [4], gives a solution under the following simplifying assumptions:

$$D_{11} = D_{22} = 0$$
$$D_{12}{}^T D_{12} = D_{21} D_{21}{}^T = I$$
$$B_1 D_{21}{}^T = D_{12}{}^T C_1 = 0.$$

These assumptions can be removed and are in computer algorithms. Since we shall not be calculating the solution by hand, we shall use these assumptions to simplify the presentation.

The H_∞ solution revolves around the two Hamiltonian matrices

$$H_c = \begin{bmatrix} A & \gamma^2 B_1 B_1^T - B_2 B_2^T \\ -C_1^T C_1 & -A^T \end{bmatrix} \tag{8.78}$$

and

$$H_f = \begin{bmatrix} A^T & \gamma^2 C_1^T C_1 - C_2^T C_2 \\ -B_1 B_1^T & -A \end{bmatrix}. \tag{8.79}$$

The technical assumptions are:

(i) (A, B_2) is stabilizable and (C_2, A) is detectable.

(ii) $D_{12}^T D_{12}$ and $D_{21} D_{21}^T$ are nonsingular.

In Doyle et al. [4], a third assumption is:

(iiia) (A, B_1) is stabilizable and (C_1, A) is detectable.

In [5], the third assumption is:

(iiib) $\begin{bmatrix} A - j\omega I & B_2 \\ C_1 & D_{12} \end{bmatrix}$ and $\begin{bmatrix} A^T - j\omega I & C_2^T \\ B_1^T & D_{21}^T \end{bmatrix}$ have full column rank for all ω.

Both (iiia) and (iiib) disallow uncontrollable modes of (A, B_1) and unobservable modes of (C_1, A) on the imaginary axis. However, (iiib) would allow open RHP uncontrollable or unobservable modes, whereas (iiia) would not. On the other hand, (iiib) states that (A, B_2, C_1, D_{12}) and $(A^T, C_2^T, B_1^T, D_{21}^T)$ shall have no transmission zeros on the j-axis, whereas (iiia) has nothing direct to say about zeros.

The theorem is the same in both Doyle et al. [4] and The Mathworks [5].

■ **Theorem 8.3**   There exists an admissible controller such that $\|T_{\mathbf{wz}}\|_\infty < \gamma$ if, and only if:

(i) H_c and H_f both satisfy the complementarity and stability conditions.

(ii) The Riccati-equation solutions P_c and P_f associated with H_c and H_f are positive semidefinite;

(iii) The spectral radius (i.e., maximum eigenvalue) $\rho(P_c P_f) < \gamma^{-2}$.

When these conditions hold, one such controller is

$$\dot{\widehat{\mathbf{x}}} = A_c \widehat{\mathbf{x}} + G_c \mathbf{y}$$

$$\mathbf{u} = -K_c \widehat{\mathbf{x}} \tag{8.80}$$

where

$$A_c = A + (\gamma^2 B_1 B_1^T - B_2 B_2^T) P_c - (I - \gamma^2 P_f P_c)^{-1} P_f C_2^T C_2 \quad \text{(8.81)}$$

$$G_c = (I - \gamma^2 P_f P_c)^{-1} P_f C_2^T \quad \text{(8.82)}$$

$$K_c = B_2^T P_c. \quad \text{(8.83)}$$

■

◆ ◆ ◆ **REMARK**

1. As $\gamma \to 0$, the solution tends to the H^2 solution. Therefore, a solution always exists for small enough γ under the technical conditions of the previous section.

2. The difference between this and standard Riccati-equation solutions is that the 12 terms of H_c and H_f may not be negative definite. If they are, then the complementarity and stability properties are satisfied.

3. The design goals are stated not in terms of $\|T_{\mathbf{wz}}\|_\infty$ but in terms of the ∞-norms of submatrices of $T_{\mathbf{wz}}$. The design process is iterative and requires attaching adjustable weight parameters to the submatrices. If the ∞-norm of a particular submatrix exceeds the specified limit, its weight is increased in the following iteration; conversely, a weight may be decreased if the ∞-norm is well below its target bound. ◆

Example 8.9   As in Example 8.8, let

$$P(s) = \frac{1}{s^2 + 0.2s + 1}; \qquad W_1(s) = \frac{s+1}{(s+.01)(s+10)}$$

$$W_2(s) = 1; \qquad W_3(s) = 2 \times 10^{-3}(s+10)^2.$$

The design objectives are:

(i) $|T_{y_d e}(j\omega) W_1(j\omega)| \le 0.1$, all ω.

(ii) $|T_{y_d u}(j\omega)| \le 1$, all ω.

(iii) $|T_{y_d y}(j\omega) W_3(j\omega)| \le 1$, all ω.

If possible, obtain H^2 and H^∞ controllers that meet those specifications.

Solution   The output of W_1, driven by e, is z_1; therefore, specification (i) is $|T_{y_d z_1}(j\omega)| \le 0.1$. We redefine z_1 by multiplying it by 10, to normalize all three specifications to an RHS of 1.

Since $z_2 = u$ and $z_3 = W_3 y$, we have

$$\|T_{y_d z_i}\|_\infty \le 1, \qquad i = 1, 2, 3.$$

From Example 8.8, the state description is

$$A = \begin{bmatrix} 0 & 1 & 0 & 0 \\ -1 & -0.2 & 0 & 0 \\ 0 & 0 & 0 & 1 \\ -1 & 0 & -0.1 & -10.01 \end{bmatrix}; \quad B_1 = \begin{bmatrix} 0 \\ 0 \\ 0 \\ 1 \end{bmatrix}; \quad B_2 = \begin{bmatrix} 0 \\ 1 \\ 0 \\ 0 \end{bmatrix}$$

$$C_1 = \begin{bmatrix} 0 & 0 & 1 & 1 \\ 0 & 0 & 0 & 0 \\ .198 & .0396 & 0 & 0 \end{bmatrix}; \quad C_2 = \begin{bmatrix} -1 & 0 & 0 & 0 \end{bmatrix}$$

$$D_{12} = \begin{bmatrix} 0 \\ 1 \\ 2 \times 10^{-3} \end{bmatrix}; \quad D_{21} = 1.$$

With the multiplication of z_1 by 10, the matrix C_1 becomes

$$C_1 = \begin{bmatrix} 0 & 0 & 10 & 10 \\ 0 & 0 & 0 & 0 \\ .198 & .0396 & 0 & 0 \end{bmatrix}.$$

Figure 8.15 shows the result of the H^2 design (MATLAB h2syn); the ∞-norm of the closed-loop system is computed to be 0.355.

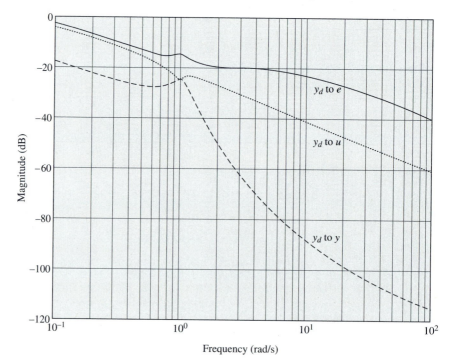

Figure 8.15 Weighted transmissions for the H^2 design

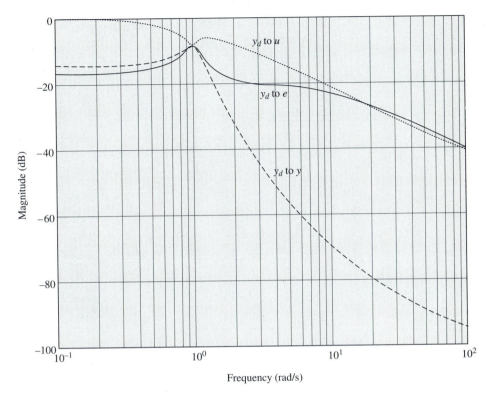

Figure 8.16 Weighted transmissions for the H^∞ design

In Figure 8.16, we display the H^∞ solution (MATLAB hinfsyn), which is really quite similar to the H^2 case. The ∞-norm of the closed-loop system is 0.336, only slightly less than that of the H^2 solution.

Example 8.10   **(Active Suspension)**

We return to the active-suspension problem of Example 2.2. Ride quality is to be measured according to the tracked air-cushion vehicle (TACV) specification [6, 7]—an upper bound on the power density spectrum of the vertical acceleration of the passenger compartment (mass M). The design objective, then, is to limit the vertical acceleration in a frequency-selective manner. At the same time, the extension of the suspension spring must also be limited; ideally, the mass M is motionless while the suspension stretches and contracts to follow the road. This is clearly impractical, as the vehicle moves over hills and valleys.

Road roughness is specified as a power density spectrum of the form A/ω^2. Thus, the power density spectrum of any output variable z in response to the input y_R is $(A/\omega^2)|T_{y_R z}(j\omega)|^2$. We may therefore consider specifications on $(\sqrt{A}/s)T_{y_R z}(s)$. We use $A = 0.25$; this corresponds to a standard deviation of the change in road level of 0.5 m over the 27 m traversed in one second by a vehicle moving at 100 km/h.

The specifications are as follows:

- The weighted vertical acceleration transmission $(.5/s)T_{y_{R}a}(s)$ shall be bounded in magnitude by the TACV specifications curve,

- The weighted transmission of the spring extension $\ell = x_1 - x_2$ shall be bounded in magnitude by 0.15 at all frequencies.

Design the active suspension control, using H_∞ methods.

Solution   The TACV specification curve is the piecewise-linear curve shown in Figure 8.20. For design purposes, we approximate it by the magnitude of the function

$$\frac{1}{W_1(s)} = \frac{.0283(s/39.6 + 1)(s/163.3 + 1)}{(s/6.28 + 1)}.$$

It is convenient to generate $y_R(t)$ as the output of a filter $0.5/s$, driven by an input w. In this way, $T_{wa}(s) = \frac{0.5}{s}T_{y_Ra}(s)$, and $T_{w\ell_1}(s) = \frac{0.5}{s}T_{y_R\ell_1}(s)$, and the frequency weighting is taken care of.

The specifications are, for all ω:

(i) $|\frac{0.5}{j\omega}T_{y_Ra}(j\omega)| \leq \frac{1}{|W_1(j\omega)|}$

(ii) $|\frac{0.5}{j\omega}T_{y_R\ell_1}(j\omega)| \leq 0.15$

We let $z_1 = W_1a$ and $z_2 = \ell_1/0.15$, so the specifications become

$$\|T_{wz_i}\|_\infty \leq 1, \qquad i = 1, 2.$$

As it stands, the problem is ill-posed, because there is nothing to limit u. For an order of magnitude on u, we note that a spring extension of 0.15 m results in a force of 450 N. We therefore write

(iii) $|T_{wu}(j\omega)| \leq 450$, all ω

Defining $z_3 = (1/450)u$, we write

$$\|T_{wz_3}\|_\infty \leq 1.$$

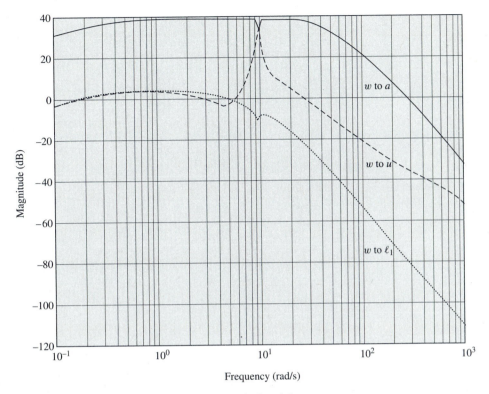

Figure 8.17 Weighted transmissions, nominal weights

We use the model developed in Example 7.11 (Chapter 7), since it already includes $y_R = (1/s)w$. With $y_R = (0.5/s)w$ instead,

$$
\frac{d}{dt}
\begin{bmatrix} \ell_1 \\ \ell_2 \\ v_1 \\ v_2 \end{bmatrix}
=
\begin{bmatrix}
0 & 0 & 1 & -1 \\
0 & 0 & 1 & 0 \\
-10 & 0 & -2 & 2 \\
720 & -660 & 12 & -12
\end{bmatrix}
\begin{bmatrix} \ell_1 \\ \ell_2 \\ v_1 \\ v_2 \end{bmatrix}
+
\begin{bmatrix} 0 \\ -.5 \\ 0 \\ 0 \end{bmatrix} w
+
\begin{bmatrix} 0 \\ 0 \\ .00334 \\ -.02 \end{bmatrix} u
$$

where $\ell_1 = x_1 - x_2$, $\ell_2 = x_1 - y_R$. The controllable-canonical realization of $W_1(s)$, driven by the acceleration, is

$$
\frac{d}{dt}
\begin{bmatrix} q_1 \\ q_2 \end{bmatrix}
=
\begin{bmatrix} 0 & 1 \\ -6467 & -202.9 \end{bmatrix}
\begin{bmatrix} q_1 \\ q_2 \end{bmatrix}
+
\begin{bmatrix} 0 \\ 1 \end{bmatrix} a
$$

$$
z_1 =
\begin{bmatrix} 228504 & 36386 \end{bmatrix}
\begin{bmatrix} q_1 \\ q_2 \end{bmatrix}.
$$

The acceleration, of course, is simply $\dot{v}_1$. Putting the pieces together,

$$
\frac{d}{dt}
\begin{bmatrix}
\ell_1 \\
\ell_2 \\
v_1 \\
v_2 \\
q_1 \\
q_2
\end{bmatrix}
=
\begin{bmatrix}
0 & 0 & 1 & -1 & 0 & 0 \\
0 & 0 & 1 & 0 & 0 & 0 \\
-10 & 0 & -2 & 2 & 0 & 0 \\
720 & -660 & 12 & -12 & 0 & 0 \\
0 & 0 & 0 & 0 & 0 & 1 \\
-10 & 0 & -2 & 2 & -6467 & -202.9
\end{bmatrix}
\begin{bmatrix}
\ell_1 \\
\ell_2 \\
v_1 \\
v_2 \\
q_1 \\
q_2
\end{bmatrix}
$$

$$
+
\begin{bmatrix}
0 \\
-.5 \\
0 \\
0 \\
0 \\
0
\end{bmatrix}
w +
\begin{bmatrix}
0 \\
0 \\
.00334 \\
-.02 \\
0 \\
0
\end{bmatrix}
u
$$

$$
z =
\begin{bmatrix}
0 & 0 & 0 & 0 & 228504 & 36386 \\
\frac{1}{.15} & 0 & 0 & 0 & 0 & 0 \\
0 & 0 & 0 & 0 & 0 & 0
\end{bmatrix}
\begin{bmatrix}
\ell_1 \\
\ell_2 \\
v_1 \\
v_2 \\
q_1 \\
q_2
\end{bmatrix}
+
\begin{bmatrix}
0 \\
0 \\
1/450
\end{bmatrix}
u
$$

$$
y =
\begin{bmatrix} 1 & 0 & 0 & 0 & 0 & 0 \end{bmatrix}
\begin{bmatrix}
\ell_1 \\
\ell_2 \\
v_1 \\
v_2 \\
q_1 \\
q_2
\end{bmatrix}.
$$

We verify controllability from the input u (MATLAB ctrb). The system is not observable from y (the weight function $W_1(s)$ is not connected to the output), but is nevertheless detectable (the modes contributed by W_1 are stable).

We check $D_{12}{}^T D_{12} = (1/450)^2$. We note that $D_{21} = 0$; we require $D_{21} D_{21}{}^T$ to be nonsingular so we resort to fictitious measurement noise, letting $y = \ell_1 + .01n$. This leads to two modifications:

$$
B_1 =
\begin{bmatrix}
0 & 0 \\
-.5 & 0 \\
0 & 0 \\
0 & 0 \\
0 & 0 \\
0 & 0
\end{bmatrix}
\quad \text{and} \quad
D_{21} = \begin{bmatrix} 0 & .01 \end{bmatrix}.
$$

Since there is bound to be some noise in the measurement of ℓ_1, this step is surely justified. The factor of 0.01 signifies measurement errors of the order of 1 cm, since n is considered to have unit norm.

Figure 8.17 shows the H^∞ design (MATLAB hinfsyn), which clearly does not meet the specifications. This leads us to ask whether the specifications can be met at all; some may have to be relaxed, or perhaps other measurements are required.

To answer that question, we multiply both z_2 and z_3 and the measurement noise by 0.0001. In effect, we allow the transmissions to ℓ_1 and u to be 10^4 times larger than specified. The results of the design are shown in Figures 8.18a and b and prove that, by relaxing the specifications and using a very low-noise measurement, the acceleration specification is indeed feasible. Note that the transmission that comes closest to violating the 0-db maximum is that from n to a; all others are well below the limit. This is an indication that the critical element may be the measurement noise rather than the actuator amplitude u or the spring extension ℓ_1.

By trial and error, we tighten the constraints. The final design has z_2 (pertaining to ℓ_1) as originally defined, z_3 (pertaining to u) multiplied by 0.01, and the noise n multiplied by 10^{-5}, as opposed to the original 0.01. Results are shown in Figures 8.19a and b. The graphs show that, indeed, the measurement will have to be very fine—unless, that is, the spectrum of n falls off rapidly beyond about 10 rad/s. The actuator will be called upon to supply more force than originally expected, especially near 8 to 10 rad/s.

Figure 8.20 shows the quantity $|\frac{0.5}{j\omega}T_{y_Ra}(j\omega)|$ and the TACV specification against which it is to be compared. The design is seen to meet the objectives, although barely.

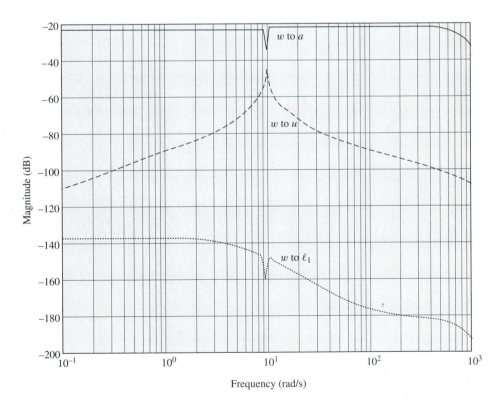

Figure 8.18a and b   Weighted transmissions, reduced constraints

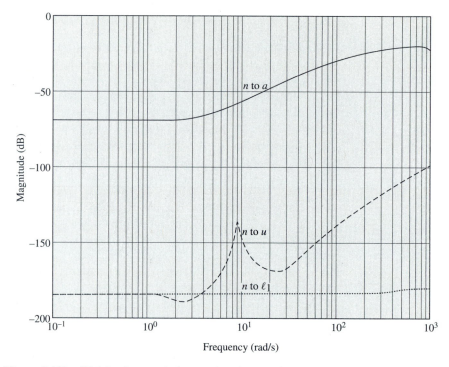

Figure 8.18b Weighted transmissions, reduced constraints

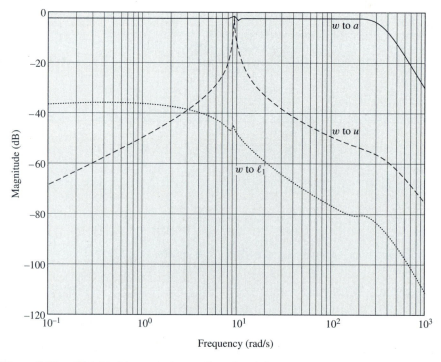

Figure 8.19a Weighted transmissions, adjusted weights

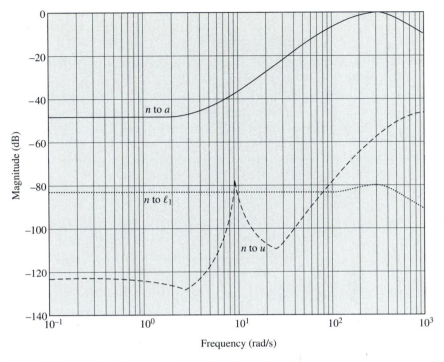

Figure 8.19b Weighted transmissions, adjusted weights

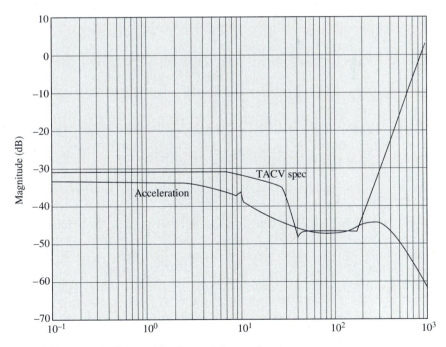

Figure 8.20 The TACV specification and the acceleration spectrum

Problems

M **8.1** For the system

$$H(s) = \begin{bmatrix} \frac{1}{(s+1)^2} & \frac{1}{(s+1)(s+2)} \\ \frac{-1}{(s+2)} & \frac{2}{(s+2)^2} \end{bmatrix}$$

a. Compute the singular values, and plot them versus frequency.

b. A controller $F = kI$ is used to control the plant in a 1-DOF feedback configuration. Show the Nyquist plots of $\det[I + L(s)] - 1$, where $L(s)$ is the loop gain, for $k = 0.1, 1, 10,$ and 100. Assess the closed-loop stability for each value of k. (You may use Bode plots instead.)

c. For stabilizing values of k, compute the singular values of $S(j\omega)$. Relate the results to those of part (b).

M **8.2** Repeat Problem 8.1 for

$$H(s) = \begin{bmatrix} \frac{1}{s+3} & \frac{-(s+1)}{(s+2)^2} \\ \frac{1}{s+1} & \frac{2}{s^2+s+2} \end{bmatrix}.$$

M **8.3** ***Drum speed control*** Repeat Problem 8.1 for the drum-speed-control problem, using the transfer functions derived in Problem 3.15 (Chapter 3). The inputs are u_1 and u_2, and the output is ω_0. Use the gain values $k = 1, 100,$ and 1000.

M **8.4** ***Blending tank*** Repeat Problem 8.1 for the blending-tank problem, using the transfer functions derived in Problem 3.17 (Chapter 3). The inputs are ΔF_A and ΔF_0, and the outputs are $\Delta \ell$ and Δc_A. Use the gain values $k = 0.1, 1,$ and 10.

M **8.5** ***Chemical reactor*** Repeat Problem 8.1 for the chemical reactor problem, using the transfer functions derived in Problem 3.19 (Chapter 3). Use ΔQ and ΔF as inputs and Δc_A and ΔT as outputs. Use the gain values $k = 10^3, 10^4,$ and 10^5.

M **8.6** For the system of Problem 8.1:

a. Calculate the norm $\|H\|_2$ using the Parseval integral.

b. From the plots of singular values, obtain $\|H\|_\infty$.

M **8.7** Repeat Problem 8.6 for the system of Problem 8.2.

8.8 For the system

$$\dot{x} = \begin{bmatrix} 0 & 1 \\ -1 & -1 \end{bmatrix} x + \begin{bmatrix} 1 \\ -1 \end{bmatrix} u$$
$$y = x$$

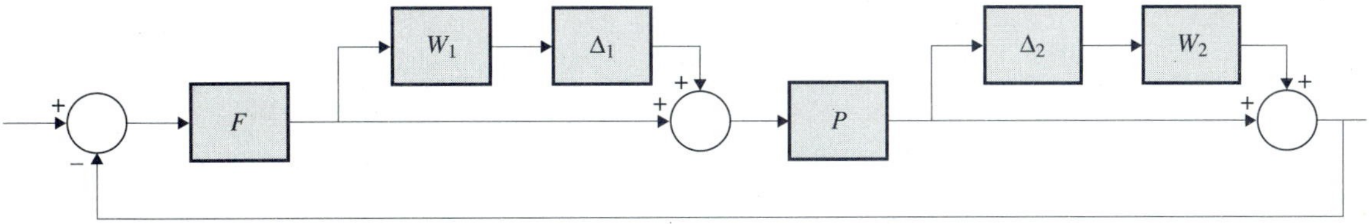

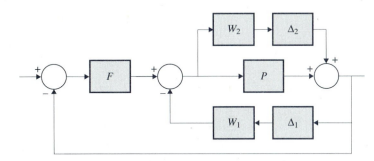

Figure 8.21 System with two uncertainties

Figure 8.22 System with two uncertainties

    a. Calculate the 2-norm, using the method based on the Lyapunov equation.

    b. Calculate the ∞-norm by hand.

8.9   Repeat Problem 8.8 for the system

$$\dot{x} = \begin{bmatrix} -1 & 1 \\ -1 & -1 \end{bmatrix} \mathbf{x} + \begin{bmatrix} 0 \\ 1 \end{bmatrix} u$$
$$y = x.$$

8.10 Figure 8.21 shows a closed-loop system with two uncertainty branches. Give the necessary and sufficient conditions for stability for:

    a. $\|\Delta_1\|_\infty < 1$, $\Delta_2 = 0$

    b. $\Delta_1 = 0$, $\|\Delta_2\|_\infty \le 1$

    c. $\|\Delta_1\|_\infty \le 1$, $\|\Delta_2\|_\infty \le 1$
    (Note that these blocks are MIMO.)

8.11 Repeat Problem 8.10 for the system of Figure 8.22.

(M) **8.12** We wish to use H^2 and H^∞ techniques to design a 1-DOF, SISO control system for the plant $P(s) = (-1)/(s^2 - 1)$. The specifications are as follows:

$$\|W_1 S\|_\infty \le 1, \qquad \|W_2 T\|_\infty \le 1, \qquad \|\frac{u}{y_d}\|_\infty \le 10$$

where $W_1(s) = 10/(s + 1)$ and $W_2(s) = .1(s + 4)$.

a. Obtain realizations for all transfer functions. (You will need to combine W_2 with another, strictly-proper block as in Example 8.8.)

b. Set up the state-space description in terms of control and disturbance inputs and measured and performance outputs.

c. Verify that the technical conditions for the H^2 and H^∞ solutions are satisfied; if they are not, add small noise inputs as required.

d. Compute the H^2 solution.

e. Compute a set of H^∞ solutions. For each one, calculate the ∞-norms of the transfer functions whose norms must be bounded. Adjust the weights on the performance outputs until the specifications are met.

(M) **8.13** Repeat Problem 8.12 with $P(s) = (-s + 1)/(s^2 + s + 1)$, all other data remaining the same.

(M) **8.14** ***Servo with flexible shaft*** A control system is to be designed for the servo of Problem 2.5 (Chapter 2). The design is to be carried out for the rigid-shaft case, with flexibility included as a modeling error.

a. With K at its nominal value, calculate $P = \theta_2/v$; calculate the same transfer function for $K = \infty$, i.e., for a rigid shaft.

b. Determine a multiplicative uncertainty frequency weight $W_1(s)$ so that $W_1(s)\Delta$ "covers" the modeling difference between nominal and infinite K, given the infinite-K model as the base case.

Using θ_2 and θ_{2d} (the set point) as measured outputs, design a control that satisfies the following:

• With $S = 1 - \theta_2/\theta_{2d}$ and $W_2(s) = 10/(\frac{s}{30} + 1)$,

$$|W_2 S| \leq 1, \qquad \text{all } \omega,$$

for the model with infinite K.

• The system is robustly stable for the uncertainty $W_1 \Delta$.

• The magnitude of $|\frac{1}{6}(v/\theta_{2d})| \leq 1$, all ω, for the model with infinite K.

Display the weighted transmission resulting from your design. (*Note:* It may be necessary to add fictitious noise inputs to the measured outputs.)

(M) **8.15** ***Drum speed control*** We wish to design a control system for the drum speed problem of Problem 2.6 (Chapter 2). The general objective is to regulate drum speed without inducing undue stresses in the two shafts. Assume the measurements to be ω_0, Δ_1, and Δ_2, (the latter two through strain gauges). The load torque T_0 is assumed to be generated by passing a signal w of unit 2-norm through the transfer function $1000/s$. The requirements are as follows:

• $|\omega_0/w| \leq 0.1$ at all frequencies.

• $|\Delta_i/w| \leq 0.1, i = 1, 2$, at all frequencies.

- $|u_i/w| \leq U, i = 1, 2$, at all frequencies.

Using an H^∞ approach, find, by trial and error, the minimum U for which all specifications are satisfied. Display the weighted frequency responses that correspond to the specifications. It may be necessary to add fictitious measurement noise to the outputs.

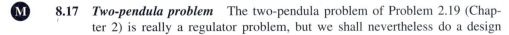

8.16 *Heat exchanger* A linearized model for the heat exchanges of Problem 2.8 (Chapter 2) is given in Problem 2.14. The two control inputs ΔF_c and ΔF_H are to be used to control the temperature outputs ΔT_{c3} and ΔT_{H3}. The interaction is to be minimized; i.e., the response of ΔT_{H3} to a set-point change ΔT_{c3d} should be small, as should the response of ΔT_{c3} to a set-point change ΔT_{H3d}. Also, the model is known to hold only to about 2 rad/min. The following specifications are given:

a. Let T_{d3} be defined by the relation $\begin{bmatrix} \Delta T_{c3} \\ \Delta T_{H3} \end{bmatrix} = T_{d3} \begin{bmatrix} \Delta T_{c3d} \\ \Delta T_{H3d} \end{bmatrix}$, and let $W_1(s) = 20/(s+1)$. Then

$$\left\| W_1 \frac{\Delta T_{c3d} - \Delta T_{c3}}{\Delta T_{c3d}} \right\|_\infty \leq 1$$

$$\left\| W_1 \frac{\Delta T_{H3d} - \Delta T_{H3}}{\Delta T_{H3d}} \right\|_\infty \leq 1$$

$$\left\| W_1 \frac{\Delta T_{c3}}{\Delta T_{H3d}} \right\|_\infty \leq 1$$

$$\left\| W_1 \frac{\Delta T_{H3}}{\Delta T_{c3d}} \right\|_\infty \leq 1.$$

b. Let $W_2(s) = 0.1(0.5s + 1)$. Then

$$\| W_2 T_{d3} \|_\infty \leq 1.$$

c. To keep the two flows down to reasonable values,

$$\| \alpha T_{dF} \| \leq 1$$

where T_{dF} is the transmission from $\begin{bmatrix} T_{c3d} \\ T_{H3d} \end{bmatrix}$ to $\begin{bmatrix} \Delta F_c \\ \Delta F_H \end{bmatrix}$.

Design an H^∞ control with as large a value of α as possible. Display (i) the weighted frequency responses corresponding to the specifications and (ii) the four transmissions, from the two set points to the two temperature outputs. (It may be necessary to add fictitious measurement noise to the measured outputs.)

8.17 *Two-pendula problem* The two-pendula problem of Problem 2.19 (Chapter 2) is really a regulator problem, but we shall nevertheless do a design for a step change in the set point x_d. The model derived in Problem 2.19 is the linearized model; use $\ell_1 = 1$ m, $\ell_2 = 0.75$ m.

We need to keep x bounded in order to have a finite "track" for the cart. At the same time, excursions from the vertical must also be kept down, to avoid nonlinear behavior. We let $T_{de} = (x_d - x)/x_d$, $T_{d\theta_1} = \theta_1/x_d$, $T_{d\theta_2} = \theta_2/x_d$, and $T_{dF} = F/x_d$. The specifications are as follows:

- $\|(0.1/s)W_1(s)T_{de}\|_\infty \le 1$, $W_1(s) = 10/(\frac{s}{6} + 1)$.
- $\|(0.1/s)W_1(s)T_{d\theta_i}\|_\infty \le 1$, $i = 1, 2$.
- $\|(0.1/s)T_{dF}\|_\infty \le \alpha$.

With x, θ_1, and θ_2 as measurements, find a design that satisfies the first two specifications while minimizing α, i.e., using the least amount of control. (Do this approximately, by trial and error.) Display the weighted transmission corresponding to the specifications. (It may be necessary to add fictitious noise to the measurements.)

(M) **8.18** *Maglev* This problem addresses the control of the magnetically levitated vehicle of Problem 2.20 (Chapter 2), in the face of guideway elevation changes z_G and wind forces F_W. A model for z_G [7] is obtained by passing a signal of unit norm through $W_G(s) = 0.01/(s + 0.5)$. The wind forces may be modeled as the output of a transfer function, $W_w(s) = 1000/(s + 3)$, driven by a signal of unit 2-norm.

The design must be such that (i) magnet gaps are kept near their nominal values, (ii) inputs are positive at all times (recall that $u = i^2$), and (iii) accelerations satisfy the requirements for passenger comfort.

Let $z_G = W_G w_1$ and $F_W = W_W w_2$. The specifications are as follows:

- $\|500(\Delta S_{Li}/w_j)\|_\infty \le 1$, $i = 1, 2$, $j = 1, 2$.
- $\|500(\Delta S_{GLi}/w_j)\|_\infty \le 1$, $i = 1, 2$, $j = 1, 2$.
- $\|.01(\Delta u_i/w_j)\|_\infty \le 1$, $i = 1, 2, 3, 4$, $j = 1, 2$.
- $\|W_1(\Delta z/w_1)\|_\infty \le 1$.
- $\|2W_1(\Delta y/w_2)\|_\infty \le 1$.

where $W_1(s) = \dfrac{(s/6.28+1)}{.0283(s/39.6+1)(s/163.3+1)}$. (See Example 8.10; $1/W_1$ is an approximation to the TACV ride specification, for vertical motion.)

a. To ascertain feasibility, assume that all state variables (including disturbances z_G and F_W) are measured, and compute a controller that meets the specifications. (It will be necessary to assume measurement noise inputs; model these by unit-norm signals passed through small gains.)

b. If part (a) yields a successful design, try a design that uses the four gap lengths ΔS_{L1}, ΔS_{L2}, ΔS_{G1}, and ΔS_{G2} as measurements.

In all cases, display the frequency responses corresponding to the specifications.

References

[1] Steward, G. W., *Introduction to Matrix Computations*, Academic Press (1973).

[2] Doyle, J. C., B. A. Francis, and A. R. Tannenbaum, *Feedback Control Theory*, Macmillan (1991).

[3] Zames, G. D, "On the Input–Output Stability of Time-Varying Nonlinear Feedback Systems," *IEEE Transcript on Automatic Control*, vol. II, pp. 228–238 (Part I) and 465–476 (Part II) (1966).

[4] Doyle, J. C., K. Glover, P. O. Khargonekhar, and B. A. Francis, "State Space Solutions to Standard H_2 and H_∞ Control Problems," *IEEE Trans. on Aut. Control*, vol. 34, pp. 831–847 (1989).

[5] The Mathworks, Inc. "Mu-Analysis and Synthesis Toolbox User's Guide."

[6] RRW Systems Group, "HSGT Systems Engineering Study: Tracked Air Cushion Vehicles," Report to OHST, U.S. Dept. of Transportation No. NECTP-219 (Dec. 1969).

[7] Atherton, D. L., P. R. Belanger, P. E. Burke, G. E. Dawson, A. R. Eastham, W. F. Hayes, B. T. Ooi, P. Silvester, and G. R. Slemon, "The Canadian High-Speed Magnetically Levitated Vehicle System," *Can. Elect. Eng. J.*, vol. 3, pp. 3–26 (1978).

Chapter

9

Sampled-Data Implementation

9.1 INTRODUCTION

In the great majority of cases, the implementation of control systems is done on computers. Typically, a design is carried out in the analog domain and "ported" into a digital implementation. This introduces two effects: sampling and quantization.

Sampling is the more important of the two and will occupy most of this chapter. It is made necessary by the fact that a digital computer requires a nonzero time interval to perform the calculations required for one time point of the system output. It follows that, to function in real time, the computer cannot do a control calculation for all time instants—hence the name "sampled-data systems." We shall assume that those sampling instants are uniformly spaced in time. This is not technologically necessary, but nonuniform sampling introduces major theoretical complications.

Quantization takes place because digital computers operate with finite arithmetic. When a system output is sampled, it is converted by roundoff or truncation to a binary value with a finite number of bits. Given the accuracy of modern digital equipment, quantization is important only in cases where extremely high accuracies are involved. We shall not treat quantization effects in this chapter; the reader is referred to Moroney [1] and Oppenheim [2] for discussions of the subject.

Figure 9.1 shows a component block diagram of a sampled-data system for a single-input, single-output (SISO) system. The analog-to-digital (A/D) block samples the system output at constant rate and gives the computer a string of digital signals. The computer performs the control calculations.

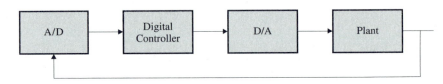

Figure 9.1 Component diagram of a sampled-data system

428

Note that the reference value is generated within the computer, perhaps entered by an operator or perhaps as the output of an algorithm. The digital-to-analog (D/A) unit converts the result of the computation to an analog signal, e.g., a value setting. It is assumed to do so in synchronization with the A/D sampling operation. The conversion is usually accomplished by producing a signal that is constant between clock pulses, with a value (properly scaled to physical units) corresponding to the output of the control algorithm.

There is much more to the topic of sampled-data implementation than is covered in this chapter; refer to Kuo [3], Franklin and Powell [4] and Aström and Wittenmark [5] for details. This chapter is slanted toward the fast-sampling case, where the design is created in the analog domain. There are cases in which it is preferable to design directly in discrete time, particularly if the sampling rate is relatively low. Examples of such systems are a paper machine, in which the weight is measured by scanning the moving web from one side to the other; some chemical processes in which analytical instruments (e.g., a gas chromatograph) require time to process a sample; and processes for which on-line instruments are not available and samples must be taken to the laboratory.

We begin with the study of the spectrum of a sampled signal and the associated phenomenon of aliasing. We then introduce the z-transform and examine the stability of closed-loop discrete-time systems. We end with a study of some common methods of translating an analog design into its digital counterpart.

9.2 SPECTRAL CHARACTERISTICS OF SAMPLED SIGNALS

9.2.1 The Impulse Samples

Figure 9.2 shows the *sample-and-hold* operation. The signal $y(t)$ is sampled at $t = kT_s$, $k = 0, 1, 2, \ldots$, and a new signal $y'(t)$ is constructed simply by holding the value until the next sampling time. We may give this operation a mathematical expression through (i) multiplication of $y(t)$ by the periodic train of impulses of unit area, $p(t)$, shown in Figure 9.3a, and (ii) passing the resulting signal

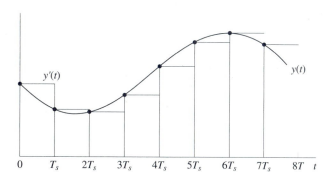

Figure 9.2   Illustration of a sample-and-hold operation

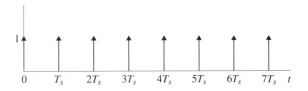

Figure 9.3a   A periodic string of unit impulses

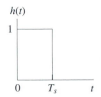

Figure 9.3b   Impulse response of a zero-order hold

through a linear, time-invariant (LTI) system whose impulse response is shown in Figure 9.3b. The process is illustrated in Figure 9.4. Multiplication by the impulse train yields

$$y^*(t) = \sum_{k=0}^{\infty} y(kT_s)u_0(t - kT_s) \tag{9.1}$$

and passage through the LTI system gives

$$y'(t) = \sum_{k=0}^{\infty} y(kT_s)h(t - kT_s). \tag{9.2}$$

Henceforth, we shall refer to sampling as multiplication by a periodic train of unit impulses. Figure 9.5 shows the symbol used to represent this operation, with T_s as the sampling period. This is never implemented in reality; it is just a convenient mathematical representation of the process. Formally, we may write the

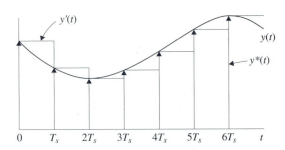

Figure 9.4   Multiplication by a periodic string of unit impulses and passing through a zero-order hold

Figure 9.5   Symbolic representation of the sampling operation

Fourier series for the pulse train $p(t)$ as follows:[1]

$$p(t) = \sum_{n=-\infty}^{\infty} a_n e^{jn\omega_0 t} \tag{9.3}$$

$$a_n = \frac{1}{T_s} \int_{-T_s/2}^{T_s/2} u_0(t) e^{-jn\omega_0 t} dt = \frac{1}{T_s} \tag{9.4}$$

where $\omega_0 = 2\pi/T_s$.

Recall that the spectrum of $e^{j\Omega t}$ is $u_0(\omega - \Omega)$, an impulse in the frequency domain. It follows that the Fourier transform $p(j\omega)$ is

$$p(j\omega) = \frac{1}{T_s} \sum_{n=-\infty}^{\infty} u_0(j\omega - jn\omega_0). \tag{9.5}$$

9.2.2 Aliasing

The sampled signal is $y^*(t) = y(t)p(t)$. Using the real multiplication theorem,

$$\begin{aligned} Y^*(j\omega) &= \int_{-\infty}^{\infty} Y(j\Omega) P(j\omega - j\Omega) d\Omega \\ &= \frac{1}{T_s} \int_{-\infty}^{\infty} Y(j\Omega) \sum_{n=-\infty}^{\infty} u_0(j\omega - j\Omega - jn\omega_0) d\Omega \\ &= \frac{1}{T_s} \sum_{n=-\infty}^{\infty} Y(j\omega - jn\omega_0). \end{aligned} \tag{9.6}$$

Equation 9.6 is illustrated in Figure 9.6. The spectrum of the sampled waveform reproduces $Y(j\omega)$ (for $n = 0$) but adds to it replicas displaced in frequency by $n\omega_0$, $n = \pm 1, \pm 2, \ldots$. These replicas are known as *aliases*.

If the spectrum $Y(j\omega)$ is band-limited to $\omega_0/2$—i.e., is zero for $|\omega| > \omega_0/2$—we see from Figure 9.6 that the aliases do not overlap. In that case, an ideal low-pass filter (or, practically, one that approaches it) will recover $Y(j\omega)$ without error. This result is due to Nyquist; if $Y(j\omega)$ is band-limited to Ω_0, then $\omega_0 = 2\Omega_0$ is called the *Nyquist rate*, corresponding to the signal $y(t)$.

---

[1]Strictly speaking, the manipulations leading to Equation 9.6 are not justified: we are playing fast and loose with impulses. The result is valid under certain mild conditions.

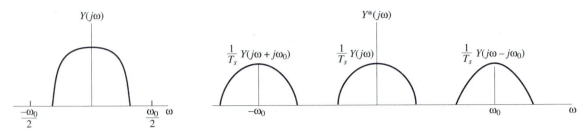

Figure 9.6 The spectrum of a signal and of the signal multiplied by a periodic string of unit impulses

If, on the other hand, the aliases do overlap, it will not be possible to recover $y(t)$ unless more than spectral information is given about the signal. The practical lesson is that we should ensure that the signal being sampled has negligible content at frequencies greater than half the sampling frequency, even if it proves necessary to filter away some of the spectrum before sampling.

Example 9.1   Let $y(t) = \cos t$. Obtain and sketch the sampled signal $y^*(t)$ and its spectrum for sampling frequencies $\omega_0 = 1$ and 4 rad/s, i.e., for $T_s = 2\pi$ and $\pi/2$ s.

Solution   The sampled signal is

$$y^*(t) = \sum_{k=0}^{\infty} \cos kT_s u_0(t - kT_s).$$

For $T_s = 2\pi$,

$$y^*(t) = \sum_{k=0}^{\infty} \cos k2\pi u_0(t - k2\pi) = \sum_{k=0}^{\infty} u_0(t - k2\pi).$$

For $T_s = \pi/2$,

$$y^*(t) = \sum_{k=0}^{\infty} \cos k\frac{\pi}{2} u_0\left(t - k\frac{\pi}{2}\right)$$

$$= u_0(t) - u_0(t - \pi) + u_0(t - 2\pi) \ldots.$$

Figures 9.7a and b show plots of the impulse area versus k. Clearly, Figure 9.7a could have been obtained by sampling a constant or, indeed, any harmonic of $\cos t$.

Since $\cos t = \frac{1}{2}(e^{jt} + e^{-jt})$, its spectrum consists of impulses of area $\frac{1}{2}$ at $\omega = \pm 1$ rad/s. For $\omega_0 = 1$ rad/s, Figure 9.7c shows the sum of $(1/2\pi)Y(j\omega)$ and the aliases centered at ± 1 rad/s and ± 2 rad/s. It is not difficult to see that the final result is a train of impulses of equal areas, $1/2\pi$, in the frequency domain. The result for $\omega_0 = 4$ rad/s is shown in Figure 9.7d.

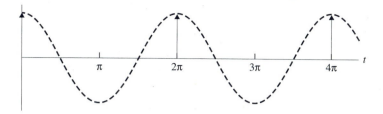

Figure 9.7a $y = \cos t$ samples with $\omega_0 = 1$ rads/s

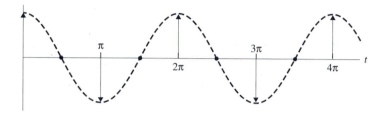

Figure 9.7b y

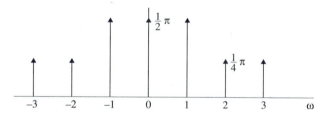

Figure 9.7c   Construction of the spectrum of the sampled waveform

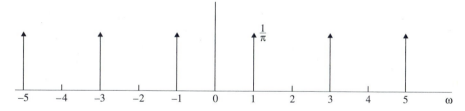

Figure 9.7d   The spectrum for the sampled waveform of Figure 9.7b

If the sampled signal is sent through an ideal low-pass filter with a bandwidth of $\omega_0/2$, the result for $\omega_0 = 1$ rad/s is a single spectral line at $\omega = 0$; the signal is constant. For $\omega_0 = 4$ rad/s, this operation recovers the original spectrum.

9.2.3 The Zero-Order Hold

As shown in Figure 9.2, the sampled signal is converted to analog form by "holding" the most recent value. This process is entirely equivalent to passing an impulse-sampled waveform through the linear system whose impulse response is shown in Figure 9.3b, known as a *zero-order hold* (ZOH). The transfer function corresponding to that impulse response is

$$H_{ZOH}(s) = \frac{1 - e^{-sT_s}}{s} \tag{9.7}$$

or

$$H_{ZOH}(j\omega) = e^{-j\omega(T_s/2)} \frac{e^{j\omega(T_s/2)} - e^{-j\omega(Ts/2)}}{j\omega}$$

$$= 2 \sin\left(\omega\frac{T_s}{2}\right) \frac{e^{-j\omega(T_s/2)}}{\omega}. \tag{9.8}$$

Thus,

$$|H_{ZOH}(j\omega)| = \frac{2}{\omega} \sin \omega\frac{T_s}{2}$$

$$\measuredangle\, H_{ZOH}(j\omega) = -\omega\frac{T_s}{2}. \tag{9.9}$$

Figure 9.8 shows the magnitude; it is seen to be low-pass in character. The phase is the same as that of a pure delay, $T_s/2$.

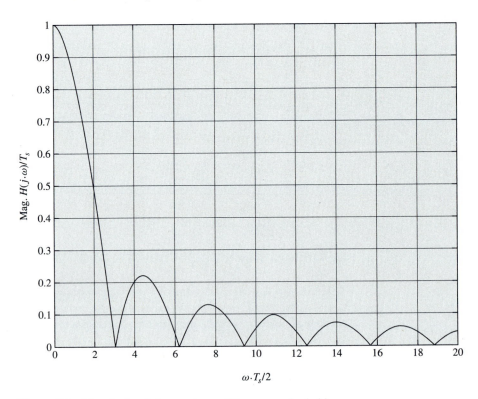

Figure 9.8   Magnitude of the spectrum of the zero-order hold

If aliasing effects are to be minimized, it is necessary that (i) the sampling frequency be at least twice as high as the highest frequency of significance in the signal and (ii) the impulse-sampled signal be low-pass filtered to attenuate the aliases. The ZOH is seen to have the desired low-pass characteristics. Other hold devices may be closer to an ideal low-pass filter (unity in the passband, zero outside) than the ZOH (see Problems 9.3 and 9.4).

9.3 THE z-TRANSFORM

An impulse-sampled signal is written as

$$y^*(t) = \sum_{k=0}^{\infty} y(kT_s)u_0(t - kT_s)$$

and its Laplace transform is

$$\mathcal{L}[y^*(t)] = \sum_{k=0}^{\infty} y(kT_s)e^{-kT_s s}. \tag{9.10}$$

If we let $z = e^{T_s s}$, then

$$\mathcal{L}[y^*(t)] = \sum_{k=0}^{\infty} y(kT_s)z^{-k}. \tag{9.11}$$

We define the z-transform of a *sequence* $\widehat{y}(k)$ as

$$\mathcal{Z}[\widehat{y}(k)] = \sum_{k=0}^{\infty} \widehat{y}(k)z^{-k}. \tag{9.12}$$

This is the so-called *one-sided* transform, which takes into account only nonnegative values of the time index k.

This definition includes Equation 9.11 as a special case: the sequence need not be generated by uniformly sampling a time waveform.

We shall use $\widehat{y}(z)$ to mean $\mathcal{Z}[\widehat{y}(k)]$; in other words, the argument will distinguish the sequence and its transform.

Example 9.2 A waveform $e^{\alpha t}$ is uniformly sampled at a sampling period of T_s. Calculate the z-transform of the resulting sequence.

Solution The sequence is $\widehat{y}(k) = e^{\alpha T_s k} = a^k$, where $a = e^{\alpha T_s}$. Then

$$\mathcal{Z}[\widehat{y}(k)] = \sum_{k=0}^{\infty} a^k z^{-k}$$

$$= \sum_{k=0}^{\infty} (az^{-1})^k$$

which we recognize to be a geometric series. Such a series converges if $|az^{-1}| < 1$, and the limit is

$$\mathcal{Z}[\widehat{y}(k)] = \frac{1}{1 - az^{-1}}, \qquad |z| > |a|. \tag{9.13}$$

◆ ◆ ◆ **REMARK**

Like a Laplace transform, a z-transform is defined only in a given region of the complex plane—in this case, the region outside the circle of radius $|a|$. ◆

◆ ◆ ◆ **REMARK**

If $|a| < 1$, the sequence tends to zero; it "blows up" if $|a| > 1$. A negative value of a is possible (although not by sampling a real exponential) and indicates a sequence of values alternating in sign. ◆

Equation 9.13 is valid for all a, real or complex. It is used as the basis from which to derive z-transforms for some of the following sequences:

(i) $\widehat{y}(k) = u_0(k)$ (discrete-time impulse):

$$\mathcal{Z}[u_0(k)] = 1$$

directly from the definition, because $u_0(0) = 1$, $u_0(k) = 0$, for $k \neq 0$.

(ii) $\widehat{y}(k) = u_{-1}(k)$ (discrete step). For $a = 1$,

$$\mathcal{Z}[\widehat{y}(k)] = \frac{1}{1 - z^{-1}} = \frac{z}{z - 1}, \qquad |z| > 1.$$

(iii) $\widehat{y}(k) = a^k \cos(\omega k)$:

$$\mathcal{Z}[\widehat{y}(k)] = \mathcal{Z}\left[\frac{1}{2}(ae^{j\omega})^k + \frac{1}{2}(ae^{-j\omega})^k\right]$$

$$= \frac{1/2}{z - ae^{j\omega}} + \frac{1/2}{z - ae^{-j\omega}}$$

$$= \frac{z - a\cos\omega}{z^2 - (2a\cos\omega)z + a^2}, \qquad |z| > |a|.$$

(iv) $\widehat{y}(k) = a^k \sin(\omega k)$

$$\mathcal{Z}[\widehat{y}(k)] = \frac{1/2j}{z - ae^{j\omega}} - \frac{1/2j}{z - ae^{-j\omega}}$$

$$= \frac{a\sin\omega}{z^2 - (2a\cos\omega)z + a^2}, \quad |z| > |a|.$$

We shall need a few of the z-transform theorems.

■ **Theorem 9.1**   **Delay Theorem** $\mathcal{Z}[\widehat{y}(k-1)u_{-1}(k-1)] = z^{-1}\widehat{y}(z)$.

Proof: $\mathcal{Z}[\widehat{y}(k-1)u_{-1}(k-1)] = \sum_{k=1}^{\infty}\widehat{y}(k-1)z^{-k}$.
Let $k' = k - 1$. Then

$$\mathcal{Z}[\widehat{y}(k-1)u_{-1}(k-1)] = \sum_{k'=0}^{\infty}\widehat{y}(k')z^{-k'-1}$$

$$= z^{-1}\sum_{k'=0}^{\infty}\widehat{y}(k')z^{-k'}$$

$$= z^{-1}\widehat{y}(z). \quad ■$$

■ **Theorem 9.2**   **Advance Theorem** $\mathcal{Z}[\widehat{y}(k+1)] = z[\widehat{y}(z) - \widehat{y}(0)]$.

Proof: $\mathcal{Z}[\widehat{y}(k+1)] = \sum_{k=0}^{\infty}\widehat{y}(k+1)z^{-k}$.
Let $k' = k + 1$. Then

$$\mathcal{Z}[\widehat{y}(k+1)] = \sum_{k'=1}^{\infty}\widehat{y}(k')z^{-k'+1}$$

$$= z\left[\sum_{k'=0}^{\infty}\widehat{y}(k')z^{-k'} - \widehat{y}(0)\right]$$

$$= z[\widehat{y}(z) - \widehat{y}(0)]. \quad ■$$

■ **Theorem 9.3**   **Complex Differentiation Theorem**

$$\mathcal{Z}[k\widehat{y}(k)] = z^{-1}\frac{d}{dz^{-1}}\widehat{y}(z).$$

Proof:

$$\frac{d}{dz^{-1}}\widehat{y}(z) = \sum_{k=0}^{\infty}k\widehat{y}(k)(z^{-1})^{k-1}$$

$$= z\mathcal{Z}[k\widehat{y}(k)]. \quad ■$$

■ **Theorem 9.4**   **Initial-Value Theorem**

$$\widehat{y}(k)\ \Big|_{k=0} = \lim_{z\to\infty}\widehat{y}(z).$$

Proof: Obvious from the definition.   ■

■ **Theorem 9.5**   **Final-Value Theorem** If it exists, the asymptotic value of $\widehat{y}(k)$ is

$$\lim_{k\to\infty}\widehat{y}(k) = \lim_{z\to 1}(z-1)\widehat{y}(z).$$

Proof: By partial fractions, this operation computes the step component of $\widehat{y}(z)$.   ■

Example 9.3   Compute the z-transform of $ka^k u_{-1}(k)$.

Solution   We apply the complex differentiation theorem. We have

$$\mathcal{Z}[a^k] = \frac{z}{z-a} = \frac{1}{1-az^{-1}}$$

$$\mathcal{Z}[ka^k] = z^{-1}\frac{a}{(1-az^{-1})^2} = \frac{az}{(z-a)^2}.$$

As in the case of Laplace transforms, the partial fraction expansion is the principal tool for the inversion of the z-transform. The following examples demonstrate this.

Example 9.4   Invert $\widehat{y}(z) = [z(z+1)]/[(z-1)(z-.5)^2]$.

Solution   We save the z from the numerator, and write

$$\frac{z+1}{(z-1)(z-.5)^2} = \frac{8}{z-1} - \frac{3}{(z-.5)^2} - \frac{8}{z-.5}$$

and

$$\widehat{y}(z) = 8\frac{z}{z-1} - 3\frac{z}{(z-.5)^2} - 8\frac{z}{z-.5}.$$

Using the results of Example 9.3,

$$\widehat{y}(k) = 8 - 6k(.5)^k - 8(.5)^k, \qquad k \geq 0.$$

Example 9.5   Invert $\widehat{y}(z) = (z+.5)/[z(z+.7)(z-.7)]$.

Solution

$$\frac{(z+.5)}{(z+.7)(z-.7)} = \frac{0.143}{z+.7} + \frac{0.857}{z-.7}$$

and

$$\widehat{y}(z) = z^{-2}\left[0.143\frac{z}{z+.7} + 0.857\frac{z}{z-.7}\right]. \tag{9.14}$$

The bracketed expression is the transform of $\widehat{y}(k) = 0.143(-.7)^k + 0.857(.7)^k$. If we apply the delay theorem twice, we obtain

$$\mathcal{Z}[\widehat{y}(k-2)u_{-1}(k-2)] = z^{-2}\widehat{y}(z)$$

so the inversion yields

$$\widehat{y}(k) = [0.143(-.7)^{k-2} + 0.857(.7)^{k-2}]u_{-1}(k-2).$$

9.4 DISCRETE-TIME SYSTEMS

By way of introduction, let us consider the first-order system

$$\dot{y} = -y + u. \tag{9.15}$$

One way of approximating this in discrete time is to use the finite-difference formula for the derivative:

$$\frac{y(t+T_s) - y(t)}{T_s} = -y(t) + u(t)$$

or

$$y(t+T_s) = (1-T_s)y(t) + T_s u(t). \tag{9.16}$$

Higher-order derivatives are approximated by higher-order differences. For instance, the finite-difference formula for $\ddot{y}$ is

$$\frac{1}{T_s}\left[\frac{y(t+T_s) - y(t)}{T_s} - \frac{y(t) - y(t-T_s)}{T_s}\right] = \frac{y(t+T_s) - 2y(t) + y(t-T_s)}{T_s^2}.$$

In general, an nth-order linear differential equation is rendered in discrete time as the nth-order difference equation

$$y(t) + a_1 y(t-T_s) + \cdots + a_n y(t-nT_s)$$
$$= b_0 u(t) + b_1 u(t-T_s) + \cdots + b_n u(t-nT_s). \tag{9.17}$$

If we now consider $y(t)$ and $u(t)$, sampled at uniform intervals T_s, Equation 9.17 becomes

$$(k) + a_1\widehat{y}(k-1) + \cdots + a_n\widehat{y}(k-n)$$
$$= b_0\widehat{u}(k) + b_1\widehat{u}(k-1) + \cdots + b_n\widehat{u}(k-n). \tag{9.18}$$

Recall that, in continuous time, the transfer function is the ratio of the output transform to the input transform, starting from rest. In discrete-time systems, initial rest means that the output was zero for $k < 0$. We assume that the input was also zero for $k < 0$. Use of the delay theorem on Equation 9.18 yields

$$\frac{\widehat{y}(z)}{\widehat{u}(z)} = G(z) = \frac{b_0 + b_1 z^{-1} + \cdots + b_n z^{-n}}{1 + a_1 z^{-1} + \cdots + a_n z^{-n}} \tag{9.19}$$

$$= \frac{b_0 z^n + b_1 z^{n-1} + \cdots + b_n}{z^n + a_1 z^{n-1} + \cdots + a_n}. \tag{9.20}$$

This is known as the *pulse transfer function*.

If $\widehat{u}(k) = u_0(k)$, the discrete impulse, then the response $\widehat{y}(k)$ is the *discrete-time impulse response*. Its z-transform is $G(z)$, because $\mathcal{Z}[u_0(k)] = 1$. As in the case of continuous-time systems, the discrete-time impulse response of a physically realizable system cannot be anticipatory. This fact forces the transfer function to be proper. To see this, consider that the long division of the ratio of polynomials of Equation 9.20 would begin with a positive power of z^* equal to i if the numerator leading term were z^{n+i}, $i > 0$, thus implying an impulse response sequence with a nonzero value at $k = -i$.

Example 9.6   Calculate the discrete-time impulse response of the system described by the difference equation

$$\widehat{y}(k) = 1.5 \, \widehat{y}(k-1) - 0.5 \, \widehat{y}(k-2) + \widehat{u}(k-1) + \widehat{u}(k-2).$$

Solution   By inspection,

$$\frac{\widehat{y}(z)}{\widehat{u}(z)} = \frac{z^{-1} + z^{-2}}{1 - 1.5z^{-1} + 0.5z^{-2}}$$

$$= \frac{(z+1)}{z^2 - 1.5z + 0.5} = \frac{(z+1)}{(z-1)(z-0.5)}$$

$$= \frac{4}{z-1} - \frac{3}{z-0.5}$$

$$= z^{-1} \left[\frac{4z}{z-1} - \frac{3z}{z-0.5} \right].$$

The impulse response is

$$g(k) = [4 - 3(.5)^{k-1}] u_{-1}(k-1).$$

As in the case of continuous time, an LTI discrete-time system can also be represented by a set of first-order equations; this time, of course, they are difference equations, written in matrix form as

$$\widehat{\mathbf{x}}(k + 1) = A\widehat{\mathbf{x}}(k) + B\,\widehat{\mathbf{u}}(k)$$

$$\widehat{\mathbf{y}}(k) = C\,\widehat{\mathbf{x}}(k) + D\widehat{\mathbf{u}}(k). \qquad (9.21)$$

This is a *state representation*, and $\widehat{\mathbf{x}}$ is the *state vector*.

It is easy to write the zero-input response of the system of Equation 9.21: with $\widehat{\mathbf{u}}(k) = \mathbf{0}$, the state equations are a simple recursion. We have

$$\widehat{\mathbf{x}}_{z_i}(1) = A\widehat{\mathbf{x}}(0)$$

$$\widehat{\mathbf{x}}_{z_i}(2) = A\widehat{\mathbf{x}}_{zi}(1) = A^2\widehat{\mathbf{x}}(0)$$

$$\vdots$$

$$\widehat{\mathbf{x}}_{z_i}(k) = A^k\widehat{\mathbf{x}}(0). \qquad (9.22)$$

If $\mathbf{v}_i$ is an eigenvector of A corresponding to the eigenvalue z_i, then, with $\widehat{\mathbf{x}}(0) = \mathbf{v}_i$,

$$\widehat{\mathbf{x}}(1) = A\mathbf{v} = z_i\mathbf{v}_i$$

$$\widehat{\mathbf{x}}(2) = A\widehat{\mathbf{x}}(1) = z_i{}^2\mathbf{v}_i$$

$$\vdots$$

$$\widehat{\mathbf{x}}(k) = z_i{}^k\mathbf{v}_i. \qquad (9.23)$$

We see that $z_i{}^k$ plays the same role in discrete-time systems as $e^{s_i t}$ does in continuous-time systems.

The zero-state response is also easily derived, by recursion:

$$\widehat{\mathbf{x}}_{z_s}(1) = B\widehat{\mathbf{x}}(0)$$

$$\widehat{\mathbf{x}}_{z_s}(2) = AB\widehat{\mathbf{x}}(0) + B\widehat{\mathbf{u}}(1)$$

$$\widehat{\mathbf{x}}_{z_s}(3) = A^2B\widehat{\mathbf{x}}(0) + AB\widehat{\mathbf{u}}(1) + B\,\widehat{\mathbf{u}}(2)$$

$$\vdots$$

$$\widehat{\mathbf{x}}_{z_s}(k) = \sum_{i=0}^{k-1} A^{k-1-i}B\,\widehat{\mathbf{u}}(i) \qquad (9.24)$$

and

$$\widehat{\mathbf{y}}_{z_s}(k) = \sum_{i=0}^{k-1} CA^{k-1-i}B\,\widehat{\mathbf{u}}(i) + D\widehat{\mathbf{u}}(k). \qquad (9.25)$$

Preceeding the transform of the state equations, we apply the advance theorem to Equation 9.21:

$$z\widehat{\mathbf{x}}(z) = z\widehat{\mathbf{x}}(0) + A\widehat{\mathbf{x}}(z) + B\widehat{\mathbf{u}}(z)$$

or

$$\widehat{\mathbf{x}}(z) = (zI - A)^{-1}z\widehat{\mathbf{x}}(0) + (zI - A)^{-1}B\widehat{\mathbf{u}}(z) \tag{9.26}$$

and

$$\widehat{\mathbf{y}}(z) = C(zI - A)^{-1}z\widehat{\mathbf{x}}(0) + [C(zI - A)^{-1}B + D]\widehat{\mathbf{u}}(z). \tag{9.27}$$

Comparing to Equation 9.22, we see that

$$\mathcal{Z}[A^k] = (zI - A)^{-1}z. \tag{9.28}$$

The matrix pulse transfer function is seen to be

$$G(z) = C(zI - A)^{-1}B + D. \tag{9.29}$$

This last equation is, with z replacing s, the same as in the continuous-time case. This implies that any software that generates $G(s)$ from (A, B, C, D) also generates $G(z)$.

To pursue the parallel, the poles of the pulse transfer functions are the eigenvalues of A in discrete time, as they are in continuous time.

Example 9.7 For the discrete-time system with $A = \begin{bmatrix} 0 & 1 \\ -.5 & 1.5 \end{bmatrix}$, $\mathbf{b} = \begin{bmatrix} -1 \\ 1 \end{bmatrix}$, $C = [1 \quad 0]$, and $D = 0$, calculate (i) the transfer function and (ii) the matrix A^k.

Solution

$$\text{(i)} \qquad (zI - A) = \begin{bmatrix} z & -1 \\ .5 & z - 1.5 \end{bmatrix}$$

$$(zI - A)^{-1} = \frac{1}{z^2 - 1.5z + .5} \begin{bmatrix} z - 1.5 & 1 \\ -.5 & z \end{bmatrix}$$

$$(zI - A)^{-1}\mathbf{b} = \frac{1}{z^2 - 1.5z + .5} \begin{bmatrix} -z + 2.5 \\ z + .5 \end{bmatrix}$$

and

$$\frac{\widehat{y}}{\widehat{u}} = \frac{-z + 2.5}{(z - 1)(z - .5)}.$$

(ii) $\quad z(zI - A)^{-1} = \dfrac{z}{(z^2 - 1.5z + .5)} \begin{bmatrix} z - 1.5 & 1 \\ -.5 & z \end{bmatrix}.$

$$= \begin{bmatrix} \frac{-z}{z-1} + \frac{2z}{z-.5} & \frac{2z}{z-1} - \frac{2z}{z-.5} \\ \frac{-z}{z-1} + \frac{1z}{z-.5} & \frac{2z}{z-1} - \frac{1z}{z-.5} \end{bmatrix}.$$

The inverse is

$$A^k = \begin{bmatrix} -1 + 2(.5)^k & 2 - 2(.5)^k \\ -1 + (.5)^k & 2 - (.5)^k \end{bmatrix}.$$

By this time, the reader will have noticed that many similarities exist between discrete- and continuous-time systems. Observability is defined the same way except that, in discrete-time systems, the observation interval is for those k greater than some positive k_1. The criteria are the same. There is a slight difference with respect to controllability, which is defined as the ability to reach to origin from any initial state. Reachability, in contrast, is defined as the ability to reach any state from any other state in a finite number of time steps. The controllability criteria of the continuous-time case are directly applicable to reachability.

9.5 SAMPLED-DATA SYSTEMS

The object of this section is to obtain the discrete-time representation of a sampled-data system using a zero-order hold to generate the plant input, as shown in Figure 9.9. We need to do this in order to construct a control system entirely out of discrete-time blocks. The controller, as a difference-equation algorithm, is in discrete time, so the part of the system that includes the ZOH, the plant, and the output sampler must be discretized. The starting point may be the transfer function of the continuous-time plant or a state representation. We begin with the latter.

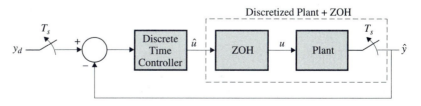

Figure 9.9 Illustration of a sampled-data system

Let

$$\dot{\mathbf{x}} = A\mathbf{x} + B\mathbf{u}$$

$$\mathbf{y} = C\mathbf{x} + D\mathbf{u}.$$

With $\mathbf{u}(t) = \widehat{\mathbf{u}}(k)$, $kT_s \le t < (k+1)T_s$, we may write

$$\mathbf{x}(kT_s + T_s) = e^{AT_s}\mathbf{x}(kT_s) + \int_{kT_s}^{kT_s+T_s} e^{A(kT_s+T_s-\tau)} B\mathbf{u}(\tau)d\tau. \tag{9.30}$$

Since $\mathbf{u}(\tau) = \widehat{\mathbf{u}}(k)$ over the interval of integration,

$$\int_{kT_s}^{kT_s+T_s} e^{A(kT_s+T_s-\tau)} B\mathbf{u}(\tau)d\tau = \left[\int_0^{T_s} e^{A(T_s-\tau')} Bd\tau' \right] \widehat{\mathbf{u}}(k)$$

where we have used the substitution $\tau' = \tau - kT_s$.

With $\mathbf{x}(kT_s) = \widehat{\mathbf{x}}(k)$, Equation 9.30 becomes

$$\widehat{\mathbf{x}}(k+1) = \mathcal{A}\widehat{\mathbf{x}}(k) + \mathcal{B}\widehat{\mathbf{u}}(k)$$

$$\widehat{\mathbf{y}}(k) = \mathcal{C}\widehat{\mathbf{x}}(k) + \mathcal{D}\widehat{\mathbf{u}}(k) \tag{9.31}$$

where

$$\mathcal{A} = e^{AT_s}$$

$$\mathcal{B} = \int_0^{T_s} e^{A(T_s-\tau)} Bd\tau.$$

These matrices are easily computed (MATLAB c2d).

The eigenvalues of the discrete-time system are those of e^{AT_s}. From Chapter 3, if $s_1, s_2, \ldots, s_n$ are the eigenvalues of A, the corresponding discrete-time eigenvalues are $e^{s_1 T_s}$, $e^{s_2 T_s}$, $\ldots$, $e^{s_n T_s}$. Since $|e^{s_i T_s}| = e^{(Res_i)T_s}$, stable eigenvalues map into discrete eigenvalues of less than unit magnitude; i.e., the left half of the s-plane maps into the inside of the unit circle in the z-plane.

If the continuous-time system is given as a transfer function, it may, of course, be realized in state form and transformed to discrete time as already described. There is an alternative way that may be simpler for hand calculation. We note that a discrete-time step, when passed through a zero order, yields a step in continuous time. It follows that sampling of the step response yields the sequence that is, in fact, the discrete step response. We then need only find the z-transform and divide by $z/(z-1)$, the transform of the discrete unit step, to obtain the pulse transfer function.

⌐⌐ Example 9.8   In Figure 9.9, the plant is $P(s) = 1/[(s+1)(s+2)]$. Calculate the discrete-time transfer function of the sampled-data plant.

Solution   The step response is the inverse transform of

$$\frac{1}{s(s+1)(s+2)} = \frac{1/2}{s} - \frac{1}{s+1} + \frac{1/2}{s+2}$$

which is

$$y_{step}(t) = \frac{1}{2} - e^{-t} + \frac{1}{2}e^{-2t}.$$

Sampling with period T_s yields

$$\widehat{y}_{step}(k) = \frac{1}{2} - (e^{-T_s})^k + \frac{1}{2}(e^{-2T_s})^k$$

whose z-transform is

$$\widehat{y}_{step}(z) = \frac{(1/2)z}{z-1} - \frac{z}{z-e^{-T_s}} + \frac{(1/2)z}{z-e^{-2T_s}}$$

$$= \frac{z[(1/2 - e^{-T_s} + \frac{1}{2}e^{-2T_s})z + \frac{1}{2}e^{-T_s} - e^{-2T_s} + \frac{1}{2}e^{-3T_s}]}{(z-1)(z-e^{-T_s})(z-e^{-2T_s})}.$$

Dividing by $z/(z-1)$ yields the discrete transfer function

$$P(z) = \frac{(1/2 - e^{-T_s} + \frac{1}{2}e^{-2T_s})z + \frac{1}{2}e^{-T_s} - e^{-2T_s} + \frac{1}{2}e^{-3T_s}}{(z-e^{-T_s})(z-e^{-2T_s})}.$$

♦ ♦ ♦ **REMARK**
The conversion from a transfer function to discrete time may be done by MATLAB c2dm. ♦

Once the conversion to discrete time has been carried out, we have a system consisting of interconnected discrete-time blocks. The algebra of block diagrams is exactly the same as in the s-domain, and the closed-loop pulse transfer function is given by the same expression as in continuous time.

9.6 STABILITY

The problem of stability in discrete-time systems is different from the continuous-time case because of the different stability domain in the complex plane; the left half of the s-plane is replaced by the inside of the unit circle in the z-plane. In principle, stability may be verified by ascertaining that all system eigenvalues are less than 1 in magnitude; however, our study of continuous-time systems showed that much more than mere eigenvalue calculations is involved.

9.6.1 The Jury Criterion

The Routh criterion is used to ascertain the number of LHP roots of a polynomial. Its counterpart in discrete time is the *Jury criterion* [3]. We begin with the polynomial

$$p(z) = a_n z^n + a_{n-1} z^{n-1} + \cdots + a_0$$

and form the two rows

$$a_n \quad a_{n-1} \ldots a_1 a_0$$

$$a_0 \quad a_1 \ldots a_{n-1} a_n$$

by writing the coefficients of $p(z)$ in forward and reverse orders, respectively.

The third row is obtained by multiplying the second row by $\alpha_n = a_0/a_n$ and subtracting from the first row. The fourth row is the third in reverse order:

$$
\begin{array}{ccccc}
a_n & a_{n-1} & \cdots & a_1 & a_0 \\
a_0 & a_1 & \cdots & a_{n-1} & a_n \\
a_n - \frac{a_0^2}{a_n} & a_{n-1} - \frac{a_0 a_1}{a_n} & \cdots & a_1 - \frac{a_{n-1} a_0}{a_n} & 0 \\
a_1 - \frac{a_{n-1} a_0}{a_n} & & \cdots & a_n - \frac{a_0^2}{a_n} &
\end{array}
$$

Note that the last element of the third row, equal to zero, is removed from its row before the fourth row is written. The next row is formed from rows 3 and 4 in the same manner as row 3 was formed from rows 1 and 2; row 6 is row 5 in reverse order with the final zero-element removed. The procedure is repeated until there are $2n + 1$ rows, with the last having only one element.

We can always make $a_n > 0$ with no loss of generality. That being the case, $p(z)$ has all roots inside the unit circle if, and only if, the leading coefficients of all odd-numbered rows are positive. If these coefficients are all nonzero, the number of negative coefficients is equal to the number of roots outside the unit circle.

Example 9.9   A plant with a pulse transfer function $P(z) = \frac{z}{(z-.5)^2}$ is controlled by a 1-DOF system with a pure-gain controller k. Calculate the range of values of k for which the closed-loop system is stable.

Solution   The closed-loop characteristic polynomial is $z^2 + (k-1)z + 0.25$. The Jury array is

$$
\begin{array}{ccc}
1 & k-1 & 0.25 \\
0.25 & k-1 & 1 \\
0.9375 & 0.75(k-1) & \\
0.75(k-1) & 0.9375 & \\
0.9375 - 0.6(k-1)^2 & &
\end{array}
$$

For stability, we require

$$1 > 0, \qquad 0.9375 > 0, \qquad 0.9375 - 0.6(k-1)^2 > 0$$

or

$$(k-1)^2 > 1.5625$$

$$|k-1| > 1.25$$

$$-0.25 < k < 2.25.$$

9.6.2 The Root Locus

The Root Locus is exactly the same in discrete and continuous time because the block-diagram algebra is the same in both cases; one is still looking to satisfy $1 + kG(z) = 0$ for a closed-loop pole. Stability requires that all poles be within the unit circle.

Example 9.10

For the system of Example 9.9, draw the Root Loci for $k > 0$ and $k < 0$, and use it to interpret the results of Example 9.9.

Solution

The two loci are shown in Figure 9.10 (MATLAB rlocus). For $k > 0$, the locus leaves the unit circle for $k = 2.25$; for $k > 0$, that gain is -0.25.

◆ ◆ ◆ **REMARK**

A strictly proper $P(z)$ under pure-gain feedback will always become unstable for sufficiently large $|k|$, because there are always as many branches going to infinity as the pole–zero difference. This is different from the continuous-time case, in which for a pole excess of 1 or 2, it is possible for all branches to remain in the left-half plane as k tends to infinity, thus retaining stability. ◆

9.6.3 The Nyquist Criterion

The principle of the argument applies to the z-plane as well as to the s-plane. This time, however, the D-contour is replaced by the unit circle, and we require all closed-loop poles to be inside, for stability. As before, infinitesimal indentations are made to avoid poles on the unit circle, and these map into circles or semicircles of infinite radii.

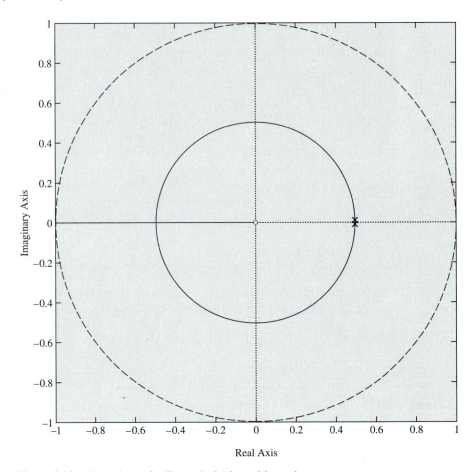

Figure 9.10a   Root locus for Example 9.10, positive gain

The map of the unit circle is $P(e^{j\omega})$, $0 \leq \omega \leq 2\pi$. Since $e^{j(2\pi - \omega)} = e^{-j\omega}$, the lower half of the unit circle ($2\pi - 0$ to $2\pi - \pi$) maps into $P(e^{-j\omega}) = P^*(e^{j\omega})$, just as the lower half of the D-contour in the s-plane maps into the complex conjugate of the map of the top half. Since $e^{j\omega}$ is periodic with a period of 2π, $P(e^{j\omega})$ is also periodic. This represents a major difference from the situation in continuous time, since the notion of roll-off is no longer meaningful. It is possible—in fact, usual—to use $z = e^{j\omega T_s}$, where T_s is the sampling period, and to plot $P(e^{j\omega T_s})$, $0 \leq \omega \leq 2\pi/T_s$. This also maps the unit circle, but ω is now related to the time base of sampled plant. The frequency $2\pi/T_s$ is half the Nyquist frequency (MATLAB dynquist).

Discrete-time Bode diagrams may also be plotted, for $0 \leq \omega \leq \pi/T_s$ (MATLAB dbode). There is no equivalent here of the low- and high-frequency asymptotic straight lines of the continuous-time plots; the Bode plot must be computed point by point.

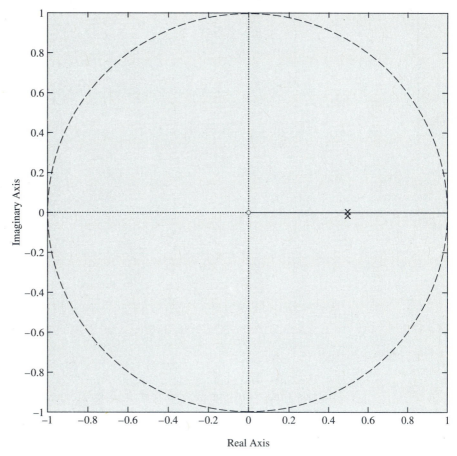

Figure 9.10b Root locus for Example 9.10, negative gain

Example 9.11 For the system of Example 9.9, obtain and interpret the Nyquist plot. Use $T_s = 1$.

Solution Figure 9.11 shows the Nyquist plot (MATLAB dnyquist). There are two open-loop poles within the unit circle ($P = 2$), and we need two closed-loop poles within the unit circle ($Z = 2$), so $N = 0$ for stability. The two real-axis crossings are at -0.444 and 4, respectively, so for stability we need

$$-\frac{1}{k} < -0.444 \quad \text{or} \quad k < 2.25$$

$$-\frac{1}{k} > 4 \quad \text{or} \quad k > -0.25.$$

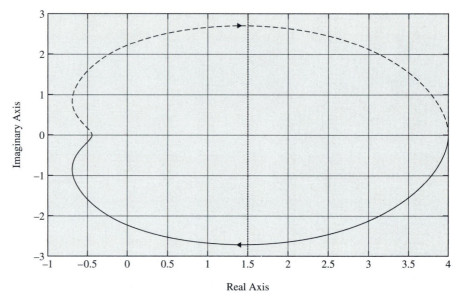

Figure 9.11   Discrete time Nyquist plot for Example 9.11

The rules for counting encirclements from Bode plots are the same as in continuous time, with the proviso that $\omega = \pi/T_s$, the highest frequency plotted, is equivalent to $\omega = \infty$ in continuous time. A real-axis crossing at $\omega = \pi/T_s$ (there will always be one) is counted as one crossing, not two.

Poles on the unit circle are dealt with by indenting to the outside. Since the phase *decreases* on the semicircle, its map is a large, clockwise semicircle (or circle, if the pole is a multiple pole). Another way to see this is to note that the contour takes a 90° right turn when entering the semicircle; hence its map must do the same. If the mapping of the entry point to the indentation is in the top half of the complex plane (phase between 0° and 180°), the real-axis crossing will take place on the far right; it will take place on the far left if the entry point maps into a point on the bottom half.

Example 9.12   For the plant $P(s) = 1/[s(s + 1)(s + 2)]$, obtain the discrete-time Bode plots for $T_s = 0.2$ s and 2 s. In each case, obtain the range of k, the gain of a pure-gain 1-DOF controller, for which the closed-loop system is stable.

Solution   Figure 9.12 shows the Bode plots. In each case, there is a pole at $z = 1$, and the Nyquist contour is indented to the right so as to include that pole. Since $P = 3$, we want $Z = 3$ and $N = 0$ for stability.

For $T_s = 0.2$, the pulse transfer function (MATLAB c2dm) is

$$P(z) = \frac{0.0012(z + 3.225)(z + 0.2299)}{(z - 1)(z - 0.8187)(z - 0.6703)}.$$

The Bode plots are shown in Figure 9.12a. We note real-axis crossings at $\omega = 1.225$ rad/s (180°, −13.1 db) and $\omega = \pi/T_s = 15.708$ rad/s (360°, −69.78 db). The pole at $z = 1$ is indented. As the indentation is entered (from below the point $z = 1$), the phase is +90 (negative of the positive-frequency plot), so the crossing is to the far right. The diagram in Figure 9.12b shows the real-axis crossings.

For $N = 0$, we require

$$0 < k < 13.1 \text{ db}.$$

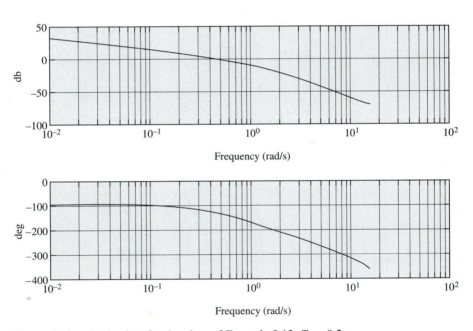

Figure 9.12a   Bode plots for the plant of Example 9.12, $T_s = 0.2s$

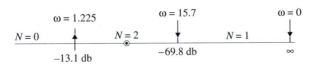

Figure 9.12b   Real axis crossings of the Nyquist plot, Example 9.12, $T_s = 0.2s$

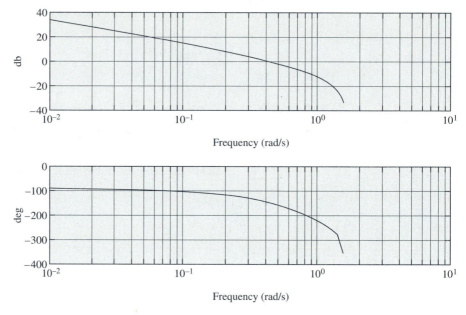

Figure 9.12c Bode plots for the plant of Example 9.12, $T_s = 2s$

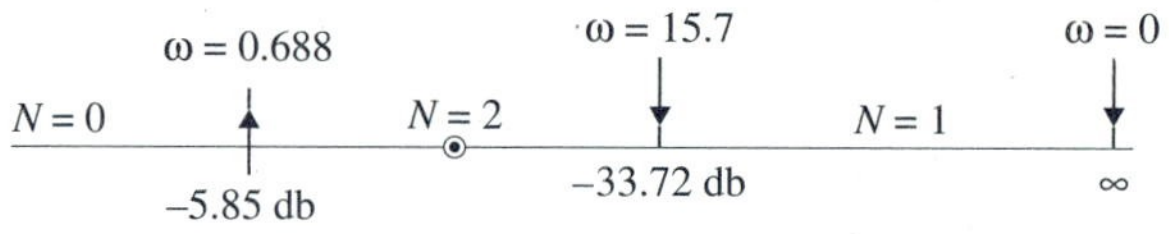

Figure 9.12d Real axis crossings of the Nyquist plot, Example 9.12, $T_s = 2s$

For $T_s = 2$, the pulse transfer function is

$$P(z) = \frac{0.3808(z + 1.1311)(z + 0.0461)}{(z - 1)(z - 0.1353)(z - 0.0183)}.$$

Figure 9.12c shows the Bode plots. They are qualitatively similar to the ones for $T_s = 0.2$. Figure 9.12d is the diagram of real-axis crossings. For stability,

$$0 < k < 5.85 \text{ db}.$$

9.7 DISCRETIZATION OF CONTINUOUS-TIME DESIGNS

9.7.1 The Case of Small Sampling Period

To consider the digital implementation of a continuous-time design, we begin with the continuous-time design of Figure 9.13a, to be approximated by the system of

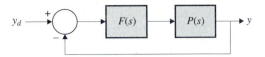

Figure 9.13a The continuous-time design to be approximated

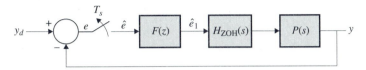

Figure 9.13b The sampled data approximation of the continuous-time design

Figure 9.13b. For simplicity, let us assume that Fourier transforms exist when they are used.

From Equation 9.6,

$$\widehat{e}(j\omega) = \frac{1}{T_s} \sum_{k=-\infty}^{\infty} e(j\omega - jk\omega_0), \qquad \omega_0 = \frac{2\pi}{T_s}.$$

The transmission in continuous time of the pulse transfer function $F(z)$ is $F(e^{j\omega T_s})$. To see this, recall that the pulse transfer function was developed by replacing e^{sT_s} with z.

There remains the transmission of the ZOH, given in Equation 9.8 as

$$G_{ZOH} = 2 \sin \left(\frac{\omega T_s}{2} \right) \frac{e^{-j(\omega T_s/2)}}{\omega}$$

and, of course, that of the plant $P(j\omega)$. The result is

$$y(j\omega) = F(e^{j\omega T_s}) G_{ZOH}(j\omega) P(j\omega) \frac{1}{T_s} \sum_{k=-\infty}^{\infty} \mathbf{e} \left(j\omega - jk\frac{2\pi}{T_s} \right). \qquad (9.32)$$

Note that $F(e^{j\omega T_s})$ is a periodic function of ω, with a period of $2\pi/T_s$, the same as that of $e(j\omega)$. Figure 9.14 shows the components of the right-hand side (RHS) of Equation 9.32 under the following conditions:

1. $|e(j\omega)|$ is band-limited.

2. T_s is small enough that the aliases of e do not overlap.

3. $|P(j\omega)|$ goes to zero as $\omega \to \infty$.

We pose the following question: Under what circumstances is the transmission from $y_d(t)$ to $y(t)$ approximately the same as in continuous time?

Reasoning from Figure 9.14, we note the following:

1. The signal e must be band-limited to half the sampling frequency so that the aliases do not overlap.

2. The open-loop transmission must be low-pass so as to attenuate all aliases save the central one around zero frequency.

3. $F(e^{j\omega T_s})$ must be approximately equal to the continuous-time compensator frequency response $F_c(j\omega)$ for $0 < \omega < \pi/T_s$.

4. $H_{ZOH}(j\omega)$ must be close to unity over the passband, to ensure that $P(j\omega)H_{ZOH}(j\omega) \approx P(j\omega)$.

If the loop transmission is low-pass, then y will be approximately band-limited. In effect, the crossover frequency ω_c must be much less than π/T_s. Aström [5] gives the following rule of thumb:

$$\omega_c T_s \approx 0.15 \quad \text{to} \quad 0.5. \tag{9.33}$$

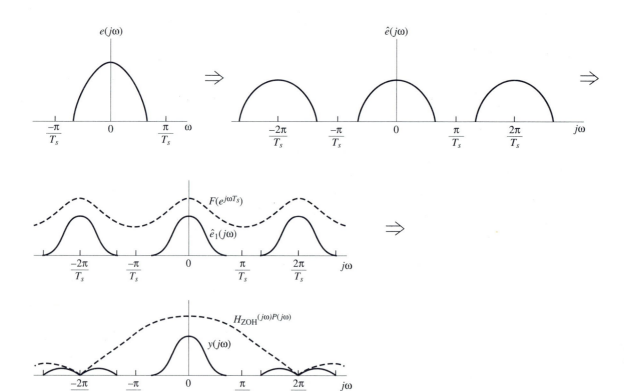

Figure 9.14 Illustrating the signal spectra in the sampled-data system

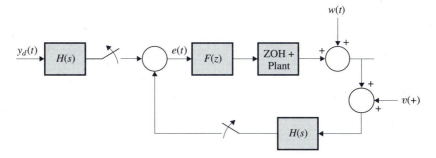

Figure 9.15

In Figure 9.15, $e(t)$ is generated by the signals y_d, w, and v; these signals may or may not be appropriately limited in bandwidth. If not, anti-aliasing filters can be used, as in Figure 9.15. Here, $H(s)$ is either an analog low-pass filter or perhaps a dedicated digital system running at a faster sampling rate than the main loop.

We conclude that, if T_s is sufficiently small and $F(e^{j\omega T_s}) \approx F_c(j\omega)$, the closed-loop system will be stable. As T_s increases, stability will be retained as long as the number of encirclements of the critical point by the discrete-time Nyquist plot does not change. This will be ensured for $T_s = T_s^*$ if the Nyquist plot does not cross the critical point for all $T_s \leq T_s^*$.

The choice of sampling period T_s can be guided by stability margins as indicators of closeness of approach to the $(-1, 0)$ point. Gain and phase margins may be specified, bearing in mind that the stability margins of the sampled-data system will always deteriorate with increasing T_s.

9.7.2 Discretization of Continuous-Time Controllers

It is natural to discretize a differential equation by replacing the derivative operation with a backward finite difference, i.e.,

$$\frac{d}{dt}y(t) = \frac{y(t) - y(t - T_s)}{T_s}$$

$$= \frac{\widehat{y}(k) - \widehat{y}(k - 1)}{T_s}. \tag{9.34}$$

In the transform variables, this is equivalent to replacement of s with $(1 - z^{-1})/T_s$ or, solving for z, replacement of z with $1/(1 - T_s s)$.

The controller is designed by replacing s with $(z-1)/zT_s$ in the transfer function $F_c(s)$. Recall from the preceding section that we need to satisfy $F(e^{sT_s}) \approx F_c(s)$ for $s = j\omega$ over as much of the range 0 to π/T_s as possible. In this case,

$$F(z) = F_c\left(\frac{z - 1}{zT_s}\right) = F_c\left(\frac{1 - z^{-1}}{T_s}\right)$$

and

$$F(e^{sT_s}) = F_c \left(\frac{1 - e^{-sT_s}}{T_s} \right). \tag{9.35}$$

Now,

$$\frac{1 - e^{-sT_s}}{T_s} = \frac{1 - 1 + sT_s - \frac{1}{2}(sT_s)^2 + \cdots}{T_s}$$

$$= s \left(1 - \frac{1}{2} s T_s + \frac{1}{3!} s^2 T_s - \cdots \right). \tag{9.36}$$

The condition $F(e^{j\omega T_s}) \approx F_c(j\omega)$ is satisfied at low frequencies, i.e., for $\frac{1}{2}\omega T_s \ll 1$. Since the approximation must hold up to frequencies where the magnitude of the loop gain is small, we require that $\omega_c T_s \ll 2$, in line with the conclusion already expressed in Equation 9.33.

The *Tustin approximation* consists in replacing s with $(2/T_s)[(z - 1)/(z + 1)]$; i.e.,

$$F(z) = F_c \left(\frac{2}{T_s} \frac{z - 1}{z + 1} \right)$$

and

$$F(e^{sT_s}) = F_c \left(\frac{2}{T_s} \frac{e^{sT_s} - 1}{e^{sT_s} + 1} \right). \tag{9.37}$$

Now,

$$\frac{2}{T_s} \frac{e^{sT_s} - 1}{e^{sT_s} + 1} = \frac{2}{T_s} \frac{-1 + 1 + sT_s + \frac{1}{2}(sT_s)^2 + \cdots}{2 + sT_s + \frac{1}{2}(sT_s)^2 + \cdots}.$$

By long division, the leading terms of the series are

$$\frac{2}{T_s} \frac{e^{sT_s} - 1}{e^{sT_s} + 1} = \frac{2}{T_s} \frac{1}{2} s T_s - \frac{1}{8}(sT_s)^3$$

$$= s \left(1 - \frac{1}{4} s^2 T_s^2 \right). \tag{9.38}$$

The condition $F(e^{j\omega T_s}) \approx F_c(j\omega)$ will hold to the extent that $\omega_c T_s \ll 2$, as before; however, since the bracket on the RHS of Equation 9.38 has no linear term, the Tustin approximation is better at low frequencies than the backward-difference approximation.

As derived, the Tustin approximation is asymptotically exact, as $\omega \to 0$. It may be scaled so that the approximation is exact at some other frequency, say, ω_1; that is known as *prewarping*. We wish to choose a scaling factor α such that

$$\alpha \frac{e^{j\omega_1 T_s} - 1}{e^{j\omega_1 T_s} + 1} = j\omega_1$$

$$\alpha \frac{e^{j\omega_1 T_s}(e^{j\omega_1 T_s/2} - e^{j\omega_1 T_s/2})}{e^{j\omega_1 T_s/2}(e^{j\omega_1 T_s/2} + e^{j\omega_1 T_s/2})} = j\omega_1.$$

Solving for α yields

$$\alpha = \frac{\omega}{\tan(\omega_1 T_s/2)}$$

and the Tustin transformation with prewarping is

$$s \rightarrow \frac{\omega_1}{\tan \frac{(\omega_1 T_s)}{2}} \frac{z-1}{z+1}. \tag{9.39}$$

Prewarping may be used to ensure a good match between $F_c(j\omega)$ and $F(e^{j\omega T_s})$ at frequencies near crossover. It is often the case that the loop gain is large at low frequencies and small at high frequencies; neither situation requires accurate approximation of the loop gain. On the other hand, the crossover region is critical, so it may be desirable to have accuracy there.

Example 9.13   (dc Servo)

In Example 6.11, a lead-lag compensator was designed for the dc servo. Obtain the discrete-time Bode plot and phase margin for the backward-difference implementation and $T_s = 0.05$ s, 0.1 s, and 0.2 s. Assume D/A conversion through a zero-order hold.

Solution   The continuous-time compensator is

$$F_c(s) = \frac{2.28(1.14s + 1)(0.63s + 1)}{(1.76s + 1)(0.177s + 1)}.$$

To convert to discrete time, we note that a factor of the form $Ts \pm 1$ transforms as

$$Ts \pm 1 \rightarrow T\frac{(z-1)}{zT_s} \pm 1 = \frac{(T \pm T_s)z - T}{zT_s}.$$

The compensator is

$$F(z) = 2.28 \frac{[(1.14 + T_s)z - 1.14][(0.63 + T_s)z - 0.63]}{[(1.76 + T_s)z - 1.76][(0.177 + T_s)z - 0.177]}.$$

For $T_s = 0.05$, the pulse transfer function of the plant (MATLAB c2dm) is

$$P(z) = \frac{(1.396e - 3)z^2 + (4.223e - 3)z + 7.6683 - 4}{z^3 - 2.224z^2 + 1.524z - 0.3004}.$$

Figure 9.16 shows the Bode plots for all three values of T_s. They are plotted against ωT_s ($T_s = 1$ in MATLAB dbode).

The phase margins (MATLAB margin) are 53.6°, 47.2°, and 35.2°, respectively, for $T_s = 0.05$, 0.1, and 0.2.

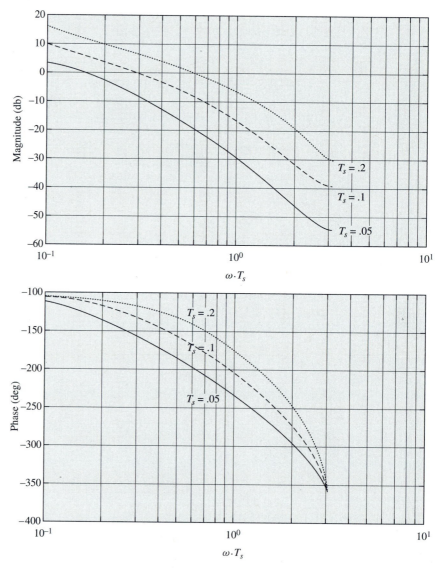

Figure 9.16 Bode plot phases for Example 9.13, for $T_s = 0.05$, 0.1, and 0.2 s

Example 9.14   Repeat Example 9.13, using the Tustin transformation with frequency prewarping. Choose the gain of the transformation so as to obtain equality between the continuous-time and discretized compensators at the crossover frequency 3 rad/s.

Solution   The design proceeds as in Example 9.13, with the discretization done on the computer (MATLAB c2dm). The Bode plots appear in Figure 9.17. The phase margins (MATLAB margin) are 55.8°, 51.8°, and 43.5°, respectively, for $T_s = 0.05$,

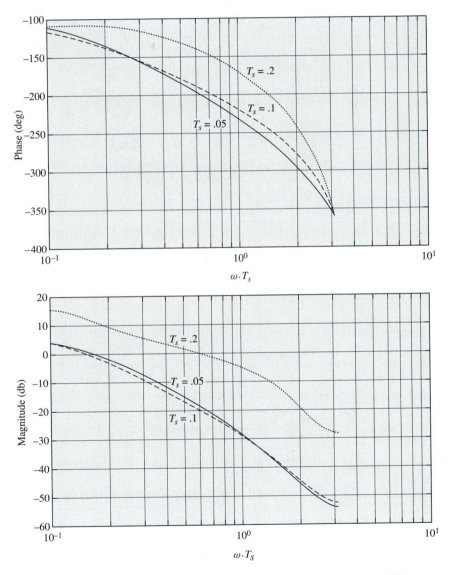

Figure 9.17   Bode plot magnitudes for Example 9.14, for $T_s = 0.05$, 0.1, and 0.2 a

0.1, and 0.2. Note that (i) the difference between $T_s = 0.05$ and $T_s = 0.1$ is much less than in the case of backward differences, and (ii) for all three values of T_s, the phase margin is greater than with the backward-difference method.

Step responses are plotted in Figure 9.18. The dotted curves are the sampled-data responses.[2]

The simplest method of discretizing a controller expressed in state form is the forward-difference method, where $\dot{\mathbf{x}}$ is replaced by $[\widehat{\mathbf{x}}(k + 1) - \widehat{\mathbf{x}}(k)]/T_s$. State equations become

$$\widehat{\mathbf{x}}(k + 1) = \widehat{\mathbf{x}}(k) + T_s A \widehat{\mathbf{x}}(k) + T_s B \widehat{\mathbf{e}}(k)$$

or

$$\widehat{\mathbf{x}}(k + 1) = (I + T_s A)\widehat{\mathbf{x}}(k) + T_s B \widehat{\mathbf{e}}(k)$$

$$\widehat{\mathbf{u}}(k) = C \widehat{\mathbf{x}}(k) + D \widehat{\mathbf{e}}(k). \tag{9.40}$$

Here, $\widehat{\mathbf{e}}(k)$ is the error signal and $\widehat{\mathbf{u}}(k)$ is the controller output, which is, of course, the plant input.

The Tustin transformation can also be used. We replace s with $(2/T_s)$ $[(z - 1)/(z + 1)] = (2/T_s)[(1 - z^{-1})/(1 + z^{-1})]$, which yields

$$1 - z^{-1}\widehat{\mathbf{x}}(z) = \frac{T_s}{2}(1 + z^{-1})[A\widehat{\mathbf{x}}(z) + B \widehat{\mathbf{e}}(z)].$$

In the time domain,

$$\widehat{\mathbf{x}}(k) = \widehat{\mathbf{x}}(k - 1) + \frac{T_s}{2} A\widehat{\mathbf{x}}(k) + \frac{T_s}{2} A\widehat{\mathbf{x}}(k - 1) + \frac{T_s}{2} B \widehat{\mathbf{e}}(k) + \frac{T_s}{2} B \widehat{\mathbf{e}}(k - 1)$$

$$\widehat{\mathbf{x}}(k) = \left(I - \frac{T_s}{2} A\right)^{-1}\left(I + \frac{T_s}{2} A\right)\widehat{\mathbf{x}}(k - 1) + \left(I - \frac{T_s}{2} A\right)^{-1}\frac{T_s}{2} B \widehat{\mathbf{e}}(k)$$

$$+ \left(I - \frac{T_s}{2} A\right)^{-1}\frac{T_s}{2} B \widehat{\mathbf{e}}(k - 1)$$

$$\widehat{\mathbf{u}}(k) = C \widehat{\mathbf{x}}(k) + D \widehat{\mathbf{e}}(k). \tag{9.41}$$

(For sufficiently small T_s, $[I - (T_s/2) A]^{-1}$ exists.)

---

[2]At this time, MATLAB has no command to compute a true-sampled-data response in continuous time. It is necessary to (i) compute the discrete-time control; (ii) fill in the control function during the "hold" time segments, using a small time step; and (iii) use ℓsim to compute the output.

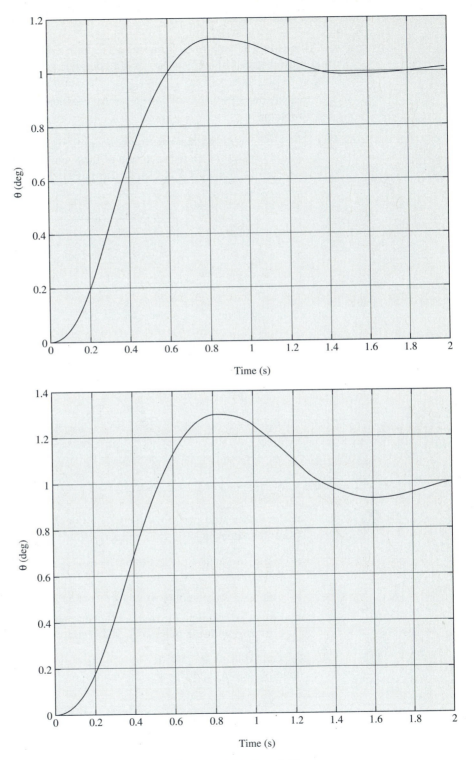

Figure 9.18   Step response for Example 9.14, $T_s = 0.05$ s

This is not in state form because of the term $\widehat{\mathbf{e}}(k)$ on the RHS. However, it is in a form that can be implemented on a computer.

Problems

9.1 A signal $y(t)$ has the spectrum shown in Figure 9.19 (the phase is zero and $Y(j1) = 0$). Sketch the spectrum of the signal obtained by impulse sampling of $y(t)$ at 1 rad/s, 1.5 rad/s, and 2.5 rad/s.

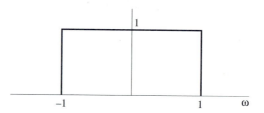

Figure 9.19 Spectrum of y, Problem 9.1

9.2 Repeat Problem 9.1 for the spectrum of Figure 9.20 (the phase is zero).

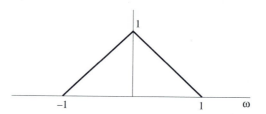

Figure 9.20 Spectrum of y, Problem 9.2

M **9.3** A first-order hold performs, for $kT_s \le t \le (k+1)T_s$, a linear extrapolation from the values at $(k-1)T_s$ and kT_s.

    a. Write the analytic expression in the time domain for the extrapolated waveforms.

    b. Calculate the response to a discrete-time impulse $u_0(k)$.

    c. From (b), calculate the transfer function and plot its magnitude and phase. Compare with the magnitude and phase of the zero-order hold.

M **9.4** Repeat Problem 9.3 for the second-order hold, where the extrapolation is done by passing a quadratic through the points at $(k-2)T_s$, $(k-1)T_s$, and kT_s.

9.5 Compute the z-transforms of the following:

    a. $\widehat{y}(k) = (0.5)^k - (2)^{k-1}$.

    b. $\widehat{y}(k) = 3e^{-k} \cos(2k + 45)$.

9.6   Compute the z-transforms of the following:

    a. $\widehat{y}(k) = (0.5)^{k-4} \sin[3(k-2)]$.

    b. $\widehat{y}(k) = k(.7)^{-k}$.

9.7   Invert the following z-transforms, assuming causal signals.

    a. $\widehat{y}(z) = [z(z+2)]/[(z-1)(z-.5)]$.

    b. $\widehat{y}(z) = 1/[z^2(z-1)(z-2)]$.

9.8   Invert the following z-transforms, assuming causal signals.

    a. $\widehat{y}(z) = [z(z+1)]/[(z-.5)^2(z-1)]$.

    b. $\widehat{y}(z) = 1/[z(z^2+z+1)]$.

9.9   Use linearity to prove the discrete convolution theorem for causal, linear, time-invariant systems:

$$\widehat{y}(k) = \sum_{n=0}^{k} h(k-n)\widehat{u}(n)$$

where $\widehat{u}$ is a causal signal.

---

*Hint   Assume $\widehat{y}(k)$ to be a linear function of present and past inputs, and invoke time invariance.

9.10 Obtain the recursive algorithm (i.e., the difference equation) corresponding to the following pulse transfer functions:

    a. $H(z) = (z+1)/(z^2+z+1)$.

    b. $H(z) = [1/z^3(z-.7)]$.

9.11 Obtain the pulse transfer function of the discrete-time system in state form, with $A = [\begin{smallmatrix} 0.3 & 0 \\ -0.5 & 0.5 \end{smallmatrix}]$, $\mathbf{b} = [\begin{smallmatrix} 0 \\ 1 \end{smallmatrix}]$, $\mathbf{c} = [1 \quad 0]$, and $D = 0$.

9.12 Repeat Problem 9.11 with $A = [\begin{smallmatrix} 0 & 2 \\ -.2 & .2 \end{smallmatrix}]$, $\mathbf{b} = [\begin{smallmatrix} 1 \\ -1 \end{smallmatrix}]$, $\mathbf{c} = [1 \quad 0]$, and $D = 0$.

9.13 Rewrite the system

$$\mathbf{x}(k+1) = A\mathbf{x}(k) + B\mathbf{u}(k-n)$$

in standard-state form, by introducing new state variables corresponding to different delayed input values. Show that the resulting A-matrix has at least n eigenvalues at the origin.

9.14 Calculate A^k, $A = [\begin{smallmatrix} 0 & 1 \\ .5 & .5 \end{smallmatrix}]$.

9.15 Calculate A^k, $A = [\begin{smallmatrix} -.5 & 0 \\ 0 & .7 \end{smallmatrix}]$.

9.16 Between sampling instants k and $k + 1$, a zero-order hold maintains the input u of a sampled-data system at a constant value $\hat{u}(k)$. Thus, $\dot{u} = 0$, $kT_s \leq t < (k + 1)T_s$. Append this to the continuous-time system state equations, and show how $\mathcal{A}$ and $\mathcal{B}$ in Equation 9.31 can be computed with software that computes e^{At}.

9.17 For the system

$$\dot{\mathbf{x}} = \begin{bmatrix} 0 & 1 \\ -1 & -1 \end{bmatrix} \mathbf{x} + \begin{bmatrix} 1 \\ -1 \end{bmatrix} u$$

$$y = [1 \quad 0]\mathbf{x}.$$

a. Compute the sampled-data state representation with $\mathbf{x}$ as the state vector, as a function of the sampling period T_s.

b. Verify that the eigenvalues of the discrete-time system are $e^{s_i T_s}$, where s_i are the eigenvalues of the continuous-time system.

c. Compute the pulse transfer function.

9.18 Repeat Problem 9.17 for

$$\dot{\mathbf{x}} = \begin{bmatrix} 1 & 0 \\ -1.5 & .7 \end{bmatrix} \mathbf{x} + \begin{bmatrix} .5 \\ 1 \end{bmatrix} u.$$

$$y = [1 \quad 1]\mathbf{x}.$$

9.19 The plant $P(s) = 0.5/[s(s + 1)(s + 3)]$ is to be discretized for $T_s = 0.5s$. Assuming a zero-order hold input, find the pulse transfer function.

9.20 Repeat Problem 9.19 for $P(s) = (2e^{-.2s})/[s(s + 2)]$.

9.21 The plant $P(z) = \frac{0.5z}{(z-1)(z-.2)}$ is to be controlled with a pure-gain controller $F(z) = k$, in a 1-DOF configuration.

a. Using the Jury criterion, calculate the range of k for which the closed-loop system is stable.

b. Sketch the Root Locus.

9.22 Repeat Problem 9.21 for $P(z) = 2/[z(z - .5)(z + .6)]$.

9.23 Repeat Problem 9.21 for $P(z) = (z - 1.5)/[z(z - .5)]$.

(M) **9.24** Compute the discrete-time Nyquist plot for the system of Problem 9.21, and compute the range of k for stability.

(M) **9.25** Repeat Problem 9.24 for the system of Problem 9.22.

(M) **9.26** Repeat Problem 9.24 for the system of Problem 9.23.

9.27 Show that the backward-difference and Tustin transformations of a proper or strictly proper $F_s(s)$ will always yield an $F(z)$ that has equal numbers of poles and zeros.

9.28 Discretize the PI and PID controller transfer functions using the backward-difference method, and write down the resulting algorithm.

9.29 Repeat Problem 9.28 using the Tustin transformation.

Ⓜ **9.30** In Problem 6.23 (Chapter 6), a lead-lag compensator was designed for the system of Problem 6.1. Discretize the controller and check stability margins for $T_s = 0.2$, 1, and 2 s. Use the Tustin transformation.

Ⓜ **9.31** In Problem 6.24 (Chapter 6), a lead-lag compensator was designed for the system of Problem 6.2. Discretize the controller and check stability margins for $T_s = 0.1$, 1, and 3 s. Use the Tustin transformation.

Ⓜ **9.32** *dc servo, simplified model* Discretize the PD compensator of Problem 6.25 (Chapter 6), ensuring a phase margin of at least 50° with the largest possible sampling period. Plot step responses for the continuous-time and sampled-data designs.

Ⓜ **9.33** *Servo with flexible shaft* Discretize the design of Problem 6.26 (Chapter 6), ensuring a phase margin of at least 40° with the largest sampling period possible. Plot step responses for the continuous-time and sampled-data designs.

Ⓜ **9.34** *Chemical reactor* Discretize the design of Problem 6.27 (Chapter 6), ensuring a phase margin of at least 50° with the largest sampling period possible. Plot step responses for the continuous-time and sampled-data designs.

Ⓜ **9.35** *Heat exchanger* Discretize the design of Problem 6.28 (Chapter 6), ensuring a phase margin of at least 35° with the largest sampling period possible. Plot step responses for the continuous-time and sampled-data designs.

Ⓜ **9.36** *Crane* Discretize the design of Problem 6.29 (Chapter 6), ensuring a phase margin of at least 40° with the largest sampling period possible. Plot step responses for the continuous-time and sampled-data designs.

Ⓜ **9.37** *Maglev* Discretize the design of Problem 6.9 (Chapter 6), ensuring a phase margin of at least 40° with the largest sampling period possible. Plot step responses for the continuous-time and sampled-data designs.

Ⓜ **9.38** *Servo, simplified model* For the state-feedback design of Problem 7.10 (Chapter 7):

a. Compute the loop gain.

---

* *H i n t*   Break the loop open at the plant input, and compute the transmission from plant input to controller output.

Compute the phase margins.

b. Discretize the design, ensuring a deterioration of no more than 10° in the phase margin with the largest sampling period possible.

(M) **9.39** *Servo with flexible shaft* Repeat Problem 9.38 for the design of Problem 7.21 (Chapter 7).

(M) **9.40** *Drum speed control* Discretize the LQ design of Problem 7.22(b) (Chapter 7). Try different sampling periods until the time response of the sampled-data system is within 10% of the continuous-time response at all times, for the test situation of Problem 7.22(b).

(M) **9.41** *Blending tank* Repeat Problem 9.40 for the LQ design of Problem 7.23 (Chapter 7), with reference to the test conditions of Problem 7.23(c).

(M) **9.42** *Crane* For the LQ design of Problem 7.24 (Chapter 7), compute the phase margin. (See Problem 9.38 for a hint.) Discretize the controller, ensuring a deterioration of no more than $10°$ in the phase margin with the largest sampling period possible.

(M) **9.43** *Two-pendula problem* Repeat Problem 9.42 for the LQ design of Problem 7.25 (Chapter 7), with $\ell_1 = 1$ m and $\ell_2 = 0.5$ m.

(M) **9.44** *Maglev* Discretize the LQ design of Problem 7.26 (Chapter 7), choosing a sampling period such that the responses for the test conditions of Problem 7.15(e) are within 10% of the corresponding continuous-time responses.

(M) **9.45** *dc servo, simplified model* A state-based controller was designed in Problem 7.57 (Chapter 7).

 a. Compute (as in Problem 9.38) the loop gain and phase margin.

 b. Discretize the controller, ensuring a deterioration of no more than $10°$ in the phase margin with the largest sampling period possible.

(M) **9.46** *Servo with flexible shaft* Repeat Problem 9.45 for the design of Problem 7.58 (Chapter 7).

(M) **9.47** *Heat exchanger* Repeat Problem 9.45 for the design of Problem 7.61. (Chapter 7).

(M) **9.48** *Two-pendula problems* Repeat Problem 9.45 for the design of Problem 7.64 (Chapter 7).

References

[1] Moroney, P. *Issues in the Implementation of Digital Feedback Compensators*, MIT Press (1983).

[2] Oppenheim, A. V., and R. W. Schafer, *Digital Signal Processing*, Prentice-Hall (1975).

[3] Kuo, B., *Digital Control Systems*, Holt-Saunders (1980).

[4] Franklin, G. F., and J. D. Powell, *Digital Control of Dynamic Systems*, Addison-Wesley (1980).

[5] Aström, K. J., and B. Wittenmark, *Computer Controlled Systems: Theory and Design*, Prentice-Hall (1984).

Index